HIMĀLAYA

HIMĀLAYA

Exploring the Roof of the World

JOHN KEAY

BLOOMSBURY CIRCUS
LONDON · OXFORD · NEW YORK · NEW DELHI · SYDNEY

BLOOMSBURY CIRCUS
Bloomsbury Publishing Plc
50 Bedford Square, London, WC1B 3DP, UK
29 Earlsfort Terrace, Dublin 2, Ireland

BLOOMSBURY, BLOOMSBURY CIRCUS and the Bloomsbury Circus logo are trademarks of
Bloomsbury Publishing Plc

First published in Great Britain 2022

Copyright © John Keay, 2022
Maps Copyright © Michael Athanson, 2022

John Keay has asserted his right under the Copyright, Designs and Patents Act, 1988,
to be identified as Author of this work

A catalogue record for this book is available from the British Library

ISBN: HB: 978-1-4088-9115-5; TPB: 978-1-4088-9114-8; EBOOK: 978-1-4088-9112-4;
EPDF: 978-1-5266-4397-1

2 4 6 8 10 9 7 5 3 1

Typeset by Newgen KnowledgeWorks Pvt. Ltd., Chennai, India
Printed and bound in Great Britain by CPI Group (UK) Ltd, Croydon CR0 4YY

To find out more about our authors and books visit www.bloomsbury.com
and sign up for our newsletters

For Nell and Amor, Coco and Indy

Contents

Maps

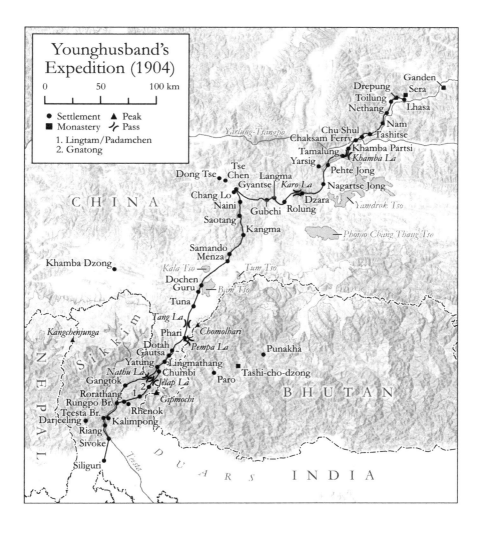

Younghusband's Expedition (1904)

0   50   100 km

- ● Settlement   ▲ Peak
- ■ Monastery   ⸘ Pass

1. Lingtam/Padamchen
2. Gnatong

Yarlung-Tsangpo

Ganden
Sera
Drepung
Toilung
Nethang
Lhasa
Nam
Chu Shul
Chaksam Ferry
Tashitse
Khamba Partsi
Tamalung
Yarsig
Khamba La
Pehte Jong
Dong Tse
Tse Chen
Langma
Gyantse
Karo La
Nagartse Jong
Chang Lo
Naini
Gubchi
Rolung
Dzara
Yamdrok Tso
Saotang
Kangma
Phouo Chang Thang Tso
Samando
Menza
Kala Tso
Tum Tso
Khamba Dzong
Dochen
Guru
Bam Tso
Tuna
Tang La
Phari
Chomolhari
Kangchenjunga
Dotah
Gautsa
Pempa La
Punakha
Yatung
Lingmathang
Nathu La
Chumbi
Gangtok
Jelap La
Paro
Tashi-cho-dzong
Rorathang
1
Gipmochi
2
Rungpo Br.
Teesta Br.
Rhenok
Darjeeling
Kalimpong
Riang
Sivoke
Siliguri

CHINA

SIKKIM

NEPAL

BHUTAN

DUARS

INDIA

Teesta

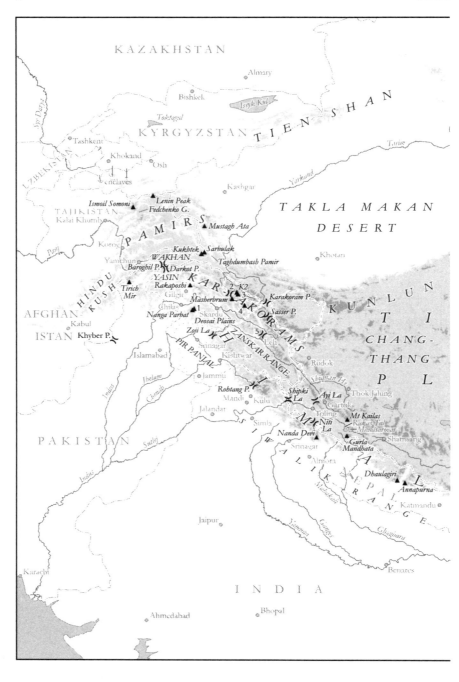

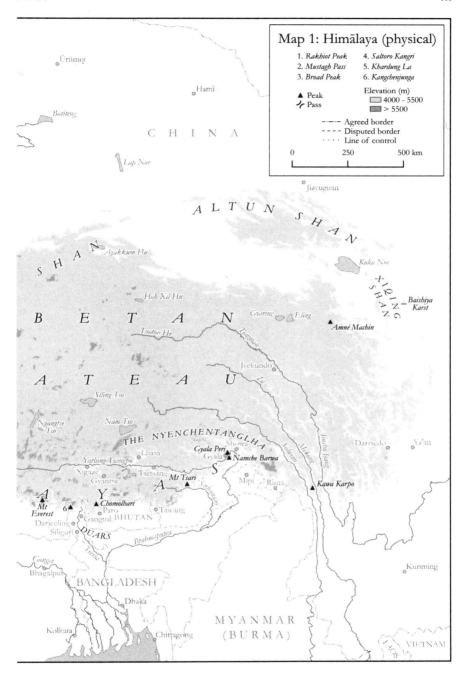

Map 1: Himālaya (physical)

1. Rakhiot Peak   4. Saltoro Kangri
2. Mustagh Pass   5. Khardung La
3. Broad Peak   6. Kangchenjunga

▲ Peak
⚲ Pass

Elevation (m)
▢ 4000 - 5500
▨ > 5500

—·—·— Agreed border
— — — Disputed border
· · · · Line of control

0   250   500 km

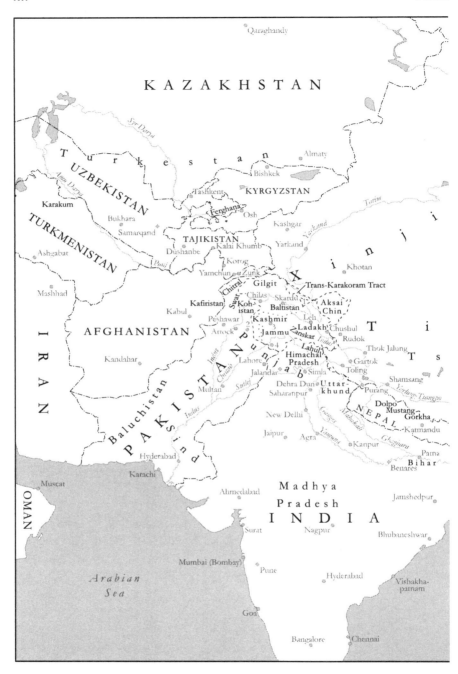

Map 2: Himālaya (human)

1. Siachen Glacier 3. Srinagar
2. Islamabad   4. Jammu

International border
—·—· Agreed
— — — Disputed

0   250   500 km

MONGOLIA

Selenga

Ider

Erdenet

Irtysh

a n g

Hami

G

Dunhuang

a n

Jiayuguan

Baotou

Beijing (Peking)

Yinchuan

C   H   s   I   N   A   Taiyuan

Amdo   Tsongkha   Jinan
Xining

(Qinghai)   Lanzhou   Zhengzhou

b   e   t
Yellow   Jyekundo   Xi'an   Huainan
(Chang'an)

a   n   g   Salween

Sichuan   Wuhan

K   h   a   m   Chengdu
Lhasa   Kongbo   Pome   Dartsedo   Ya'an
Xigaze   Gyala   Showa   Yangtze   Chongqing   Nanchang
Gyantse   Tsetsang   Pemako   Changsha   Pingxiang
Sik-   Mipi   Rima
kim   BHUTAN   Arunachal   Guiyang   Guilin
Paro   Tawang
Darjeeling   A   s   s   a   m   Yunnan
Cooch Behar   Kunming
BANGLA-
DESH   Dhaka   Nanning   Guangzhou
(Canton)
Chittagong   MYANMAR   Hong
Kolkata   (BURMA)   VIETNAM   Kong
(Calcutta)   Hanoi
Naypyidaw   LAOS

Bay of   Vientiane   Mekong
Bengal   Rangoon   South China
THAILAND   Sea
Bangkok
CAMBODIA

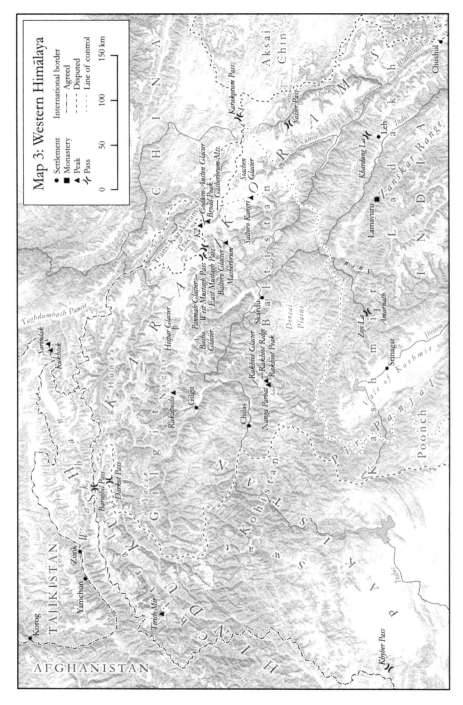

Map 3: Western Himālaya

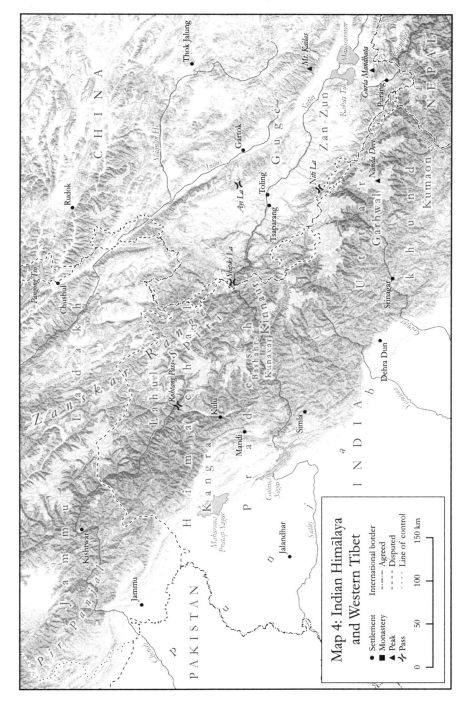

Map 4: Indian Himālaya
and Western Tibet

- Settlement   International border
■ Monastery   ------ Agreed
▲ Peak   ---- Disputed
⅄ Pass   ···· Line of control

0   50   100   150 km

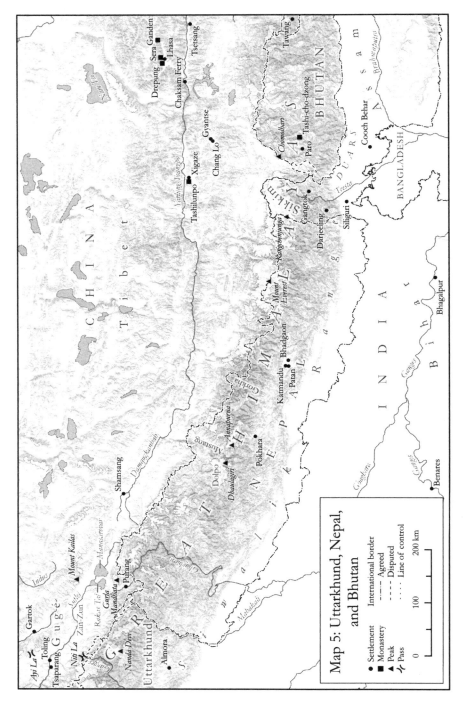

Map 5: Uttarkhund, Nepal, and Bhutan

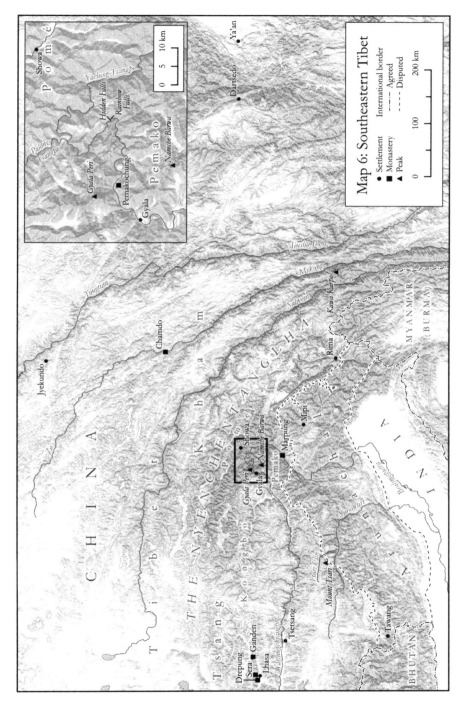

Map 6: Southeastern Tibet

Settlement • Peak ▲
Monastery ■
International border
----- Agreed
----- Disputed

0   100   200 km

Preface

History has not been kind to Himālaya. Those acquainted with our spherical planet's most spectacular protuberance have been tempted to appropriate it. Buddhist India claimed Himālaya from the south, Islam put down roots in its western approaches, Mongols and Manchus rode in from the north, and from the east an irredentist China continues to engross what it prefers not to call Tibet. Empires have here collided, cultures clashed. Surveyors staked out frontiers, hunters decimated the wildlife and mountaineers bagged the peaks. Machinery is now gouging out the minerals.

A jumble of borderlands and buffer states on the map, Himālaya is quite unlike, say, Amazonia, Australia or most of the world's other distinctive ecozones. Another of them, Antarctica, when threatened by international rivalries in the 1950s, was demilitarised and apportioned into national study zones by mutual agreement. Himālaya, being no less fragile and just as globally significant, would benefit from a similar consensus. But it's unlikely to emerge. Antarctica has no history and no native Antarcticans; Himālaya has much history and many native Himalayans.

The most we can hope for is a wider recognition of the region's physical integrity as the only high-altitude ecozone. The clashing of cultures, ideologies and empires on the 'roof of the world' has long been matched by a comparable variety of competing reasons for taking Himālaya seriously. But in trying to do justice to all these interests, this book may seem as disjointed as the Himalayan skyline. It can't be helped. Transitioning from geology and palaeontology to mysticism and mountaineering by way of archaeology, monastic warfare and exotic commodities cannot be expected to be seamless. To provide a thread of continuity I have instead focused on the human component in Himalayan studies, the personalities and their adventures, and I have sometimes taken liberties with the chronology. Biography enlivens geography; and in Buddhist societies no one is likely to take issue with a narrative which leaps across improbable distances and incorporates demanding flashbacks. For what follows I plead the nature of the subject, plus the indulgence of the mountains and all who love them.

Prologue

The common anglicised pronunciation is Hĭmălayă. But in recent years there has been a tendency among superior folk to say Hĭmālĭyă or Hĭmălĭyă … The sum of it all is that Himalaya is a Sanskrit word, and there is no doubt about the correct Sanskrit pronunciation. The English equivalents of the vowel sounds are these: Hi- as in 'him', -mā- as in 'father', -la- and -ya as in 'fur' or French 'le'.

Sir Geoffrey Corbett (1929)[1]

Visitors to Darjeeling are urged to rise early. From this ramshackle city slung across once lush hillsides at the northern apex of the Indian state of West Bengal the best chance of a close encounter with the world's most spectacular urban backdrop comes at first light. Even then you may be disappointed. Kangchenjunga (Kangchendzonga), our planet's third highest mountain, is like the tiger in the forest – ever out there but not readily spotted. Or more accurately, it's ever *up* there. Instead of straining towards some pearly peak on a distant horizon, you need to crane backwards. The mountain is overhead; and though the snows of its summit are billowed in cloud, should it deign to disrobe, be prepared for a vision. You're standing beneath an apotheosis, a shy, radiant deity. Even empire-builders never forgot their first sighting of Kangchenjunga.

It is Friday 19 June 1903. Dawn comes up grey and unpromising. The monsoon has just broken and the rain is said to have been 'coming down in cataracts'. This inconvenience fails to deter a small group of sightseers gathered outside the town's Rockville Hotel. The mountain is anyway invisible from the narrow lanes round the Rockville, and here at this unearthly hour there is another attraction. Horses are being saddled, servants berated, last-minute orders issued. By the time the newly appointed commissioner for Tibetan frontier affairs emerges from the hotel, all is ready. A short figure with a droopy moustache, receding hairline and not much in the way of small talk, the forty-year-old Captain Francis Edward Younghusband is dressed, beneath his dripping oilskins, in 'Marching order' – 'breeches, gaiters, brown boots, flannel shirt, khaki coat and forage cap'.[2] The captain mounts, his escort forms up and the cavalcade clatters off. They head north for Sikkim and Tibet. As the rain hammers down, someone is heard to call out 'Good luck.' With half a dozen Himalayan passes to cross, a near-polar winter in prospect and 650 kilometres of the world's most hostile terrain, they need more than luck.

Four months later the captain, now promoted to colonel, is back in Darjeeling. Luck has failed him; Tibet is no-go. Not for the first time – but the last if Younghusband has anything to do with it – an expedition angling for permission to visit the Tibetan capital of Lhasa has been rebuffed. The mission was supposed to have put relations between British-ruled India and Chinese-claimed Tibet on an amicable basis. A previous convention had promised reciprocal trade, border demarcation and respect for one another's sovereign territory. But though the Chinese had signed this document, the Tibetans had not (and, according to the British, had the Tibetans signed it, the Chinese would not; collusion supposedly underlay the rejection). Lhasa had therefore felt entitled to remove the new boundary markers, claim grazing rights in their vicinity, nullify cross-border trade and return unopened any letters thought to be protesting against these actions.

Younghusband had counted on gaining Tibetan compliance by himself infringing the convention. With an escort of 500 mainly Indian troops plus countless servants, porters and pony-men, his mission had crossed the British-protected state of Sikkim and clambered up a pass 5,200 metres above sea level (asl) to set up camp at Khampa Dzong on Tibetan soil. There they had indeed been met by a Tibetan delegation. But the delegates were thought to be of inferior rank and neither

Khampa Dzong, the fortress well inside Tibet where Younghusband tried to open negotiations in 1903

empowered nor disposed to negotiate. Younghusband insisted that in the face of such intransigence his orders were to advance. The Tibetans insisted he must withdraw; only when his mission had vacated Tibetan territory might even talks about talks be sanctioned.

The stand-off had lasted from June till October. Younghusband hadn't budged. The rains eased off and the thermometer began to plummet. Tibetan troops were reportedly mobilising to contest any further advance. A couple of Sikkimese subjects, who were entitled to British protection and were in fact British spies, were arrested and maltreated by the Tibetans. But that was only half the story.

> These insults would never have given rise to the despatch of an expedition if the Tibetans had not added injury to them by their dalliance with Russia [wrote a well-informed correspondent of the London *Times*]. As it was, there was nothing else to do but intervene, and that speedily.[3]

Speedily enough, orders had been drafted for a more forceful 'intervention' well before Younghusband, leaving his escort at Khampa Dzong, rode back into Darjeeling. He was immediately summoned to Simla, the summer capital of the raj, to consult with Lord Curzon, the British viceroy. With more troops and greater firepower, the Younghusband 'diplomatic mission' of 1903 was about to be reconfigured as the Younghusband 'military expedition' of 1904. At the time the Tibetans and the Chinese rightly called it an 'invasion'; subsequently even supporters would concede it was an 'armed incursion'. The weaponry being readied included mountain artillery and the latest in death-dealing machine guns. 'Whatever happens, we have got / The Maxim gun and they have not,' quipped a Hilaire Belloc character in the only known maxim about the Maxim.

For an elevated wilderness as apparently impregnable, unproductive and physically demanding as Himālaya's heartland, Tibet has attracted a surprising number of invaders. Many came from the north and east – Mongols, Jurchen, Manchus, Maoists and currently Han immigrants. Equally enticing were Himālaya's outlying regions. Both Kashmir and Nepal had been repeatedly overrun from the south – by Rajputs, Afghans, Mughals, Dogras – while, from the same satellite kingdoms, invasions of Ladakh and Tibet itself had been mounted.

As early as 326 BCE, Alexander the Great, when marching a detachment of his Macedonian veterans from northern Afghanistan to India, seems to have notched up conquests in Kafiristan (Nuristan), Chitral and Swat, valleys where Himālaya's western fringes serve as India's north-west frontier. Along the same frontier but more recently, British expeditions in the 1890s had fought their way into the mountain kingdoms of Hunza and Chitral (both now in Pakistan). Younghusband himself had been involved in these latter-day ventures, refuting Chinese claims to Hunza in the Karakorams as a political agent and reporting on the British occupation of Chitral in the Hindu Kush on a journalistic assignment. As an explorer and collector of military intelligence, he was already famous for exploits beyond India's frontiers in the Pamirs and Eastern Turkestan (Xinjiang). But perhaps Younghusband's most useful discovery had been made in Chitral where, in 1894, while travelling the length of that valley in the company of a globe-trotting MP, he had found his future patron.

*Francis Edward Younghusband, the soldier-explorer
responsible for the 1904 invasion of Tibet*

Captain Francis Younghusband and the Hon. George Nathaniel Curzon MP appeared to have little in common. Younghusband, born into a military family in India, had belied his slight stature by winning cross-country races at school and going straight into the army. Shy, introspective and prone to embarrassing revelations of a simplistic fervour, he seems to have embraced exploration as an escape from subaltern camaraderie. Curzon, on the other hand, the heir to a barony, had garnered prizes at Eton and Balliol and been elected president of the Oxford Union. A coveted fellowship of All Souls and magisterial works on Persia and Central Asia had followed. Few would doubt, least of all Curzon himself, that as a scholar and highly articulate parliamentarian he was destined for the highest office. His appointment in 1899 as the youngest ever viceroy seemed preordained and his six years of forceful rule would generally be reckoned to mark the apogee of the British raj.

But while Curzon winced at Younghusband's naivety and Younghusband crumpled under Curzon's interrogation, neither ever doubted their mutual commitment. In the face of criticism over an assignment like Tibet, Curzon could count on Younghusband's expansionist instincts while Younghusband could bank on Curzon's unwavering support. Fired by a sense of imperial mission, both believed that 'Providence or the laws of destiny had called Britain to India for [in Curzon's words] "the lasting benefit of the human race".'⁴ The benefit was thought self-evident and amply justified a degree of compulsion. But British rule in India was constantly under threat, both from within and from without, and nowhere more so than on its long continental frontier in Himālaya. Here, beyond the mountains, another empire, that of tsarist Russia, was invoking Providence and the laws of destiny to justify extending its own lasting benefits to the human race.

It was an old story. Anglo-Russian rivalry in Inner Asia had been smouldering since Napoleonic times. As an Asia-wide 'cold war' punctuated by diplomatic flurries and unexpected troop movements, the stand-off was likened by one Russian foreign minister to 'a tournament of shadows'. Earlier an Irish cavalryman had called it 'a great game' – until, that is, his luck ran out when he was confined in a pit of snakes in Bukhara prior to being executed. In the mid-nineteenth century the Uzbek capital had been as much a forbidden city as Lhasa, its emir being just as isolationist as the Dalai Lama. But by the 1870s most of Central Asia had been overrun by the Russians. 'In 1865 the great walled city of Tashkent submitted to the tsar. Three years later it was the turn of Samarkand and Bukhara, and five years after that ... the Russians took Khiva.' Peter Hopkirk quotes Fridtjof Nansen's calculation that the tsarist frontier was advancing on India at the rate of 55 square miles (142 square kilometres) a day. In 1900, '2000 miles [3,200 kilometres] separated the British and Russian empires in Asia. By the end of [the century] this had shrunk to a few hundred [miles], and in parts of the Pamir region to less than twenty.'⁵

In the process 'the Great Game' acquired its definite article, though it was not given initial capitals until the mid-twentieth century. It entered popular usage without them when Kipling adopted the term in his much loved *Kim*, published in 1901. In that same year it was reported in St Petersburg that envoys from Tibet were being entertained in the Russian capital. They bore gifts and overtures from the young Dalai Lama and were accompanied by a man long resident in Lhasa whom the British identified as Aghvan Dorzhiev.

Aghvan Dorzhiev, the Buriat Mongol whom Younghusband
and Curzon suspected of being a Russo-Tibetan go-between

To those, like Curzon and Younghusband, of a decidedly bullish disposition, this news was as the reddest of red rags. Dorzhiev was portrayed as an unscrupulous adventurer and master of intrigue. It was not his first Russian visit and he was actually a Russian subject. But as a Buriat (Siberian) Mongol, he was a Buddhist by birth and had a genuine interest in Tibetan scholarship. He may have encouraged the Dalai Lama to challenge those of his ministers who favoured good relations with the British, and he may genuinely have believed that Tibetan Buddhism would be safer under the protection of Tsar Nicholas II, among whose subjects were several other Buddhist peoples, than under the newly crowned King Edward VII.

Curzon chose to believe that Dorzhiev had actually drawn up a Russo-Tibetan treaty; Younghusband was convinced that St Petersburg was supplying Lhasa with modern arms; and both men credited reports of a cossack detachment standing by to rush to Tibet's aid whenever required. But neither of them, nor seemingly anyone else, appreciated that the Game was pretty much over. An Anglo-Russian agreement about the always contentious Afghan frontier had finally been implemented and, in the Far East, Russian expansion was even now being brought to a shuddering halt. On 8 February 1904, just as the new Younghusband expedition edged out of Sikkim into Tibet, the imperial Japanese navy was staging what reads like a rehearsal for Pearl Harbor as it took the Russian fleet unawares at anchor at Port Arthur (Lüshun).

A year later, by when members of the Younghusband expedition were happily souvenir-shopping in Lhasa, Japanese troops swept through Korea and pushed the Russians out of southern Manchuria (Liaoning). Growing unrest within the tsarist empire, plus another naval debacle, then sealed the modern era's first ever defeat of a European power by a resurgent Asian nation. The Russo-Japanese War cost an estimated 150,000 lives yet it barely outlasted the British invasion of Tibet. Tsar Nicholas survived, though only by conceding reforms that prepared the way for the 1917 revolution. If Aghvan Dorzhiev had indeed masterminded a Russo-Tibetan treaty, the tsar would never have been in a position to honour it.

None of this news made much impression on British India. By November 1903 Younghusband was in Simla being briefed by Curzon. The new expedition was to have British troops as well as more Indian regiments, more and bigger guns, and a supply chain capable of supporting around 2,000 combat troops over vast distances and for several months. But according to a supposedly definitive statement issued by the Cabinet in London, the objectives had changed. The 'sole purpose' of the expedition was now that of 'obtaining satisfaction' for the recent border transgressions. To this end Younghusband might proceed as far as Gyantse, about halfway to Lhasa. He would withdraw as soon as 'reparation is obtained'; no permanent mission was to be left in Tibet and no part of the country occupied. Nor was there any mention of a treaty or future relations. 'In view of the recent conduct of the Tibetans' (besides the arrest of the two Sikkimese, Lhasa stood accused of an 'act of overt hostility' when some Nepali yaks had been seized), the expedition no longer had a diplomatic purpose. It was simply a punitive action.[6]

All of which was meant to placate the Russians, with whom the London government was anxious to stay on good terms, and reassure the British public, whose suspicions of imperial adventures had lately been heightened by the Jameson Raid in South Africa and a second Boer War. Curzon, on leave in London when the expedition began its advance in December 1903, suddenly found it politic to eat his words. As he explained in a letter to Younghusband, it was no longer wise to play up the idea of Russian intrigues. 'We are advancing not because of Dorjieff, or the mission to the Czar, or the Russian rifles in Lhasa but because of our Convention shamelessly violated, our frontier trespassed upon, our subjects arrested, our missions flouted, our representations ignored.' In terms of the expedition's *raisons d'être* it was back to square one. The invading force would need to show restraint – unless and until, Curzon added, 'it is converted by hostile acts into a military expedition'.[7]

Reassuring the British public proved more straightforward than placating the Russian bear. It simply meant keeping them on side by providing an informative newsstream. The 1903 mission to Khampa Dzong had been shrouded in secrecy, but the 1904 expedition was to be a more public affair. Behind the advancing troops a telegraph line was being hastily erected, postal runners plied daily back to Darjeeling, and officers were allowed to bring along their cameras and sketching materials. Captain Walton was in charge of the natural history collection, a young geologist, Henry Hayden, was on loan from the Geological Survey, and Lieutenant Ryder from the map-making Survey of India. Austine Waddell, an elderly doctor, headed the medical team and, on the strength of previous acquaintance with the Buddhist traditions of Sikkim and Darjeeling, doubled as the expedition's cultural attaché. Waddell helpfully lists at least five officers who were filing reports or writing articles for the press. In addition, accredited correspondents from *The Times*, *Daily Mail* and Reuter's vied for front-line stories and marvelled at the strange land in which they found themselves. All, like Younghusband, wrote letters home and many, including Dr Waddell, subsequently produced detailed and handsomely illustrated accounts of the whole expedition.

From this mass of reportage, there emerges yet another incentive for the whole undertaking. Besides regulating frontier relations with the Tibetans, exacting satisfaction for their 'insults' and countering Lhasa's supposed overtures to tsarist Russia, the expedition would generate its own dynamic. What Waddell calls the 'Land of Mystery', 'the mystic land of the lamas', was exercising its allure, drawing the intruders onwards. In the thin air it seemed possible to do anything, go anywhere. Dispelling Tibet's 'forbidden' reputation and exploring its romance became ends in themselves.

The clarity of the atmosphere sharpened contours and heightened their coloration. The remotest horizon looked attainable. Kangchenjunga and Chomolhari signposted the Lhasa road; the high passes served as its milestones. Mirages swam into view – or perhaps they were hallucinations induced by the oxygen deficiency? The local way of life reeked with the mystique of antiquity. Beyond the furthest ridge there was always another attraction – a mythic Shambhala, Elysian slopes abounding in game, an unclimbed peak, an uncharted river. For the scholar there were texts to be studied, languages to be learned; physical anomalies awaited the geologist, exquisite primulas rewarded the botanist. In a topsy-turvy world where eternal truths were forever within reach yet never quite grasped, where the gods were interchangeable with demons, and where empire had somehow preceded statehood, Himālaya would give as good as it got, hypnotising the visitor, inveigling the invader.

'It is reckoned eight days from Darjeeling to Chumbi, but, riding all day and most of the night, I completed the journey in two,' recalled the *Mail*'s Edmund Candler. 'Newspaper correspondents are proverbially in a hurry. To send the first wire from Chumbi I had to leave my kit behind and ride with poshteen [long sheepskin coat] and sleeping bag tied to my saddle.'

Younghusband, after again setting off from Darjeeling's Rockville Hotel, was a month ahead of Candler. Having crossed the Jelap (Dzelap) La, the colonel and his escort were already encamped in the Chumbi valley, a wedge of Tibetan territory between Nepal and Bhutan. There Candler aimed to join them. Hastily packing a Gladstone bag, the *Mail*'s man had taken the train from Calcutta to Darjeeling and then splashed out 300 rupees on a pony 'of the most modest pretensions' for the onward journey. The expense was justified not just to catch up with the expedition but to see off the competition.

> I was racing another correspondent. At Rungpo I found that he was five hours ahead of me, but he rested on the road, and I had gained three hours on him before he left the next stage. Here I learned that he intended to camp at Lingtam, twelve miles [19 kilometres] further on … I made up my mind to wait outside Lingtam until it was dark, and then to steal a march on him unobserved.[8]

Which of his fellow correspondents Candler was so keen to overhaul he doesn't say. It could have been the urbane Perceval Landon of the London *Times*. Landon, an Oxford graduate, gave 'foreign correspondent' as his profession; Candler, a Cambridge classicist, embraced journalistic assignments as a break from teaching in Darjeeling. But Landon was slow off the mark in 1904. At this stage Candler's rival must have been Henry Newman, the assistant editor of a Calcutta newspaper who was now representing Reuter's news agency.

Across Sikkim, Candler rode on through the night. Nocturnal pony-trekking was a novelty. The swish and smell of the changing vegetation hinted at progress; 'uphill and downhill could only be distinguished by the angle of the saddle'. Some charcoal-burners told him he was only half an hour behind the competition. 'But I believed no one. Wayside reports were probably intended to deceive me, and no doubt my informant was his [Newman's] unconscious confederate.' As the moon rose, the 300-rupee pony collapsed from exhaustion. 'I rested a few hours, bought a good mule, and pressed on.'

Mess mates of the expedition's 'Mounted Infantry'. Bottom right is Eric 'Hatter Bailey and immediately behind him the Daily Mail *correspondent Edmund Candler*

Breakfast came courtesy of the 8th Gurkhas encamped at Gnatong. 'I met a subaltern with a pair of skates. He showed me to the mess-room.' The skating subaltern looks to have been 'Hatter' Bailey, sportsman and fitness fanatic, who commanded a unit of the contradictory-sounding 'mounted infantry'. Comprised of Gurkha and Sikh foot soldiers who'd lately been given a crash course in horsemanship, the mounted infantry would prove invaluable for scouting and skirmishing. Candler would mess (that is, share a field kitchen and meals) with their officers. Newman mostly had to fend for himself, while Landon, a friend of both Kipling and Curzon, was sufficiently well connected to mess with Younghusband and the headquarters staff.

Now on a borrowed pony, in the afternoon of Day Two the man from the *Mail* crested the Jelap Pass to the fluttering of prayer flags and the shrill whistling of red-billed choughs.

> Behind and on both sides was a thin mist, but in front my eyes explored a deep narrow valley bathed in sunshine. Here, then, was Tibet, the forbidden, the mysterious … Far to the north-east Chumulari (23,930 feet [7,294 metres]), with its magnificent white spire rising from the roof-like mass behind, looked like an immense cathedral of snow. Far below on a yellow hillside hung the Kanjut Lamasery above Rinchengong. In the valley beneath lay Chumbi and the road to Lhasa.[9]

Descending to the upper part of the valley, Candler wearily picked his way through a knacker's yard of dead transport animals. The precipitous terrain, the sub-zero temperatures and an outbreak of anthrax were taking a heavy toll on the expedition's supply train. A later report would itemise the total expenditure on quadrupeds as: mules 7,096 (of which 910 died), bullocks and buffaloes 5,372 (1,091 deceased), yaks 4,466 (4,114 deceased, mainly due to the anthrax), ponies 2,668 (1,200 deceased) and camels 6 (no survivors). This was in addition to '10,091 coolies', or porters, of whom 88 were listed as 'casualties', and the two experimental 'zebrules'. Half zebra and half donkey (or Clydesdale according to Dr Waddell), the last were intended as gun transports but, even when not loaded, were having to be hauled up the scree with ropes; 'the men hated them' – which no doubt prejudiced their chances of survival. Both died.[10]

Now nearing the end of his ride, Candler marvelled how easily the cliffs that commanded the trail might have been used to ambush the expedition. From such vantage points 'a few riflemen might annihilate a column with perfect safety and escape into Bhutan before any flanking movement could be made ... yet miles of straggling convoy are allowed to pass daily with the supplies necessary for the existence of the force ahead'.[11]

Instead the Tibetans seemed to have placed their faith in drystone walls thrown across the valley but now abandoned. There were two outside Yatung, the cobbled border town that was supposed to host the Anglo-Tibetan trade mart. Its Customs House was home to a Chinese official who, like most of China's customs men at the time, was British, and the town's only 'trader' was actually a stiff-skirted Presbyterian missionary, also British. Both would join the expedition. Ranald Parr, the customs man, put his Chinese to good use as a rather voluble liaison officer with Beijing's intermediaries, while Annie Royle Taylor, the doughty Scots missionary, offered Dr Waddell her services as a nurse.

From Yatung it was only 5 kilometres to the open dust bowl, flanked with sombre firs and strewn with tents and rough shelters, which the British were optimistically calling New Chumbi. Here the bulk of the troops that now constituted the mission's 'escort' would spend the next three months. Their commanding officer, an unloved brigadier-general called James Macdonald, welcomed reinforcements, organised his supply train and sat out the worst of the winter within easy reach of firewood. Younghusband and his staff had already moved on. Candler was consoled by having at least pipped Reuter's Henry Newman: 'I reached [New] Chumbi on the evening of January 12, and was able to send the *Daily Mail* the first cable from Tibet, having completed the journey from Darjeeling in two days' hard riding.' New Chumbi was not quite the picnic Candler's readers might have supposed. Officers lived in 'primitive dens' dug out of the ground, walled with boulders and roofed with fir branches. 'I write at an operating table after a dinner of monal (pheasant) and yak's heart.' But Chumbi was infinitely preferable to Phari, 40 kilometres to the north, let alone the 'desolate hamlet' of Tuna, another 25 kilometres, where Younghusband was spending a winter under canvas in temperatures as low as minus 25 degrees Fahrenheit (minus 31 degrees Celsius).

Phari had two claims to fame. Its massive *dzong*, whose walls and bastions reared bleakly from a dead-flat plain, was the most formidable fortress on the whole Gyantse road, and the 'rabbit-warren' of hovels nestling in its shelter was what Landon would declare 'the filthiest town on earth'. Here accumulations of offal and sewage blocked the main street to a depth that obliged the householder to dig a tunnel through them to reach his front door. The stench was 'fearful', especially when the midday sun thawed the yellow ice into an 'iridescent' slime from which protruded the hair and horns of yak skeletons and the half-decayed corpses of long-dead dogs. 'It must be seen to be believed,' sniffed Landon.

Candler was no less disgusted by the fort. Like the stone barriers across the trail, it had been evacuated by the Tibetans ahead of the advancing mission. Lest the Tibetan garrison return, the British had to occupy it. Candler was one of those unfortunate enough to be billeted in it. It was said to have been built in the 1600s and never mucked out since. The dirt was here the dust of ages which a blizzard-wind redistributed faster than eighty sweepers could remove it. Everyone was coughing. Whatever they touched was coated in grime. At breakfast Candler's companions thrust their chapatis in their pockets to keep them clean and covered their yak meat with soup plates 'making surreptitious dives at it with a fork'. At dinner in the evening the same company was quite unrecognisable. 'Ruffianly-looking bandits in a blackened, smut-begrimed room, clad in wool and fur from head to foot, bearded like wild men of the woods', the officers of the mounted infantry swigged their rum in the dark and choked on the smoke from a fire of smouldering yak-dung.[12]

'All the while I was in Phari I forgot the mystery of Tibet,' says Candler. The people were just too savage, too smelly, their country too cold, too grubby. It was quite inconceivable that men who lived their entire lives in such squalor could be responsible for the intricate frescoes barely visible on some of the fort's walls. Yet one could not deny that the Tibetans had 'a settled dogma and definite convictions about things spiritual and natural that are not easily upset'.

> Perhaps before we turn our backs on the mystery of Tibet we
> will realise that the lamas despise us as gross materialists and
> philistines – we who are always groping and grasping after
> the particular, while they are absorbed in the sublime and the
> universal.[13]

With such unfashionable sentiments, Candler took leave of Phari in March. Brigadier-General Macdonald with the bulk of the expedition's guns and troops had just vacated their winter quarters to join

Younghusband at Tuna. 'Groping and grasping after the particular', Candler went forward with them. Tibetans were reported massing 12 kilometres ahead near the Hot Springs (Chumi Shengo) of Guru. A new wall there blocked the road to Gyantse, and defensive sangars (breastworks) overlooked it. All needed to be cleared.

The expedition appears to have been relieved to be on the move again after three months of inaction and was confident of a Tibetan withdrawal. 'Most of us thought that the Tibetans would fade away in the mysterious manner they have, and build another futile wall further on.'

In a last-minute parley Younghusband is heard repeating that his orders are to advance. The Tibetan commander says he will try to restrain his men from firing. Candler is still convinced they won't risk a fight.

> The morning of the 30th [March] was bitterly cold. An icy wind
> was blowing and snow was lying on the ground. I put on my
> thick sheepskin for the first time for three months, and I owe my
> life to it … No one dreamed of the sanguinary action that was
> impending.[14]

On orders from Macdonald, Gurkha and Sikh detachments climb to the flanking sangars to disarm and disperse their occupants. Candler is among the onlookers watching through field glasses. The defendants offer no resistance; in good humour they rejoin their comrades behind the wall. The artillery and the Maxims now command the whole site. The British congratulate themselves on having defused the situation. Field glasses are returned to their cases. According to Landon, 'the incident was regarded as practically over'.[15]

Dismounting, Candler uses his saddle as a desk to scribble a wire to Fleet Street: 'the Tibetan position has been taken without a shot being fired'. Younghusband reports to London along the same lines. The road to Gyantse and on to Lhasa beckons. It remains only to deprive of their weapons, mostly antiquated matchlocks and rusty heirlooms, the perhaps 1,500 Tibetans penned behind the wall. The Sikhs entrusted with this task encounter some resistance. 'But not a single shot was fired,' notes Dr Waddell. 'If there had been any firing,' adds Candler, 'I would not have been wandering about by the Tibetan flank without a revolver in my hand.' He will later admit that the oversight was sheer folly.

Candler is down. Two or more Tibetans clad in homespun stand over him, hacking away with knives and swords. The attack was so sudden that the first man was upon him before he had time to dig inside his sheepskin for his revolver. His assailant held aloft a sword in both hands as he charged. Candler ducked and rugby-tackled him but was dragged to the ground in the process. 'Trying to rise, I was struck on the temple by a second swordsman, and the blade glanced off my skull.'[16] Now face down on the ground, he relies on his heavy sheepskin for protection, using his hands to ward off blows to the head. Seven sword cuts will be found in his poshteen as well as the twelve found about his person. Bullets are finally flying, mostly from the Lee–Metfords of the Sikhs.

> After a time the blows ceased; my assailants were all shot down or had fled. I lay absolutely still for a while until I thought it was safe to raise my head … Seeing no Tibetans nearby in an erect position, I got up and walked out of the ring between the rifles of the Sikhs.[17]

It would be claimed by Landon and others that it was the Tibetan commander who fired the first shot. It blew off the jaw of a Sikh soldier who had grabbed the Tibetan's bridle, and was supposedly a signal to his troops to open fire. Candler disagreed. 'The Lhasa general must have fired off his revolver after I was struck down'; seeing his men being disarmed, the man probably despaired and panicked; the first shots must then have come not from the Tibetan but from the Sikhs who had opened fire on Candler's attackers.

This mattered in light of the carnage that ensued. 'From three sides at once a withering volley of magazine fire crashed into the crowded mass of Tibetans. It was like a man fighting with a child,' thought Landon. 'The issue was not in doubt, even from the first moment … Straight down the line of fire lay their only path of escape. Moved by a common impulse, the whole jostling one against another with a curious slow thrust, they set out with strange deliberation to get away from this awful plot of death.'[18]

Candler was by now having his wounds dressed in a field station. Looking up he witnessed the rout and could hardly believe his eyes.

> They were walking away! Why, in the name of all their Bodhisat[tva]s and Munis [holy men], did they not run? There was cover behind a hill a few hundred yards distant, and they were exposed to a devastating hail or bullets from the Maxims and rifles, that seemed to mow down every third or fourth man. Yet they walked!
> It was the most extraordinary procession I have ever seen … They were bewildered. The impossible had happened. Prayers, and charms, and mantras, and the holiest of their holy men, had failed them … They walked with bowed heads, as if they had been disillusioned in their gods.[19]

The disillusionment was not entirely one-sided. 'I got so sick of the slaughter that I ceased fire,' recalled the commander of the Maxim gunners.[20] The untrained Tibetans with their cumbersome matchlocks and assorted blades stood not a chance against regular troops with

Some of the 628 Tibetans mowed down at Guru

breech-loading rifles, machine guns and shrapnel-firing cannon. British indifference to the modern euphemism of 'proportionality' was evident in the casualties: of the Tibetans, 628 lay dead and 222 wounded; on the British side, three were slightly wounded, three more seriously wounded (including Candler) and there were no fatalities whatsoever. 'It was all over in about ten minutes,' according to Waddell.[21]

'Perhaps no British victory has been greeted with less enthusiasm than the action at the Hot Springs,' thought Candler. His mess mates, who enjoyed the occasional pursuit of Tibet's wildlife, prided themselves on notions of fair play and sportsmanship. This was neither. 'Certainly the officers … had no heart in it. After the first futile rush the Tibetans made no further resistance. There was no more fighting, only the slaughter of helpless men.'[22] He thought it could all have been avoided if the disarmament had been explained to the Tibetans and more time allowed. He was not persuaded that an early demonstration of overwhelming firepower would convince the enemy of the futility of further resistance. And nor did it.

Three months later Candler found the dead of Hot Springs still lying where they had fallen. 'One, shot through the shoulder in retreat, had spun as he fell facing our rifles. Another tore at the grass with futile fingers through which a delicate pink primula was now blossoming.' Candler himself had fared better. Stretchered painfully back to Darjeeling, he was now recovering well, having lost only his left hand to the surgeon's knife. But he was still liable to be declared unfit to rejoin the expedition and therefore approached Gyantse with caution. His timing was perfect. Reaching the town on 12 July, he learned that the move on Lhasa would begin on the 14th.

In Candler's published account Henry Newman, the Reuter's man, assumes responsibility for the narrative during the author's absence. Gyantse, a large town by Tibetan standards, had been reached on 11 April. There had been other one-sided engagements on the way, but with less carnage. Next day Gyantse's great cliff-top fort had been surrendered. The people of Gyantse 'seemed friendly and brought in large quantities of supplies', says Newman. Poplars rustled and irises bloomed in the spring sunshine. Younghusband informed London that, in this favoured part of Tibet, all resistance was at an end. There being no need to occupy the fort, the mission took up quarters in the more salubrious residence of a local aristocrat called Chang-lo. General Macdonald, however, opted to retire all the way back to New Chumbi. It was supposedly to reduce the strain on the now over-extended supply column. With him went his staff, the larger guns and about half the troops.

This was a risky move; and two weeks later it was compounded by another. To clear a concentration of Tibetans who were reported dug in at the next pass, most of the remaining troops, plus the remaining artillery, were sent ahead up the Lhasa road. Younghusband's orders were to negotiate at Gyantse and proceed no further. The Karo Pass was over 5,000 metres asl and at least 70 kilometres away. However strongly held, it posed no threat to the British in Gyantse or to their communications. It was an obstacle only if the expedition was indeed heading for Lhasa.

As Curzon expected, but to the government's consternation, Younghusband now appears to have been writing his own script: either he was aiming to provoke the Tibetans into a show of defiance that would justify his advancing on the capital, or as Lord Ampthill, the stand-in viceroy, put it, 'he is going off the rails'. It was probably a bit of both. 'The unfastened power that Younghusband enjoyed at Gyantse brought out an unpleasant and disturbing side of his character,' says his biographer.[23] His letters became peppered with racist and sectarian slurs; it was the 'cringing' and 'filthy lecherous' lamas and their 'pope' (the Dalai Lama) who were egging on the lay authorities to fight and reject negotiations; they were worse than Dorzhiev and his pro-Russian backers.

The assault on the Karo La entailed climbing to 5,640 metres asl and was promptly declared 'the highest skirmish in military history'. Again the defenders were routed, but only after fierce fighting and some of the expedition's first fatalities, including that of a British officer. Tibetan losses were just as disproportionate as at the Hot Springs. Meanwhile, back in Gyantse the first of numerous night attacks on the expedition's now hopelessly under-manned Chang-lo compound very nearly succeeded. A prolonged siege ensued, during which the Tibetans, having reoccupied the great fort, bombarded Chang-lo with home-made cannonballs fired from the long-barrelled matchlocks known as jingals. They were surprisingly accurate up to about 2,000 metres, and 'the courage now shown by the enemy altered all our previous conceptions of the fighting qualities of the Tibetans', says Newman.

Given the uncertain military situation, Lhasa was more reluctant than ever to negotiate, while London's grudging support for the

Loading mules on the uncontested Tsangpo ferry at Chaksam

expedition was beginning to look distinctly shaky. To stop the rot, Younghusband fired off a flurry of telegrams and dashed back to New Chumbi for consultations. He now urged an immediate advance on Lhasa and authorisation to stay there until a treaty was signed; as well as Macdonald and his troops, he also needed more reinforcements, and if these demands were not conceded, he would resign. That settled it. The prospect of explaining away the resignation of a resolute and apparently successful leader while his well-publicised expedition was in peril deep inside Tibet appalled both London and Simla. As Curzon and Younghusband had anticipated – not to say engineered – the expedition was directed to march on the forbidden city.

The addition of more mounted infantry, more heavy cannon and another 2,000 infantrymen cleared the supply trail back to Chumbi and soon turned the tables at Gyantse. The besieged became besiegers as the Chang-lo defenders joined their rescuers to invest the fort. Outlying villages, the nearby monastery and then the great *dzong* itself were pounded and stormed in the fiercest fighting of the whole campaign. It was also the most decisive. Following the three-month standstill and this second Gyantse surrender, Tibetan resistance was indeed 'at an end'. Hence the resumption of the advance on Lhasa just as Candler unexpectedly rejoined his colleagues in mid-July.

Instead of retracing his ride through Sikkim, Candler had this time entered Tibet by what he calls 'the Bhutan route'. It saved him from being jostled from the overworked trail across the Jelap La and 'it avoided running the gauntlet of the medicals' (who might declare a notably 'handicapped' non-combatant a liability). An added attraction was that the route took in Kalimpong and the border town of Pedong. Here lived the seventy-eight-year-old Père Desgodins, a long-bearded Methuselah, who had spent the last forty years in Himālaya probing the Tibetan frontier from various parts of China, Assam, Bhutan, Uttarkhund and Ladakh, all in a fruitless endeavour to reach his place of work as prospective 'Vicar Apostolic of the Roman Catholic Mission to Western Tibet'. According to Candler, Desgodins knew Tibet and the Tibetans better than anyone; he might make an excellent 'curé de Lhassa', he thought, and he was certainly a fund of information. The visit may have been purely social, but the timing suggests otherwise, as do Candler's subsequent plans.

The expedition had never been simply an Anglo-Tibetan affair. About a third of the British troops were in fact Nepali Gurkhas who relished the chance of renewing hostilities with their Tibetan neighbours. Nepalis, Sikkimese, Ladakhis and Bhotias (hill people from the formerly Nepali territories of Garhwal and Kumaon) made up the bulk of the supply column, while as allies and intermediaries the British called on the good offices of the royal houses of Nepal, Bhutan and Sikkim. The expedition was always an all-Himalayan affair and at this moment the British were counting heavily on Bhutan's de facto ruler to broker, if not a treaty, an uncontested advance on Lhasa. Using his personal rapport with the Dalai Lama, Bhutan's Tongsa Penlop duly obliged. The expedition proceeded over the Karo La with little opposition, and none at all on the three-day march round the azure inlets of the Yamdrok Lake.

The last obstacle looked to be the Tsangpo–Brahmaputra, the great artery of southern Tibet. Sighting the river from the last pass, Candler was reminded of the Thames. 'In the heart of Tibet we had found Arcadia.' Homesteads were dotted among the cornfields and shaded by trees, 'real trees, not dwarfish and stunted in the struggle for existence but stately and spreading'. After the tundra of Tuna and the dirt of Phari, 'everything is homelike'.

> I would have stayed there. Lhasa was only four marches distant,
> but to me, in that mood of almost immoral indolence, it seemed

that this strip of verdure, with its happy pastoral scenes, was the most impassable barrier that Nature had planted in our path. Like the Tibetans, she had menaced and threatened us at first, then she turned to us with smiles and cajoleries, entreating us to stay, and her seduction was hard to resist.[24]

Crossing the river proved little more than a formality. The Tibetans made no attempt to oppose it and had obligingly left the two capacious ferries in good working order. 'The whole force was across in less than a week.'

The only disappointment came in an order from Simla countermanding plans for a small party to return to India by following the river downstream. At the time it was generally accepted that Tibet's Tsangpo must be the upper reaches of India's Brahmaputra; but where the former became the latter and how, while plunging from a height of 3,000 metres to 300, it managed to cut right through, or under, the main Himalayan range remained one of Himālaya's greatest mysteries.

Enthusiastically supported by Younghusband, Captain Ryder, the expedition's surveyor, believed the presence of a British expedition in the heart of Tibet was the perfect opportunity for mapping the river's unknown course; Lieutenant Bailey of the mounted infantry was always game for a challenge; and Candler had surely picked the brains of Père Desgodins, who had once attempted to enter Tibet by working upstream from Assam. No doubt there were other volunteers as well, but it was not to be. Simla thought it far too risky. Ryder and Bailey were needed for a safer excursion through western Tibet, and the one-handed Candler could well have proved a liability. 'It was a great disappointment. I had counted on it as early as February [so before his mauling at the Hot Springs] and had made all preparations to join it.'

That left just Lhasa. On 3 August, as the advance guard drew near to the city, Dr Waddell was reminded of the Crusaders straining for their first sight of Jerusalem – until, that is, a glimpse of the Potala, or 'palace of the Buddhist pope', reminded him of the Vatican.[25] For Landon too, 'the hour teemed with a fierce interest of a kind that no man will perhaps ever feel again'. 'I took off my smoked glass spectacles,' he says, then he obligingly pulled out all the literary stops. 'There at last it was, the never-reached goal of so many weary wanderers, the home of all the occult mysticism that still remains on earth. The light waves of mirage dissolving impalpably just shook the far outlines of the golden roofs and dimly seen white terraces. I do not think any of us said very much. Life seemed very full: but the fact of achievement seemed remote and impossible. Still, there it was. There was Lhasa.'[26]

And there Landon would have been content for it to remain. Closer acquaintance, as when next day Younghusband staged a ceremonial procession through the city, invariably resulted in disappointment. The 'holy city' was as filthy as Phari and, according to Landon, only the Potala was 'really worthy of all the rumour and glamour and romance with which in the imaginings of man it has been invested'. The mission's reception was also disappointing. The Dalai Lama had already fled the city, Dorzhiev had disappeared off to Mongolia with him, Lhasa harboured no Russians, and there were no more Russian rifles in the state armoury than there were Lee–Metfords, most of both being locally made copies.

Younghusband would nevertheless insist on his treaty. It was duly signed within the Potala on 7 September. Until it was later modified by London and he himself censured, the colonel basked in the general approbation and felt vindicated over a job well done. The euphoria

Monks blow radung, *giant trumpets, to salute the Dalai Lama in his Potala palace at Lhasa*

induced one of his 'Road to Damascus' moments. On an early-morning ride into the mountains 'there came upon me what was far more than elation or exhilaration ... I was beside myself with an intensity of joy ... And with this indescribable joy came a revelation of the essential goodness of the world. I was convinced past all refutation ... that men at heart are divine.'[27] 'That single hour on leaving Tibet was worth all the rest of a lifetime.'[28]

In a mind as tender as Younghusband's, Himālaya's virgin peaks and stark contrasts awoke a surge of compassion and a streak of mysticism. It was as if the heights were calling him, uplifting him literally and spiritually. The scale of the Karakorams had so 'enthralled' him that he had asked 'how could I ever be little again?' But Tibet was both epiphany and finale. He would never return to the country nor lead another military expedition. He had found, if not God, then a noble goal which he called the 'Creative Spirit', a 'Universal Divinity'. After a spell as British resident (representative) in Kashmir, he would resign from government service and devote the rest of his life to ethereal matters ranging from the World Council of Churches to extraterrestrial encounters and the scaling of that slippery ladder to the stars known as Mount Everest.

For Edmund Candler, Himālaya's revelation was nothing like as lofty. It came as they were approaching Lhasa. Candler had left the banks of the Tsangpo–Brahmaputra reluctantly and, still in a mood of 'almost immoral indolence', now contemplated giving the city a miss altogether.

> Tomorrow, when we enter Lhasa, we will have unveiled the last mystery of the East. There are no more forbidden cities which men have not mapped and photographed. Our children will laugh at modern travellers' tales ... For now that there are no real mysteries, no unknown land of dreams, where there may still be genii and mahatmas and bottle-imps, that kind of literature will be tolerated no longer. Children will be sceptical and matter-of-fact and disillusioned, and there will be no sale for fairy stories any more.
>
> But we ourselves are children. Why could we not have left at least one city out of bounds?[29]

The mystique had indeed been dispelled. Yet in Himālaya as a whole, mysteries aplenty remained – scientific as well as esoteric, linguistic, geographical, archaeological and natural.

1

An Orogenous Zone

Of Geology and Geognosy we know enough: what with the
libraries of the Werners and the Huttons, what with the ardent
genius of their disciples, it has come about that now, to many a
Royal Society, the creation of the world is little more mysterious
than the cooking of a dumpling.

Thomas Carlyle (1838)[1]

Heading west across Tibet and Ladakh for all of 1,500 kilometres, the
upper Indus suddenly has second thoughts. Rather than expire in the
deserts of Inner Asia like its not too distant cousins the Oxus (Amu) and
Jaxartes (Syr), the river begins to buckle, then, on entering what is now
Pakistan, to plunge abruptly south. Tibet's other watercourses do much
the same; few so-called Himalayan rivers originate in the Himalayas.
The Tsangpo–Brahmaputra, the Satluj, the Ganges and its tributaries,
the Mekong – nearly all have their sources north of the mountains. And
there, penned back on the Tibetan plateau, they spend their infancy,
sloshing beneath the eaves of the Great Himālaya like rainwater in a
gutter till a duct is found down which to rampage through the mountain
barrier on to the plains of South Asia. In other words, and to the
consternation of border-demarcating surveyors in the nineteenth and
twentieth centuries, the Asian watershed here runs not along the line of
the loftiest peaks but considerably to the north of them. Imperceptible
to anyone unequipped with a theodolite, this actual watershed has few
features of value to the surveyor, yet its apparent misplacement north of
the mountains is not without consequence. It has been blamed for one

of the nineteenth century's worst natural disasters; and from that same misplacement geophysicists have drawn an unexpected conclusion: the rivers here predate the mountains. Inner Asia was draining to the south long before the mountains got in the way, long before there were any mountains.

So the Indus's sudden change of direction is no fluke of nature. It's a reversion to its original pre-Himalayan direction of flow. Ahead there lies some leisurely irrigation of the Punjab's cereal crops and a warm-water welcome to the Arabian Sea, but only after the river has cut clean across the main Himalayan axis at one of its highest points. The challenge demands a change of tactics. Instead of birling along a rock-strewn bed under big Tibetan skies, the grey-brown flood boils through steep-sided gorges and dark chasms, writhing and thrashing as it forces a passage round the obstruction that is Diamer, or Nanga Parbat, the 8,126-metre 'Naked Mountain' which anchors the western extremity of the Great Himālaya.

The mountains of the Great Himālaya are not just younger than the rivers but younger than most other mountain ranges. Seen from afar, with the snows of their dragon's-back skyline snagging the clouds, they epitomise permanence and eternity; but geologically speaking they are neither permanent nor eternal. 'As old as the hills' is here not that old. The main Himalayan ranges arose later than other youngish mountains like the Andes and Rockies, while compared to the weathered humps of, say, the Appalachians or the Grampians they rank as newborns. Mount Everest is so young it's still growing – by an estimated 2 to 4 centimetres a year. The deeply buried rocks of which it is composed have been melting and metamorphosing for 50 million years – and counting, for the process is still ongoing, as witness the great earthquake in Nepal in 2015. Yet the spectacular skyline that is the Himalayan profile is scarcely half as old. It dates from 'the height of Himalayan mountain-building' about 20 million years ago, says Oxford's Mike Searle. 'The combination of uplift of the rocks with erosion, combined with the recent glacial weathering, has produced the wondrous Himalayan landscape we see today.'[2] Altitude in mountains is often indicative of youth. Age, on the other hand, is more reductive. It pares the peaks, gnaws at their rocky flanks and slowly chamfers their jagged ridges into rounded saddles. Were it not for erosion's timeless attentions, Mount Everest's increment of 2 to 4 centimetres a year might be 6 to 8.

Something similar is going on right across the bleak and elevated zone which, for want of any other accepted designation, this book is calling Himālaya. In a 1930s volume entitled *Sketch of the Geography and Geology of the Himalaya Mountains and Tibet* Sidney Burrard and Henry Hayden (respectively superintendents of India's Trigonometrical and Geological Surveys, and Hayden being the man who had served the Younghusband expedition as geologist) declared that all the high country between the alluvial plains of India and the deserts of Central Asia constituted 'one geographical whole'. Few have quarrelled with this, for so indeed it does. But borders, politics and scholarship have taken little account of its essential integrity, while the Burrard–Hayden suggestion that the region be known as 'High Asia' has gone unheeded.

On physical maps this High Asia/Himālaya is the big white void at the heart of the Asian continent; sometimes it's so spattered with purple-to-chocolate splodges it looks like a bad case of cartographic bruising. Stretching from Tajikistan and Afghanistan to Burma (Myanmar) and the western provinces of China, Himālaya is best defined by its average height of over 3,000 metres (or nearly 10,000 feet) above sea level (asl). Not surprisingly it includes a lot of mountains – the high Hindu Kush, the treeless Pamirs, the once jade-rich Kun Lun, the peaks and glaciers of the Karakoram, the lake-spattered Trans-Himālaya ranges of Tibet, and of course the 2,500-kilometre-long swag of the Himalayas themselves together with their ruckled extremities. To all such features the map confidently awards a static prominence, and that too is deceptive. The region is one of the most seismically active in the world with an earthquake a week not uncommon in north-west Pakistan and several a day reported in Afghan Nuristan (formerly Kafiristan). Summer brings avalanches, rainfall triggers landslides, winter obliterates trails. Glaciers retreat and occasionally advance, rivers change course, roads revert to rock slides and whole lakes quietly evaporate.

Lakes can also be destructive, nowhere more so than in the vicinity of the upper Indus. The classic case was in 1841, a year of retribution if ever there was one. That January, not far from the source of the river in western Tibet, a few thousand Chinese troops, ably assisted by the severities of a Tibetan winter and an altitude of over 5,000 metres (about 15,000 feet), wiped out an invading army commanded by General Zorawar Singh on behalf of the Sikh kingdom of Lahore. Simultaneously, on the far side of the Khyber Pass the English East India Company's Army of the

Indus was being annihilated in the course of what the British called the First Afghan War. And all the while, though little remarked at the time, on the Indus itself the same winter was plotting the destruction of yet another expeditionary force.

This third catastophe would be backdated to December 1840 when a landslide sliced off a chunk of the Hattu Pir, a well-known declivity by which the Kashmir–Gilgit trail traversed a spur of Nanga Parbat. The detached mountainside ended up in one of the Indus gorges with the inevitable consequence – an almighty dam. Lodged in the gorge like a cork in a bottle, the compacted mass of rock, rubble and ice grew to a height of 150 metres and held the river in check for six months. In April, from near the blockage, the rajahs of Gilgit and Astor warned of impending disaster. They were ignored. So were reports of the river having slowed to a trickle downstream in the plains. By June the gorge had become a lake, reportedly 300 metres deep and 'nearly 40 miles long [so that it] reached almost to Gilgit town'.

No one seems to have witnessed the actual bursting of the dam; even the exact date is uncertain. But two trustworthy accounts agree in saying that it was on or about 1 June and that the lake emptied in a day, with 'the immense volume of water rushing down to Attock'.[3] Notable for its Mughal fort and its occasional bridge of boats, Attock, 400 kilometres downstream from the dam, commanded the river's debouchure on to the plains of Punjab. It was also where, on the stifling afternoon in question, a contingent of the Lahore army commanded by the Sikh general Sham Singh Atarewala happened to be encamped beside the dried-up bed of the river. By chance, at the crucial moment Sham Singh himself along with a small detachment of troops was up in the hills chasing a troublemaker. They were the lucky ones.

According to Ashraf Khan, a *zamindar* or landowner in nearby Tarbela, the first anyone knew of the advancing flood was an ominous mumbling sound.

> We began to exclaim 'What is this murmur? Is it the sound of
> cannon in the distance? Is it the bellowing of Gundgurh [a local
> monster]? Is it thunder?' Suddenly someone cried out: 'The river's
> come.' And I looked and perceived that all the dry channels were
> already filled and that the river was racing furiously in an absolute
> wall of mud … it was a horrible mess of foul water, carcases of

soldiers, peasants, war-steeds, camels, prostitutes, tents, mules, asses, trees and household furniture, in short every item of existence jumbled together in one flood of ruin ... About 500 [of Sham Singh's] troops were at once swept to destruction; just as a woman with a wet towel sweeps away a legion of ants, so the river blotted out the army of the raja.[4]

With the Indus running some 20 metres above its highest-ever recorded level, the floodwaters backed up to where the Kabul River joins it. From there, all the countryside 'to within ten miles [16 kilometres] of the city of Peshawar' disappeared beneath the flood along with 'four towns and some twenty villages'. The devastation was total, the fatalities uncountable. Dr Hugh Falconer, who was the first to submit a report on 'the recent cataclysm of the Indus', could recall nothing remotely comparable. It was surely 'one of the most remarkable natural catastrophes hitherto recorded as having occurred on the continent of India, or anywhere else, in the deluge way'.[5]

Falconer, superintendent of the government's botanical garden at Saharanpur (north of Delhi), was one of only a handful of Europeans who had some first-hand experience of the upper Indus. Four years previously he had visited Baltistan, a dirt-poor and sparsely populated reach of the river due north of Kashmir. It lay well upstream of the gorges and the fatal dam, but Dr Falconer had got to learn of the river's reputation and actually measured its summer discharge. Shuddering to imagine the consequences of the 1841 dam-burst, he assumed that its trail of destruction would reach all the way to the sea. He also supposed the dam was the work of a glacier advancing across the path of the river. He was wrong on both counts, although glacial blockages were indeed of frequent occurrence.

In 1833 a Karakoram glacier had nosed across the bed of the Shyok, an upstream tributary of the Indus, resulting in a similar lake and a similar cataclysm, albeit on a lesser scale. Baltistan was still recovering from that 1833 devastation when, seven years later, there came a repeat performance. Major Alexander Cunningham, who visited Baltistan at the head of a border commission in 1847, made careful enquiries about this second glacial blockage of the Shyok. It lasted six weeks and the dam had burst, he was told, on 1 June 1841. That was of course the day on which the great Indus dam at Hattu Pir was thought to have burst.

Cunningham's informants may have been confusing the two blockages; alternatively it could be that the rush of water from the dam-burst on the Shyok dislodged the much bigger dam downstream at Hattu Pir. Such a domino effect could not be ruled out. There would be ten more Indus floods over the next few decades, six being attributable to the sudden emptying of glacial lakes and four to the yielding of landslide dams.

Earthquakes, too, could interfere with the river's course. Kenneth Mason, a twentieth-century Himalayanist and surveyor, thought it was probably a local earth tremor that had sent that chunk of Nanga Parbat slithering into the Indus gorges in the first place.[6] Even today the devastation downstream may be traced through the gorges, though not in the plains. There, and rather ironically, the 250 square kilometres of prime Punjabi farmland flooded in 1841 have, as of 1976, been reflooded, this time by the reservoir of the Tarbela Dam, 'the largest earth-filled dam in the world'. What with the seasonal extremes and all this unexpected rearrangement of the terrain, even frequent travellers can struggle to identify their whereabouts in Himālaya. Trees have occasionally been reported on the treeless Pamirs; earthquakes erase whole villages in Nepal; unsuspected Shangri-las are swiftly overrun by peace-seeking hedonists; and Beijing's munificence in the matter of infrastructure – highways, railways, runways, pipelines, 'restricted areas' and ever more hydroelectric dams – challenges the comprehensive coverage promised even by Google Earth.

Maps play catch-up as best they can. What they can't show is how Himālaya, with or without human intervention, is as much geology as geography, as much ongoing process as finished product. The process in question is called orogeny or orogenesis, 'mountain-making', and nowhere on earth is it more in evidence than in Himālaya. Like the Indus River and the Karakoram glaciers, the hills and the high plateaux are themselves on the move. They've been jostling one another – upthrusting here, down-wearing everywhere – for 50 million years and look set to continue for another 50 million. This is our planet in action. The action ranks as the most dynamic geological phenomenon with which mankind's belated and as yet comparatively brief presence on earth has coincided. It's our privilege to be witnesses, albeit fleetingly.

And it is only recently – very recently in the great scheme of things – that scientists have gained some understanding of what's actually happening.

An unforeseen casualty of this research has been the classroom certitude that Himālaya's mountains are the world's tallest. They are not. The truism proves to be untrue. Though soaring to an unrivalled height above sea level, they're dwarfed by mountains that are largely below sea level. In the middle of the Pacific the volcanoes of Mauna Loa and Mauna Kea rise only 4,000 metres (13,000 feet) above the Hawaian shoreline. It could be more or less in the case of Mauna Loa because it's still active. But in both cases all that's visible is the apex of a much bigger mountain beneath the water. From where the waves break against the islands' tousled flanks it's another 13 kilometres down to where their nursery slopes meet the seabed, itself here much depressed by their weight. From tip to toe that makes 17 vertical kilometres (55,000 feet). Everest, at less than 9 kilometres (29,000 feet), is only just over half as high as these mermaid mounts in the mid-Pacific.

Such revelations came courtesy of the Cold War. In the 1950s, as a prelude to the clandestine deployment and tracking of one another's nuclear-armed submarines, the superpowers embarked on the mapping of the oceans' seabeds. Using a variety of probes – dredging, satellite sensing from space and sonic soundings by surface ships, submarines and underwater drones – the mapping continues; deep-sea drilling for hydrocarbons, rare metals and precious minerals can be expected to follow. The ocean floor promises to be as bountiful as any continental El Dorado and proves just as spectacularly endowed with natural wonders. Already more volcanoes, both active and extinct, have been discovered in the depths than are to be found on dry land. And though only a few of them rise to sea level, their positioning, often in a line or arc across the ocean floor, has afforded corroboration of the faulting and fracturing to which the earth's outermost crust is subject. It was the study of this crustal activity, now known as plate tectonics, that would substantially resolve the mystery of mountain-making, of orogeny.

Describing a similar arc, the map shows the east–west alignment of the Himalayan ranges bulging south, or outwards, as if sagging under the weight of the Tibetan plateau. Whether this could really be the case intrigued nineteenth-century pioneers, men like Joseph Dalton Hooker and Henry Haversham Godwin-Austen, and alerted them to other possible clues to the mountains' genesis. Hooker, the greatest of Himalayan botanists who would succeed his father as director of London's Kew Gardens, was a close friend and colleague of Charles Darwin. To Darwin, Hooker dedicated the *Himalayan Journals* of his three-year sojourn in India, and it is these journals that place the thirty-three-year-old botanist on the Donkhya Pass in Sikkim on 7 November 1849. From there Hooker sketched the view north-west over Tibet, taking care to frame the scene with some typically jagged geology of upended rock strata and teetering slabs.

Of no less interest was Hooker's discovery on the same 18,000-foot (5,000-metre) pass of an outcrop of fossil-bearing limestone. Limestone is formed from sediment consisting of the pulverised shells and skeletons of millions and millions of minute marine organisms. Mostly coccoliths, these organisms constitute a calcium-carbonate plankton and each is so small it's been estimated you need 400 million just to cover a £1 coin. When deposited in unfathomable quantities on the seabed and left to lie there for a very long time, the sediment agglomerates and is compacted into limestone rock. Hooker would have known of limestone as the material of which the white cliffs of Dover and the chalk downs of southern England are composed. It had no right to be cropping up on a 5,000-metre Himalayan pass – unless, that is, the pass had once somehow been submerged. Yet Hooker had no doubt that it was indeed limestone. He even found a small coin-shaped fossil embedded in it which he tentatively identified as a nummulite.

'If proved to be that genus [a nummulite],' noted H. H. Godwin-Austen after reading Hooker's account, '[it] would indicate the Eocene age of the formation, with all that would mean.'[7] At the time fossils provided the most reliable way of dating any rock strata in which they happened to occur. Not until the discovery of the properties of radioactive isotopes in the early twentieth century would fossil-dating be superseded by the much more accurate radiometric dating. Because the occurrence of single-cell nummulites had already been dated to the epoch known as the Eocene ('New Life'; it ran from 56 to 35 million years ago), Godwin-Austen's cryptic 'with all that would mean' meant that, even within the foreshortened geological time frame accepted in the mid-nineteenth century, the Himalayas were indeed young.

As a multi-purpose natural scientist Godwin-Austen was intrigued. He was best known as the first surveyor to have made a close reconnaissance of the mountain called K2. In his honour that shy Karakoram peak, second only to Everest in height, had been renamed Mount Godwin-Austen – and would soon be unnamed when the presumption of attaching a foreigner's moniker to someone else's mountain began to be questioned. But lest none of this provide sufficient grounds for fame, the only biography of Godwin-Austen bears the unlikely title of *The K2 Man (and his Molluscs)*.

Molluscs comprise a vast division of invertebrate life forms including squids, octopuses, snails and slugs as well as oysters, mussels and clams (the unusual use of brackets in the book's title looks like a typographical nod to such bivalves). Molluscs meant as much to Godwin-Austen as mountains. He collected and wrote about them at length, becoming the leading authority of his day. At least one was named in his honour and in this case the name would stick. But for the most part molluscs live in or near water, especially seawater; they often have shells and are not generally to be found up mountains or slithering across high-altitude deserts. Godwin-Austen was therefore even more excited than Hooker when, in the course of a foray into western Tibet in 1862, he stumbled upon both snail and clam shells on the barren shores of the azure Pangong Lake. They were so plentiful he could scoop them up by the handful. It confirmed his supposition that the lake, big and brackish as it was, must once have been much bigger and saltier and could even have been part of an inland sea.

Limestone and the fossils of both marine and freshwater molluscs have since been found from one end of Himālaya to the other. The women of landlocked Ladakh, which is about as far from the sea as anywhere on earth, sew cowrie shells and bits of vermilion coral alongside the silver rupees and turquoise pendants in the lappits of their elaborate headdresses. Endowed with magical powers, ammonites (or saligrams), another species of marine mollusc from the hills, find eager takers among India's miracle-workers and are particularly associated with the worship of Lord Vishnu. The Gandak River, a tributary of the Ganges, 'is known in Nepal as the Saligrami because it brings down such large quantities of *Saligrams*', reported M. S. Krishnan, superintendent of the Geological Survey of India in the 1950s.[8] When in the 1960s Himalayan mountaineers anticipated the first lunar landings by returning from their summit conquests with rock samples, analysis duly revealed the fossilised presence of yet more minuscule marine life forms.

The discoveries made by Hooker, Godwin-Austen and others on the fringes of Tibet in the mid-nineteenth century remain as eloquent as any in the history of geology. To the Pulitzer-winning American writer John McPhee they say it all. In the course of his blockbusting *Annals of the Former World*, a five-book geological road trip across the United States made in the 1980s and 1990s, McPhee offers the thought that, were he obliged to reduce his 700-page odyssey to a single sentence, 'this is the one I would choose: the summit of Mount Everest is marine limestone'.⁹

Mike Searle, Oxford's professor of earth sciences, would agree. He enlarges on Himālaya's fossil phenomena in his *Colliding Continents*:

> Rocks from the summit of Everest still preserve tiny crinoid ossicles, the fossilised stems of sea-lilies that lived in a shallow, warm, tropical sea 400 million years ago. Fossil remains of tiny primitive horses, hippos, and palm trees are preserved in the 5-kilometre-high barren desert plateau of Tibet, showing that major climate and altitude changes have occurred. Ancient human remains have been found in the Siwalik deposits along the southernmost ranges of the Himalaya. There is even a suggestion that the migration of Tibetan people south into the Sherpa lands of Nepal, Sikkim, and Bhutan may have preceded the final stage of the rise of the Himalaya during the Holocene [the post-Eocene epoch which runs from 11,000 years ago to the present].¹⁰

But how come? How did the palms and hippos find their way into Tibet? How did the limestone and the fossils get up there? How did the mountains get up there? Himālaya's back story proves long and rocky, a bit like John McPhee's stop–start progress from one exposed rockface to the next across the width of North America. It does lead, eventually, to the sun-drenched Pacific islands and the Himalayan heights, but not before braving the ice cap of the world's largest island and not until two centuries of analysis and guesswork had elapsed since its cloistered beginnings in the academic salons of eighteenth-century Saxony and Scotland.

Like Thomas Carlyle in the quotation at the head of this chapter, the easy-riding John McPhee, as he heads into the sunset along Interstate

80, assesses the competing claims to be the founding father of modern geology of two eminent savants: Abraham Gottlob Werner of Freiburg and James Hutton of Edinburgh. Writing in the 1830s, Carlyle seems to have reckoned the honours even; the two men had jointly convinced their eighteenth-century peers that the creation of the world owed little to divine agency, in fact was 'little more mysterious than the cooking of a dumpling'. Carlyle, as so often, was being satirical. Though Werner and Hutton were no more atheists than he, it was as well to anticipate the scriptural fundamentalists who might rubbish any presumption of a plausible creation that conflicted with the Book of Genesis.

McPhee, never without a weathered geologist in the passenger seat beside him, is better informed. In his *Annals* – which might well be subtitled 'Rocks and the Art of Reading Them' – he brings the reader up to speed on the modern standing of geology's pioneers. Neither Werner nor Hutton, it seems, was a geologist in today's sense. Werner was nearest; he called himself a geognost, a term cognate with geognosy, 'earth knowledge'. Werner's great achievement lay in describing and classifying the different formations of rock strata to be found in and around Freiburg. Ill health meant he never left Saxony, but his taxonomy of rocks and of their successions of strata was so universally admired that '[his] impact on the half century of geological endeavour beginning in 1774 was unquestionably greater than Hutton's'.[11] Werner also aired the idea that the earth might be a lot older than the few thousand years allowed by most of his contemporaries, or even than the 6,000 years supposedly allowed in Genesis. The laying down of all those rock formations and strata was the work of many millennia, he thought, or even longer, 'perhaps a million years'.

James Hutton never declared himself even a geognost. His doctorate was in medicine and his inheritance made him a gentleman farmer. But he was also an intuitive genius with a wide range of other interests, acute powers of observation and the great good fortune to be a native of Edinburgh, a city as cluttered with crags as with churches and at the time the scene of an extraordinary efflorescence of philosophical and scientific speculation known as the Scottish Enlightenment. Within a cobbled stroll of Hutton's doorstep, spectacular rock formations reared from the close-cropped sward; and for a receptive audience he had only to turn to illustrious neighbours like David Hume and Adam Smith (philosophy/political economy), William Cullen and

Joseph Black (chemistry) and John Playfair and James Hall (geology). Hutton's prose being decidedly indigestible, it was John Playfair's posthumous *Illustrations of the Huttonian Theory* that would establish Hutton's reputation; and it was James Hall's public experiments which often rendered that theory intelligible to a wider audience. Thus, to demonstrate how, if horizontal pressure were applied to flat strata of slightly viscous rock, the strata would bend, fold and otherwise contort, even being forced upwards as mountains, James Hall in 1812 regaled the Royal Society of Edinburgh with a do-it-yourself experiment.

> He piled several pieces of cloth … on a table, placed an unhinged
> wooden door and some weights over them … and then moved
> two vertical boards at the ends of the pile toward each other by
> hammering them with a mallet. In consequence the pile shortened,
> the weighted door was gradually raised, and the cloth strata developed
> folds very much resembling those of [the] Berwickshire [coast].[12]

As an explanation of how mountains might be formed, this was pretty crude. But it was consistent with the then prevalent idea that the earth was slowly cooling and, in the process, buckling and contracting. For this, the example given was that of a baked apple, the shrinkage of the earth's outer crust being likened to the wrinkling of the apple's skin and the earth's core being thought capable, given the evidence of earthquakes and volcanoes, of exerting the forces needed to mould and manipulate rock.

Werner's stratigraphic studies, instead of endorsing this contraction idea, favoured a dynamic role for the receding waters of a one-time world ocean, with mountain ranges being fashioned along the ocean's shores like sand-ridged tidelines. This placed him in the ranks of the so-called Neptunists (because they emphasised the earth-shaping effects of seas and tides), while Hutton, confident that sufficient heat to mould and move rock would be found within the earth's core, was more nearly aligned with the so-called Vulcanists. Of volanoes as such Hutton had little to say, and he couldn't of course prove the existence of his supposed 'subterranean fire'. But the presence of such a heat-source was supported by the higher temperatures encountered deep underground by coal miners, and it accorded well with Hutton's big idea – or, as per the title of his seminal work, with his *Theory of the Earth*.

Hutton's theory held that the earth should be understood as a self-sustaining and constantly self-renewing 'system', a bit like a living organism. As a farmer he was well acquainted with soil enrichment and as a doctor he had written on the circulation of the blood. Just so, he conceived of the whole earth as a life-supporting 'mass of matter', some of it originally organic – animal (such as molluscs) or vegetable (such as coal) – but most of it mineral (such as stone and rock). In the course of time, rock got eroded by the elements into grit and tilth. It was a slow process, barely perceptible within a single lifetime and made up of what he famously called 'little causes, long continued'. The same elements – wind and rain, frost and flood – then conveyed this soil to the fields; and by river, glacier, drain or deluge the soil was eventually conducted down to the sea. There currents and tides spread what was now seaborne sediment over the ocean bed. The pressure at great depth plus that subterranean heat source compacted the sediment into differing rock strata; and these, folded and contorted, were eventually uplifted to form the continents, islands, deserts and mountains that we know as terra firma.

This whole cycle – of erosion, deposition, distribution, submersion, consolidation, reassembly and elevation – could be regarded as a self-perpetuating continuum activated by observable forces of nature. Hutton likened the cycle to the revolutions of the planets. The consistency of planetary movements left no doubt in his mind 'that there is a system by which they are intended to continue those revolutions'. Likewise it was to be expected that there existed a system by which earthly matter was destined to be continually recycled. Today's earth and its mountains should therefore be seen as the product of a succession of previous earths and their mountains, and they in turn will form subsequent earths and their mountains. The Himalayas of our maps are not, then, the first Himalayas, nor are they the last. They're the product of a sequence of birth and rebirth – in fact of reincarnation, not unlike the karma-regulated life cycles to which Himālaya's Buddhist and Hindu inhabitants subscribe. Landslides and Indus floods are just part and proof of the 'system'. The destruction they wreak in the here and now is laying the foundations of worlds to come.

On the strength of this far-sighted and influential vision, most of today's geologists, including those consulted by McPhee, make James Hutton the undisputed 'founding father of modern geology'. Hutton himself claimed only to have formulated a plausible hypothesis; it was for others to substantiate or reject it. But his theory did in fact give geology a new and enduring direction. By the end of the nineteenth century Wernerians and Neptunists would be outdated oddities, while Huttonians and Vulcanists had become mainstream.

The debate took the best part of a century not because Hutton died before it got under way, nor because his prose was so uninviting, but because the implications of his theory were so disturbing. For in addition to finding a new direction (that is, his system of continuous renewal) he had found the time required for it – literally. Werner had suggested 'perhaps a million years' for the age of the earth. Hutton ventured nothing in the way of a 'perhaps' age but insisted that the natural laws governing his 'system' must have been uniform throughout the earth's existence – and throughout that of all the earths that preceded it and would remain so throughout all the earths that would follow it. In effect he was claiming an unlimited time frame. The age of the earth was so great it defied measurement. As he put it in the last line of his *Theory*'s much quoted final sentence, 'the result, therefore, of our present enquiry, is that we find no vestige of a beginning, no prospect of an end'.

Undaunted by the challenge of quantifying eternity, and quite unknown to Werner or Hutton, others had reached much the same conclusion at least 1,500 years earlier. Like Hutton's theory, India's ancient chronologies posit cycles of creation, destruction, repose and recreation. Composed of four *yuga* and known collectively as a *kalpa* (or *kappa* in Buddhist texts), each of these cycles lasts 8,649,000,000 years. When one is complete, another begins, and 'there is no end to this process, nor purpose other than the Lord's *lila*'.[13] *Lila* is a term meaning sport or spectacle, and the Lord in question is the supreme deity Brahma (sometimes in the guise of Vishnu). In effect the cosmic cycle goes on repeating itself for as long as it pleases their Lordships.

So in Indic cosmology both time and space are in ample supply. For the benefit of less numerate initiates, spatial ideas are often conveyed by reference to one of the many versions of the mandala, the cosmic diagram, while time is often measured in terms of movement. The shortest interval is a blink of the eye (about 1/15th of a second), and the longest is the *kalpa*, which India's *Mahabharata* epic describes as lasting for as long as it takes to flatten the Himalayas by relieving them of one grain of sand a day. The Buddhist *Saṃyutta Nikāya* pursues the same idea but with a bit more circumstantial detail:

> It is as if, O priest, there was a mountain consisting of a great rock, a league in height, without break, cleft, or hollow, and every hundred years a man were to come and rub it once with a silken garment; that mountain consisting of a great rock, O priest, would more quickly wear away and come to an end than a world-cycle [a *kalpa*].[14]

Here was a time span more generous even than that contemplated by Hutton. But it too operated through the action of 'little causes, long continued' – just like those observed by Hutton in the landscapes of Britain. The dripping of an icicle, say, or the trickle of a runnel of wind-blown sand might be more plausible than a once-a-century caress with a silken garment, but if continued for the best part of eternity, all were well capable of fashioning mountains and cutting mighty gorges.

Though long entertained in the East, in Europe the idea that the earth's beginning was infinitely more remote even than Werner had proposed threw wide the shutters on a new and breathtaking perspective. Not everyone was impressed. As the leading physicist of his day William Thomson, Lord Kelvin, adduced evidence for a slowing-down in the rotation of the earth that limited its potential age to 200 million years, a figure which he later revised down to a mere 25 million. Geologists and evolutionists were less cautious. As palaeontologists, palaeobotanists, palaeogeographers, palaeoclimatologists and so on they have been pushing back the palaeo-horizon ever since. Darwin in *On the Origin of Species* (1859) pushed too hard. He worked out an age of 306,662,400 years for the exposure of the sandstone and clay formations of the Weald near his home in Kent, and like Kelvin he was wrong. In current estimation the Weald is not half that old.

Darwin and Kelvin notwithstanding, the trend towards an older earth more in line with Indo-Buddhist cosmology has been sustained. Roughly speaking, the blinding light at the far end of our earth's long time tunnel has been pushed back nearly half a billion years in every decade of the past century. Hence estimates for the total age of the earth now stand at around 4,500,000,000 years (4.5 billion or just over half a full *kalpa*) and geologists have come to rival astrophysicists in their dab handling of unwieldy values. The same cannot be said for the rest of us. As McPhee puts it, 'numbers do not seem to work well with deep time. Any number above a couple of thousand years – fifty thousand, fifty million – will with nearly equal effect awe the imagination to the point of paralysis.'[15]

Because so little is known of the longest and deepest of all geological time's named eras, even homely analogies with the calendar or the kitchen clock may not be very helpful. Thus, taking the calendar year as a microcosmic yardstick for the earth's entire time span, the earliest and most enigmatic geological era 'runs from New Year's Day till well after Halloween'. Known as the 'pre-Cambrian' and extending from 4.5 billion years ago to around 0.5 billion, it leaves precious little time for anything else. As a result earth's best-known epiphanies and extinctions come thick and fast in an end-of-the-year last gasp. 'Dinosaurs appear in the middle of December and are gone the day after Christmas,' says McPhee. 'The last ice sheet melts on December 31st at one minute before midnight, and the Roman Empire lasts five seconds.'[16] It is as if life on earth, and especially that of *Homo sapiens*, is but a doodled afterthought or a decorative flourish as peripheral to the earth's story as the roots of grasses and trees are to the great mass of our planet's vast and inhospitable core.

With time in superabundance and no longer acting as a constraint on speculation, the late nineteenth and early twentieth centuries witnessed a spate of important new theories about the earth and its constituent elements, especially mountains. In mainland Europe geologists had long been looking to the Alps for the challenges and insights afforded Hutton by the Scottish hills. In particular, Swiss and French alpinists studied glaciation and found, in the imperceptible advance of Alpine glaciers, an example of how substances that seemed as solid and static as rock could in fact be in motion. Hutton's ideas about sediment fusing, or granite flowing, could be right. Here, too, lay an explanation for the puzzling presence of free-standing rocks and bus-sized boulders in pastures several days' march from, and sometimes above, the scarps from which they had clearly been detached. Evidently glaciers had once been very much larger and thicker and had distributed their embedded rocks very much further. Their passage was attested by the scarring of the surfaces over which they had passed and by the stony embankments that had once been their moraines. At times in the not so distant past, glacial ice caps like that of today's Antarctica had been so extensive as to cover the entire world. Climatologists called these white-out eras 'ice ages'. There had been several and they had changed the face of the earth.

The Face of the Earth, or *Das Antlitz der Erde* in the original German, was the title chosen by Eduard Suess for four colossal volumes published between 1883 and 1904. A prodigious polymath who, though born in London, spent his long working life in Vienna, Suess studied the Alps specifically for clues to how mountains were formed. He also read up extensively about the findings of other geologists all over the world. With the water take-up by ice-age glaciation being responsible for lowering sea levels – and the melting of glacial ice for rising sea levels – Suess, like Werner, saw the oceans as a powerful formative force. He explained the earth's crust beneath the ocean floor as having sunk because it was denser and heavier than that of the lighter and more buoyant dry-land masses. And noting how mountain ranges often mirrored the sweep of the nearest coastline (for example, the Rockies/Andes running parallel to the Americas' Pacific coast) he explored the idea of a compensatory relationship between the two. Thus the east–west depression that is the Mediterranean basin is seen as being matched by the east–west mountain continuum of the Pyrenees, Alps and Balkans to the north of it.

This led him to postulate something similar for Himālaya. There was of course no Mediterranean Sea lapping at the Himalayan foothills, but Suess, familiar with the reports of travellers like Godwin-Austen and Hooker, agreed there must once have been such a sea. He called it Tethys after a water goddess in Greek mythology, and he gave it generous proportions. It had covered, he determined, not only most of what is now the Indian subcontinent but also much of what is now the Indian Ocean, Australia, South America, the Middle East and Africa. It was this Tethys Sea or Ocean, moreover, that separated the two supercontinents into which, 250 million years ago, the world's dry land had apparently been divided: on the one hand, an elongated Eurasia stretching unbroken from North America to Mongolia and Tibet that he called Laurasia, and on the other a southern land mass comprising all those lands whose later global positioning the waters of Tethys were pre-empting and which he called Gondwanaland.

Tethys and Gondwanaland, a superocean and a supercontinent – these were Suess's twin legacies and both have contributed mightily to the understanding of orogeny. They still feature prominently in histories of the earth, although somewhat modified by later research. In the 1980s a good case was made for an earlier and still bigger Tethys. Evidence for this 'Palaeo-Tethys' has been found in Himālaya's northernmost ranges, the Kun Lun and Pamirs, and has greatly complicated their geological interpretation. On the other hand Godwin-Austen's Pangong Lake, like most other Tibetan lakes, is now thought not to be a Tethyan relic. It resulted from a glacial dam, like those that continue to block the Indus's tributaries but which in this case acquired permanence when precipitation diminished and sand encroached. Genuine relics of Tethys are larger and include the Mediterranean, Black and Caspian seas.

Gondwanaland, or sometimes just Gondwana, would prove to be an even more revolutionary idea than Tethys. As nineteenth-century geologists and naturalists dispersed around the globe in the wake of Europe's empire-builders, striking similarities in the rock strata, the fossils and even the living flora and fauna found in very different parts of the world had been noted. For instance tillite rock formed from the sediment left by glaciers was identified on the baking escarpments of central India. This was odd in itself, but odder still were reports that more tillites of identical composition were to be found in the semi-desert Karoo of South Africa. Similarly lemurs were thought unique to Madagascar, yet a species of lemur was reported in India. And likewise several of the early reptiles that had scuttled across terrains thousands of kilometres apart appeared to be of the same species.

Suess read up on the occurrence of all these different life forms and rock formations. The examples were too similar and numerous to be explained by coincidence and yet too distantly distributed to support suggestions of natural transference (in the case of life forms) or of one-time contiguity (in the case of rock sequences). In particular Suess concentrated on the fossils of an extinct fern-like tree called *Glossopteris* which had been discovered in abundance in India and other parts of Asia as well as in Africa, South America and Australia. For such *Glossopteris*-bearing rock formations he borrowed the term Gondwana as adopted by geologists in India, the Gonds being one of India's indigenous peoples and Gondwana the name by which history knew their homeland in Madhya Pradesh.

In the first years of the twentieth century Suess's Gondwanaland supercontinent was awarded yet another continent. Polar explorers like Captain Robert Falcon Scott extracted rock samples in Antarctica that contained more examples of fossilised *Glossopteris*. A Gondwanaland that had included Antarctica as well as India had obviously experienced major climate change as well as piecemeal relocation. But Suess was at a loss to explain either convincingly. A 'contractionist', he still subscribed to the 'baked-apple theory' of the earth – as a cooling, shrinking core within a crumpling outer crust. Faults could open in the crust, mountains could be pushed up, ocean floors be pulled down and sea levels advanced or retracted, but in the contractionist's reading all such movement was essentially vertical. Any horizontal, or lateral, movement of the earth's crust would be incidental and quite incapable of despatching entire continents to long-haul destinations in another hemisphere.

Nor was it likely, as had been proposed, that these errant land masses had once been connected by chains of islands, or 'land bridges', which might have facilitated the transference of species. A principle known as isostasy (whereby equilibrium was maintained between buoyant continents and the deeper, denser oceanic floor) ruled it out. 'With isostasy it would have been physically impossible for portions of the Earth's crust to have risen and fallen so dramatically.' Thus the idea of crustal plates oscillating up and down 'became difficult to sustain'. Likewise the idea of a steadily cooling earth was becoming unsustainable at a time when radioactive heat was found to be convecting within the earth's core. 'Altogether the homely analogy of an apple with a shrivelling skin now seemed misleading.'[17]

It was time, high time, to think again. The twentieth century was shrinking the world without reference to either its contracting core or baked apples. From China and then Russia, revolution was about to trouble Himālaya's borderlands. From British India the early years of the new century had brought Younghusband's military invasion of Tibet, much far-from-innocent exploration of the trans-Himalayan lands and the first shaky advances in Himalayan mountaineering. Aeroplanes were taking to the empty Himalayan skies, horseless vehicles to the switchback mule trails. And still the mountains awaited a satisfactory explanation of how they had got there in the first place.

2

War of the Plates

Whence this creation has arisen; whether he founded it or did not; he who in the highest heaven is its surveyor, he only knows, or else he knows not.

Rig Veda[1]

Probably no symbolic diagram is as richly freighted with ritual and mystic associations as the Indo-Buddhist mandala. In Hindu tradition the mandala's geometry serves as a stylised map of the whole universe, its concentric rings have suggested a template for inter-state relations and its multiple compartments are perfect for accommodating a pantheon of deities. All creation muscles in on the Hindu mandala; seas are awash with dairy products; an ocean of salt water necessitates one of sweet water (or 'treacle' according to Macaulay); indeed the whole construction is hostage to domestic imagery – the cosmic egg, the unfolding lotus blossom, the succulent mango, the layered coverings of the coconut. The mandala of Buddhism, on the other hand, is less appetising but more esoteric. Its creation is itself a ritual and its contemplation an aid to attaining enlightenment. As the visual equivalent of the *Om mani padme hum* mantra, its circle-within-a-square-within-a-circle is inescapable throughout Himālaya, from the remotest Ladakhi chorten (stupa) to Lhasa's Potala palace.

Yet, despite the differences, the basic idea of the mandala is the same for both Hindus and Buddhists – the expression of hierarchy through spatial relationships and symmetries. Typically both versions have at their centre the mythical supermountain known as Meru (or Sumeru). Mount

Meru is the axis of the universe; and by way of a reminder of its colossal three-dimensional reality, the mandala's concentric circles may be read as the mountain's contour lines. Like Mount Olympus, Meru's upper slopes are for the nearly emancipated and the gods. There is in fact a whole stack of heavens, each offering the comforts appropriate to the stage of enlightenment achieved by its inmates, and all temporary because rebirth awaits even the most favoured. One example, and perhaps the most enticing, is the curiously unBuddhist Heaven of Transferring Pleasures, so called because here 'subjective desires are at once transformed into objective pleasures, thoughts as well as wishes being creative forces'.[2]

On the mountain's lower slopes are located the Himalayas; they constitute Meru's foothills. Below (or around) them the four continents are disposed in the four quarters. In Hindu cosmology the southern continent is the one known as Jambudwipa, the 'land of the rose-apple tree', which the more prosaic among us know as earth. Lower still, in the subterranean depths of the great mountain, come the nether worlds. Many are conventional hells, each meting out the punishments due for particular transgressions. In one, for instance, Buddhists who have slaughtered a yak get slaughtered by a yak. In another, the condemned rend one another apart with metal claws and are then destroyed by fire 'but revived by a cool wind in order to undergo further torture'; in yet another, they're tossed in a lake of flames 'and pierced with iron spikes when they attempt to escape'. These are just the hot hells. There are also eight cold hells offering various forms of chilblain, frostbite and brittle dismemberment. Three such cold hells go by the telltale names of Atata, Hahara and Huhuva, these being the shivering utterances of which their cold-convulsed inmates are alone capable.[3]

For South Asia's sun-blessed peoples frigidity holds as many horrors as fire; for Europe's winter-accustomed peoples the gelid is less forbidding. It was a better understanding of Alpine glaciers and their earth-shaping propensities that led Eduard Suess to speculate on how continents might have been formed, oceans distributed and mountains uplifted; and though Suess was sometimes wrong, it was seldom for lack of evidence.

Just so, the behaviour of ice was again to prove instrumental in nudging men of science away from the static geography of the maps and into the dynamic wonderworld of drifting continents and plate tectonics. But it was not the ice of Himālaya or Antarctica that was now responsible; rather was it the freezing void indicated by the third great white splodge on the physical map of the world. Less an ice cap than an ice carapace, this was the glacial shield, up to 3 kilometres deep, beneath which lay the subcontinent-sized island misleadingly known as Greenland.

Alfred Wegener, the German balloonist, explorer and physicist who championed the idea of continental drift

Again, too, it was neither a Himalayanist nor a dedicated geologist who made the breakthrough. Alfred Lothar Wegener, the fifth child of the evangelically inclined supervisor of a Brandenburg orphanage, would be attracted to many disciplines, geology not being foremost among them. As a scientist Wegener would be best known for his work in meteorology, astronomy and astrophysics. Initially the structure of the earth interested him less than did the ether above it, and it was while studying the layering of the atmosphere that he first tasted fame. In 1906 he and his brother set a world distance record for 'joint balloon flight'. Their aerial marathon lasted over fifty-two hours and took them from Berlin north to Jutland and out over the Baltic before depositing them, frozen stiff and barely able to stand, in a field near Frankfurt. 'It was a major event in the early history of aeronautics,' says Wegener's biographer, 'and ... gave him a distinct achievement as an explorer before he had even gone anywhere.'[4]

The travel deficit was quickly redressed. Later in 1906 Wegener left for the Arctic as scientific supernumerary on a Danish expedition to the north-eastern tip of Greenland. The expedition lasted two and a half years and included grisly overwinterings in a one-room shed amid the ice floes on the island's east coast. In retrospect it served Wegener as an induction into the connoisseurship of cold hells. A quiet and imperturbable twenty-seven-year-old at the time, he fended off the depression induced by the endless storms, the twenty-four-hour darkness and the minus 40 degree temperatures by single-mindedly dedicating himself to his mostly meteorological tasks. Tobacco and coffee sustained him, his strength of purpose was matched by an iron physique, and if not entirely humourless, neither was he inclined to levity.

This was the first of Wegener's three great Greenland expeditions and, though mainly spent on the island's coastal fringe, was much the longest. It ought also to have served as a warning. Twice he himself nearly drowned. The leader of the expedition along with two of his fellow Danes did perish – in an ill-judged bid to contest the findings of their Arctic rival, the American Robert Peary. As a polar explorer Wegener is sometimes ranked alongside Peary, sometimes with Amundsen and Scott. Were it not that, at the time, Greenland afforded better employment opportunities than Antarctica, and were it not for the international community's cold-shouldering of German scientists during and after the Great War, the acknowledged pioneer of 'continental drift' might have been reckoned the twentieth century's greatest innovator-explorer.

It was after the first Greenland expedition but before the second in 1912–13 that Wegener engaged with Suess's idea of continents having shifted. 'The first notion of the displacement of continents came to me in 1910 when, on studying the map of the world, I was impressed by the congruency of both sides of the Atlantic coasts, but I discarded it at the time because I did not consider it probable.'[5] A year later he returned to it. He now likened the 'congruency' (or, in plainer English, the 'neat fit') between the western seaboard of Africa and the eastern seaboard of South America to that between two bits of roughly torn newspaper; if the torn edges fitted and if the lines of print ran on from one to the other, they must belong to the same sheet and have once been conjoined. Similarly, two distant land masses whose rock strata and fossilised life forms matched as neatly as the newspaper's lines of print must likewise have been conjoined and subsequently sundered. It was all so blindingly obvious that Wegener could hardly claim the idea as his own. Others, from Abraham Ortelius and Francis Bacon in the seventeenth century to the still living Eduard Suess, had also pored over the maps and been impressed by that 'congruency'. The American Frank Bursley Taylor actually went into print on the subject in 1910, the year in which Wegener conceived his own 'notion'. Wegener claimed he knew nothing of Taylor's work at the time – and probably didn't.

More certainly, the model and the metaphor for continental displacement were indeed Wegener's own and were derived from his experiences in Greenland. The drifting apart of Africa and South America he would liken to the movement of 'bits of a cracked ice floe in water'.[6] And once adrift, such continental land masses achieved an equilibrium which he thought 'similar to a floating iceberg in water'.[7] He returned again and again to the iceberg analogy.

On a sledging excursion to the northern tip of Greenland in 1907 he had skirted the ice cap that covers the whole interior of the island. (On his second expedition he would cross it, and on his third his mortal remains would be consigned to it.) About 1,000 kilometres wide and rising to a height above sea level of over 2,000 metres, the Greenland ice cap is second only to that of Antarctica. But what intrigued Wegener was the way in which the snouts of its peripheral glaciers plunged from the ice cap straight into the ocean. Travelling by sled along the icy margin between these glacial juggernauts and the heavings of the frozen sea, he witnessed a clash of the elements that called to mind the dawn of creation. 'The ice begins to groan and ache. The tide is coming in. The pulse of the sea still penetrates this icy armour [the sea ice].'[8]

One glacier was over 20 kilometres wide and trundling seawards at the rate of 30 metres a day. Its progress was actually visible to the naked eye. Another he observed descending to the shoreline, bulldozing its way beneath the sea ice and there 'calving'. This was the accepted terminology for when a section of glacier broke off to become a free-floating iceberg. Like a gigantic torpedo, it then 'burst up through the newly formed [sea] ice, sending radial cracks in every direction with thunderous reports'.[9] Wegener was transfixed. Might this be how the detachment of continents came about? True to the cliché, icebergs revealed only their tips; mountains, too, were believed to be dwarfed below ground by even bigger root bowls of even denser rock; and no doubt something similar was true of continents. Might the buoyancy of icebergs in water provide a clue to how, given a 'sea' of viscous rock melt beneath the earth's crust, land masses and mountains could be rendered mobile? At the very least it was all highly suggestive. What Wegener had originally called continental 'displacement' was soon being generally referred to as continental 'drift'.

But such analogies, like his claim to precedence, could easily be contested. What was different about Wegener's promotion of the displacement hypothesis was that he was so convinced of it that he made it his own and boldly proclaimed its relevance to the widest possible spectrum of scientific opinion. 'This book is addressed equally to geodesists, geophysicists, geologists, palaeontologists, zoogeographers, phytogeographers [specialists in plant distribution apparently] and palaeoclimatologists,' he announced in the foreword to the fourth edition of his *The Origin of Continents and Oceans*.[10] The list mirrored the spread of evidence – geological, geophysical, climatic, botanical and zoological – which he marshalled in support of his contention. Like Suess, he was a voracious reader of scholarly papers and an adept in deploying other people's findings in support of his own.

Drawing together all the available evidence, he made the case for the African and South American continents having been so prised apart, or displaced, that the basin left between them had been filled by the nascent Atlantic Ocean. Over the long course of geological time, then, land masses did move horizontally, and oceans too came and went. But his suggestion that this had something to do with the *Polflucht* ('flight from the poles') triggered by the centrifugal force of the spinning earth was unconvincing. He couldn't really explain how continents got displaced and nor could anyone else; the mechanism, though often presumed to depend on the energy released by radioactivity, is still not fully understood. Wegener trusted his theory was sufficiently persuasive regardless of why and how displacement occurred. Yet, without such an explanation, the theory was clearly incomplete and far from conclusive. It proved robust enough only to provoke a half-century of impassioned debate which would comfortably outlast Wegener himself, rumble on through the Second World War and ripple on into the more science-savvy 1960s and 1970s.

Wegener had survived the First World War thanks to the non-combatant status conferred by a spell in the trenches that left him wounded in the neck and arm. A family man by then, he had eventually found a congenial post-war position as professor of meteorology and geophysics at the University of Graz in Austria. Sadly neither the war wounds, the professorial chair nor the devotion of loved ones deterred him from leading his third great expedition. An all-German affair this time, it reached Greenland in 1929 with 98 tonnes of equipment including propeller-driven sleds. The main objective was to conduct observations throughout the long Arctic winter of 1930–1 from an igloo-like encampment called *Eismitte* ('Mid-Ice'), because it was established halfway across the great ice cap.

With the thermometer falling to minus 61 degrees Celsius, on 1 November 1930, his fiftieth birthday as well as All Hallows, Wegener and a companion set off from this *Eismitte* to return to the coast. They left by dog sled (the motorised vehicles had already departed), and when the dogs could go no further, they took to skis. The distance was about 500 kilometres, the conditions a pitch-dark white-out and the timing a month later than Wegener himself had declared the last safe date for ice-cap travel. It was one cold hell too many. Neither Wegener nor his companion was ever seen alive again. With no radio operator at *Eismitte*, it was six months before anyone even realised they were missing.

What became of Wegener's companion, a Greenlander called Rasmus Villumsen, remains a mystery. His trail simply expired on the ice cap; but he had evidently outlived his leader, for Wegener's body would be found tied in a sleeping bag and reindeer skins, carefully buried beneath the snow, and the spot marked by his crossed skis. Medical opinion would suppose death the result of a heart attack, the consequence of excessive strain from physical exertion and the extreme conditions.

The world's press heaved a belated sigh over yet another intrepid pioneer lost to the frozen wastes; Germany mourned its greatest polar explorer; and if none of the paeans said much about his scientific breakthrough, Else, his widow, ensured that continental displacement was not entirely forgotten. Declining offers to bring her husband's body home, she insisted it be left where it was, deep-frozen into the Greenland ice cap, thence in the fullness of time to be borne by glacial cortège down to the coast for committal to the deep from a calved iceberg adrift in the pack ice.

In sparing Wegener the scholarly ridicule that followed, death may have done him a favour. Continental drift was not going down well at the time. 'In life and in death, he was a target of scorn,' says John McPhee, readying a veritable thesaurus to lob in Wegener's direction. 'His idea provoked gibes, jeers, sneers, derision, raillery, burlesque, mockery, irony, satire, and sarcasm, but it could not be ignored.'[II]

One problem was that Wegener had kept changing his ground. Conceived after the first Greenland expedition, published in 1915 after the second, and much expanded and amended in the three editions that appeared between the second and last Greenland expeditions, *The Origin of Continents and Oceans* was proving as elusive a target as the erstwhile balloonist responsible for it. Bits of the original argument had simply been jettisoned, while new evidential ballast had been freely incorporated in response to objections. Wegener had insisted that, in the case of the Greenland subcontinent, westward displacement was still ongoing and could actually be measured by astronomical observations. But this claim had to be abandoned when the instruments, procedures and calculations used were all found to be at fault. Instead of Greenland heading away from Scandinavia at Wegener's cracking rate of 14–28 metres per year, or even the 11 metres of another calculation, it was not going anywhere far or fast enough to be convincingly measured using the data and techniques available. Yet it was what he called 'this mathematical proof', more even than all the geological and biological data, that had convinced Wegener of continental drift in the first place.

Less disastrously, the 'congruency of continents' was early changed to the congruency of their continental shelves; being submerged, these were harder to argue about. The continental shelves in turn would eventually be supplanted by tectonic plates; and these, though they underlaid continents, shelves and even the mid-ocean floor, would seldom prove to be coterminous with any of them. There was a lot more to the displacement of continents than Wegener had supposed.

As the motherland for these drifting continents/shelves, Wegener had adopted Suess's supercontinent of Gondwana. But with an important addendum: in the 1922 edition of his book, he included a

Gondwana prequel. Called Pangea (Pangaea or Pangaa, 'All Earth'), this colossus of a land mass had comprised all our planet's terra firma until some 250 million years ago, at which point the encroaching Tethys had divided it into the supercontinents of Gondwana and Laurasia. Mysteriously, though, mention of Pangea disappeared from later editions of *The Origin of Continents and Oceans* as completely as it had from his maps. 'Wegener just left it out,' says a latter-day critic.[12] None of which proved or disproved its existence. A better understanding of the ocean floor would eventually bear out Wegener's Pangean proposition, while the subsequent proliferation of other fancifully named palaeo-supercontinents – Lemuria, Rodinia and so on – may be taken as an endorsement.

It is as if geology were intent on trumping its own revelations. Just as James Hutton's self-sustaining world presumed a cycle of former and future worlds, each with its own Himālaya, so Wegener's Pangea is now credited with 'all-earth' predecessors and the prospect of 'all-earth' progeny. When supercontinents split, their component parts go their separate ways. But, like formation dancers moving to the laggardly music of time, they disperse, advance, retire and rotate, only to reunite eventually for another Pangea, and then start all over again.

The break-up of Wegener's Pangea looks to have left its mark on Himālaya. Oxford's Mike Searle lists three major plate collisions prior to the current one (the India–Asia crash of the last 50 million years). The earliest of these former collisions appears roughly contemporaneous with Wegener's Pangea.

> They are known as the Kun Lun, Qiangtang [Changthang], and Lhasa terranes [a terrane being a detached section of a crustal plate that collides with and adheres to another plate], and from stratigraphic and palaeontological ages we know they collided with and accreted to Asia approximately 250 million years ago, 200 million years ago and 160 million years ago respectively. The final collision, that of India with Asia … occurred around 50 million years ago [and probably won't be 'final'].[13]

Nor is all this cyclical choreography the exclusive prerogative of science. For consistency and duration, the geophysicists' Pangean cycle is rivalled by that offered in India's Vedic Puranas. There too, after every 306,720,000 years (that being the length of a single day in the life of Lord Brahma), 'a dissolution of the universe occurs'. In this case all that has existed is consumed by fire, leaving just a vast ocean on which the weary Lord Brahma reposes. Ensconced on his lotus bed, His Lordship sleeps through the long Brahma night (another 306 million years), 'at the close of which he creates anew', says the *Vishnu Purana*. The earth is reclaimed from the depths and set afloat 'like a mighty vessel'; the ground is levelled and apportioned; and 'He who never wills in vain, created, by his irresistible power, those mountains again upon the earth which had been consumed at the destruction of the [previous] world.'[14] So up pop the Himalayas once more, just as Hutton had surmised, and it's business as usual for another 306,720,000 years.

Much of the mockery and bemusement that greeted Wegener's *The Origin of Continents and Oceans* was prompted by the book's graphics. Typical was a vertical sequence of three maps, each featuring an elliptical globe, that appeared in the 1922 edition to illustrate the progression (i) from Pangea to Gondwana, (ii) from Gondwana to the separation of Africa and South America, and (iii) from the African separation to the detachment of Antarctica, Australia and India. To the modern eye Wegener's grainy images of a womblike world look like an expectant mother's ultrasounds, with the foetal Pangea extending its limbs a bit further in each frame. Wegener had rejected the usual Mercator map projection found in the atlases because it segmented the southern polar region where the continents had once fitted so neatly together. But his preference for the less familiar 'equal area' projections merely confused readers and emboldened critics.

The problem, of course, was that spherical surfaces, however projected, don't lend themselves to faithful reproduction on the flat. Moreover, because most of us are accustomed to the two-dimensional projections, we find it hard to appreciate a theory based on lateral movement around a three-dimensional sphere. Wegener urged sceptics to spend more time studying a classroom globe. His father-in-law, who was also his one-time mentor and colleague, was said always to carry about his person a small pocket-size globe, possibly on a key-ring, for handy reference whenever Alfred raised the subject of continental drift.

If the maps were confusing, the text of Wegener's book strove to make things clearer. As it appeared in the 1929 edition, his basic contention was simple enough. It holds good even in terms of today's plate tectonics:

> Antarctica, Australia and India, up to the beginning of the Jurassic [so 145 million years ago], lay alongside southern Africa and formed, together with it and South America, a single large continent partly covered by shallow water. This split up into separate blocks … and the sub-blocks drifted away in different directions …

Wegener's text then makes one of its rare and no less prescient references to Himālaya. Coming from someone who never set foot in Asia, it's a bold application of his hypothesis and at first sight not an entirely convincing one:

> In the case of India the process [of splitting into blocks and drifting off into the blue yonder] was somewhat different. Originally [India] was joined to Asia by a long strip of land, mostly under shallow water. After the separation of India from Australia, on the one hand, and from Madagascar on the other, this long junction zone became increasingly folded by the continuing approach of present-day India to Asia; it is now the largest folded range on earth, i.e., the Himalaya and the many other folded chains of upland Asia.[15]

Here Wegener muddied the waters of Tethys by making them shallow enough for a long and barely submerged 'strip of land'. His objective may have been to appease those of his critics who still swore by the idea of intercontinental 'land bridges'. Yet subsequent findings would authenticate his supposition: the floor of Tethys had been traversed by the ridges and island arcs of ancient crustal faults.

It was these sea-floor features that first felt the impact of our planet's most dramatic ongoing event. Detached from Africa, and then from Madagascar and Australia, the land mass that would become the Indian subcontinent ground its way north, runkling the seabed and obliterating obstacles. The speed of this advance – possibly 16 centimetres a year – is testimony to its inexorable force. When, 50 million years ago and somewhere in the vicinity of the equator, it first came into contact with the almost static Asian (or Laurasian) plate, its leading edge was not just a chunk of Africa or Gondwanaland but crustal souvenirs acquired along the way. Naturally these included marine scoopings from the Tethys sea floor and naturally these scoopings included seams of the usually white sedimentary rock formed from vast accumulations of coccoliths that is limestone.

So Himālaya is evidence of continental drift, and a continental collision explains why Himālaya is where it is. Neither, though, explains how Himālaya got to be quite so high, nor how the limestone found its way to the top of Mount Everest. For that the world would have to wait while Wegener's theory was ridiculed, discredited and then grudgingly accommodated among the more sophisticated considerations that govern plate tectonics.

For the latest in the way of graphics the internet is the place. There the nearest thing to Wegener's sequence of 'ultrasounds' is a succession of brightly coloured diagrammatic maps posted in 1999 by a Swiss Himalayanist called Pierre Dèzes. These, too, show India's northward progress from east Africa to southern Tibet, although as of mid-2018 the images in question appear to have been taken down and replaced with a single monochrome rendering. Happily the thesis they accompanied is still there. 'Tectonic and Metamorphic Evolution of the Central Himalayan Domain in South-East Zanskar' was never intended for the general reader. Dèzes was addressing fellow geologists at the University of Lausanne and did so convincingly enough to be awarded a doctorate. But anyone with experience of the mountain citadel that is Ladakh's southernmost province of Zanskar will be tempted to persevere with his text, and especially so if they happen to have noticed the prominence awarded 'Zanskar' in Dr Dèzes's maps.

The maps, which appear to have been derived from previous studies of the earth's palaeomagnetism, show a recognisably triangular India taking leave of Madagascar and, twisting slightly, inching north on its 6,000-kilometre journey in 25 million years. Only one location on the future subcontinent is marked. Sri Lanka is there but not identified and likewise Bengal; worth mentioning too might have been the ancient rock escarpments known as the Deccan Traps with their telltale tillites betokening a glacial past. But no, the only place named on the maps posted by Dèzes is a rather random spot of no apparent importance on India's leading edge that is boldly identified as 'Zanskar'.[16]

Zangs being the Tibetan word for 'copper', 'Zan(g)s-skar' is thought to mean the 'land of copper'. Dr Dèzes mentions this without comment, but it may have prompted an unexpected contention: that, before the break-up of Gondwana, this 'land of copper' was subject to 'a major tecto-magmatic event'. Dated to about 500 million years ago and known as 'the late Pan African Event', it saw a marked intrusion of granite into a wide swathe of Gondwanaland stretching from what are now Africa and Arabia to Australia and Antarctica. The nature of the event, says Dèzes, 'is still poorly understood'. It may or may not have been purely tectonic. But 'evidence for it is frequently observed within Himālaya' and was duly collected by Dèzes and his team in Zanskar.[17] Zanskar can thus be considered as having been a geologically recognisable entity before India parted company with Gondwana. Indeed it can be said to retain geological traces of its pre-Asian provenance to this day. Cradled in its bosom, Himālaya still hugs a sliver of what was once Africa.

However all this may be – and Dèzes is suitably cautious – even non-geologists find much that is challenging in Zanskar's orography. In a maze of twisting valleys and dark gorges the granite of the Great Himālaya range here brushes up against the limestone of the Zanskar range. The Zanskar range parallels the Indus on its passage through Ladakh, with the even higher Ladakh range, an extension of the Karakorams, visible beyond it. Or so the maps say. But on the ground such is Zanskar's chaos of lateral spurs and random ridges, such their gradients and their bare-rock savagery, that the logic of ranges and rivers is often hard to discern.

No one confronted by these contorted scarps could possibly question the instability of Himālaya or the one-time plasticity of its rocks. In colours as bright as confectionery, a seam of ochre traverses a cliff face of burnt umber; close up, the white and yellow veins of a splintered boulder tie themselves in knots like ribbons; and from the heights, waterless rivulets of slate and shale trickle down to a spreading fan of purple gravel. The vertiginous is here the norm, the horizontal the exception. Buckling, backtracking and always aslant, the geological strata dip, swell and even somersault across every exposed surface. The thin air positively zings with lithospheric tension. The gorges are being squeezed, the mountains twisted. Pocked cliffs promise 'caverns measureless to man'. Higher still, plastered into a cleft, a glint of white betrays a cliff-face retreat. The black substance lolling from a ledge like liquorice is also rock. And rock-cut too is the ziggurat that teeters in a sunless grotto. Its jumble of stone beams and upended slabs could once have been an acropolis.

The trail glints with mica, boots scrunch on quartz. A prickle of vegetation is sufficient excuse for a halt. Were it not for the profound silence and the absence of drag lines and juggernauts, Zanskar could be one giant quarry, an open-cast mine churning out novelty aggregates. And in a sense it is. Whether copper-rich, African or neither, it's certainly being worked. Mountains are being uplifted, rocks rendered and their product removed. Zanskar is where continents collide and tectonic muscles flex.

'With regard to the lithosphere [the earth's outer crust], it's a good place to sit and watch the plates move. It's a moment in geography that does your thinking for you.'[18] Or so writes John McPhee, though not of Zanskar. Anticipating the end of his transcontinental road trip, McPhee has flown into San Francisco. On the agenda are briefings with Princeton geologist Kenneth Deffeyes and an encounter with the San Andreas fault. Deffeyes has no trouble with a bit of Africa ending up in Himālaya. The American West Coast, he says, is itself the result of at least three such unexpected intrusions. Sutured on to the American continent like mudpacks, it is these immigrant land masses that have come together to form what we call California.

> In each instance a great island had closed up a sea and hit into
> America – just as India hit Tibet … Fossils from the mid-Pacific
> have been found here in the [American] West, and [so have]
> limestones that lithified a thousand miles south of the equator.
> Formations in California have alien fossils with cousins in the rock
> of New Guinea.[19]

Running roughly south-east to north-west, the several strands of seismic activity that constitute the San Andreas fault follow the boundary between two of these foreign accretions. The fault is what's called transform, or strike slip, because one side (the Pacific plate) is moving north while the other side (the North American plate) is heading south. It's a glancing clash of passing plates rather than a head-on collision of continents. And the vantage point near the Golden Gate Bridge from which McPhee chooses to 'watch the plates move' is where California and the San Andreas part company. After 800 kilometres of anxious acquaintance, the fault forsakes dry land, spills down a cliff, runs out aross a ledge and plunges beneath the Pacific breakers. There it continues its plate-parting function, but not as a landlubbing fault: as a deep-ocean trench.

It was the discovery of deep-ocean trenches in the 1960s that did more than anything to rehabilitate Wegener's continental-drift theory and so pave the way for plate tectonics. Attitudes to continental drift had always had as much to do with national prejudice as with science. While American academics remained hostile to German scholarship and wary of anything to do with what they called the 'mobilism' of continents, in Europe and parts of the British Commonwealth support for Wegener's displacement theory had gathered some badly needed momentum.

Initially this was thanks to a puzzling phenomenon called Apparent Polar Wander. APW, as it was known to the initiated, suggested that anomalies in the earth's magnetism could best be explained by the magnetic poles having a tendency to roam. The poles didn't just wander about but could even reverse their polarity and had done so more than once. At a given moment in the history of the earth the North Pole could be in Florida, at another in Australia. Rock, as it cooled and solidified, acquired some of this magnetism; its inclination (or deviation from the true pole) was that which prevailed at the time the rock was formed; and this magnetic legacy was retained within the rock ever after as 'remanent magnetism'.

It followed that once these vagaries of the poles had been plotted through time, it should be possible to calculate the age of a rock from the angle of its magnetic inclination. Or would it? In other words, was it really the poles that had gone walkabout? Or was it the rocks and the continents to which they belonged? In McPhee's words, 'Either the Equator had gone to Minnesota, or Minnesota had gone to the Equator.'[20] Birbal Sahni, the Indian palaeobotanist who'd worked on the distribution of *Glossopteris*, may have been the first to suggest using this palaeomagnetism to test Wegener's mobilism. Matching the magnetisms in different parts of the world as they worked backwards through time, Sahni and others successfully retraced India's life story right back to its calving from Gondwana. In effect remanent magnetism had provided 'the first solid physical evidence for continental drift'.[21] The equator could have gone to Minnesota but more probably it was Minnesota that had gone to the equator.

APW was thus discredited, but only to make way for RPW (Real Polar Wander). The magnetic poles did move, although nothing like as far as APW had suggested. Yet even minor variations in their position were of vital importance in cartography, telegraphy, geodesy, navigation and aviation. Keeping track of the movement of the magnetic pole became a turn-of-century obsession. Wegener might have been reassured to learn that, of the rather few German contributions to the exploration of Himālaya, the most notable were those of the Schlagintweit brothers in the 1850s and of Wilhelm Filchner in the 1930s. The expeditions of both were principally concerned with conducting far-flung magnetic surveys designed to monitor polar wander; and where better than across the unfrequented and electrically undisturbed terrain of, respectively, the western Himalayas and China/Tibet?

While continental drift, or mobilism, was at last gaining some traction in the rest of the world, in the US it was as despised as ever. American geologists clung fast to the 'fixist' beliefs of Suess. They wanted nothing to do with Wegener's mobilism. How could continents possibly drift apart? After thirty years the mobilists themselves still hadn't an answer. And what, if anything, could mobilism tell us of mountain-making? By 1950 Wegener's theory was as its lowest ebb in America. When taught at all, it was as an object lesson in bad science. Teutonic enthusiasm for grand ideas had led Wegener to neglect the scientist's foremost duties of collecting and publishing data, assembling an argument, anticipating every possible objection and inviting review. His theory was not just incredible; it was wrong because the methodology was wrong.

Aware, perhaps, that with a name like Hess his own ancestry might prompt similar suspicions, in 1960 the chair of Princeton's geology department prefaced his deep-sea findings with a catalogue of caveats: very little was known about the world's oceans, said Harry Hess; their origins were a matter of conjecture, their history was obscure, and their present structure could be understood only if some rather big suppositions were made. Hess knew of the perils of what he called 'geopoetry' but hoped to counter them by candidly admitting his own speculative assumptions. These did not include 'mobilism'. He had already seen enough of the ocean floor and read enough about remanent magnetism to be persuaded that Wegener was right. 'The general picture' derived from palaeomagnetism was 'sufficiently compelling', he wrote. True, Hess's ideas about sea-floor spreading did not entirely corroborate continental drift. 'The continents do not plow through oceanic crust impelled by unknown forces,' he declared; 'rather they ride passively on mantle material as it comes to the surface at the crest of the [mid-ocean] ridge and then moves laterally away from it.'[22] But this was a quibble. Land masses did indeed move, and he had some idea of how. As of 1960 continental drift in the US may be said to have come in from the cold; and since it was an American who'd opened the door, national pride was assuaged.

Hess was not just an American, nor just an academic. He was a rear admiral in the US navy with a distinguished record of wartime service in the Pacific. There he had commanded a vessel equipped with sonar for detecting Japanese submarines; and when not so engaged, he had continued to use the sonar to scan and map the ocean floor. The results were unexpected and had been confirmed by similar surveys elsewhere, including beneath the Atlantic.

On average, ocean floors were around 5 kilometres lower than dry land. They were nearer the crustal action and exhibited more recent evidence of it. Mountain ranges stretching up to 25,000 kilometres bisected the ocean's basins and, like stretch marks, faithfully preserved the boundaries of, for instance, the African and South American plates. In the vicinity of these mountain ridges, or 'rises', Hess detected evidence of new sea floor being expelled from crustal vents as molten rock and flowing outwards; hence his best-known discovery, that of 'sea-floor spreading'. Some of these fault lines were so long that they 'hit the coastline and could be traced into strike-slip faults such as the San Andreas fault in California'.[23] Others, instead of extruding new sea floor, gobbled up old sea floor. Known as deep-ocean trenches, these plunged to depths of 12 kilometres, which is about as low as Hawaii's mermaid mounts are high. And that was only the beginning: 'The [old] seafloor goes down four hundred miles [644 kilometres] after it goes into the trenches. On the way down, some of it melts, loses density, and – white hot and turbulent – rises toward the surface of the earth, where it emerges as volcanoes …'[24] In the Pacific, Mauna Loa and Mauna Kea were the obvious examples. But Hess also mapped extended chains of flat-topped excrescences which he called guyots and identified as former volcanoes, albeit now extinct and decapitated by the ocean's currents.

Boosted by these underwater insights into the relationship between crustal faults and mountain-building, the research that led on to plate tectonics started to pick up speed. All that was needed was an understanding of the earth's crustal composition (the plates) and of their structure (the tectonics). Answers to both came almost simultaneously and from opposite sides of the Atlantic. In 1967 Dan McKenzie and Rob Parker at Cambridge published 'The North Pacific: An Example of Tectonics on a Sphere'; and also in 1967 astrophysicist W. Jason Morgan, one of Hess's Princeton protégés, published 'Rises, Trenches, Great Faults and Crustal Blocks'. Neither party acknowledged a debt to the other. Morgan's forte lay in spherical trigonometry and his brainwave consisted in spotting that fracture zones were disposed across the Pacific concentrically. In other words they shared a common pole. It was somewhere in Siberia but the important thing was that, since all shared it, they must belong to the same plate. Plates could therefore be bigger than the continents and even than the oceans that they underlaid.

McKenzie and Parker deployed similar spherical mathematics. They concentrated more on the behaviour of plates but also dealt with the formation of faults, ridges and trenches. Like Morgan they thought there might be twenty or so plates and they stressed their rigidity. They called them blocks or paving stones, and they named their findings 'the paving-stone theory of world tectonics'. Whatever the name and whoever was first, 1967 was a milestone year in the history of science. The ever quotable McPhee puts it well: 'As has happened only twice before in geology – with Abraham Werner's neptunist system and James Hutton's *Theory of the Earth* – the theory of plate tectonics has assembled numerous different phenomena into a single narrative. Where plates separate, they produce oceans. Where they collide, they make mountains.'

Others have come up with even more extravagant claims for plate tectonics. Cherry Lewis, biographer of Arthur Holmes, the geophysicist who championed continental drift in Britain, ranks the theory of plate tectonics as 'one of the great unifying theories of all time, alongside Darwin's Theory on the Origin of Species and Einstein's General Theory of Relativity'.[25] Henry R. Frankel, doyen of American science writers, draws the same parallels: what evolution was to the natural sciences, and what relativity and quantum mechanics are to physics, plate tectonics are to the earth sciences. Together, these three theories determine our current understanding of the physical world.

How Wegener's continental drift was subsumed into plate tectonics is the subject of Professor Frankel's monumental *Continental Drift Controversy*. Frankel, of the University of Missouri, Kansas City, devoted thirty-five years to his definitive work. Continuing the practice of affording geological publications geological proportions, it runs to four volumes, each of 400–600 pages, and was published in 2012. In hardback all the volumes are of identical size and design but with a colour photo of a different geological phenomenon on the cover. When neatly piled on top of one another in order of publication, their spines call to mind geological strata as exposed by, say, a road cutting. The stratigraphy of a mountain, one might suppose, would be something like this, with the earliest layers/volumes at the bottom and later layers/volumes ranged in order above them.

Alas, though, this is not how the strata stack up; in fact the exact reverse is just as likely. As noted by Godwin-Austen, the limestone on Hooker's Sikkimese pass was quite young while the rock above and around it could be quite old. And the same goes for the limestone at the summit of Everest.

Most Himalayan peaks are composed principally of either granite or gneiss, rocks hard enough to withstand prolonged exposure to those 'little causes, long continued' that pare away more friable sedimentary rock. In the case of granite, durability is achieved when molten magma from within the earth's mantle is extruded and cooled under pressure. As an igneous rock, then, granite denotes a process as much as a substance; and over time the process may be repeated with reliquefied granite being intruded into younger rocks.

Gneiss is the result of a similar but less extreme process known as metamorphism; this requires somewhat lower temperatures and less pressure over a shorter time span. Though not strictly speaking an igneous process, it relies on the same combination of inner earth forces. And thus it happens, with an irony that will come as no surprise, that Himālaya's frozen realms are the product of a subterranean inferno, and earth's highest peaks consist of materials forged in its deepest depths. The coldest of hells is fashioned in the hottest of hells.

The uplifting of all this rock is the result of tectonic collision. In the absence of the Pacific's all-devouring deep-ocean trenches, the Tethys Sea offered nothing to detain the advancing Indian plate and left it no alternative but to slam head-on into the Asian plate. As it sought a way under, over or around this obstruction, its leading edge was so compressed and folded that, north to south, the earth's crust was shortened by around 800 kilometres. Combining both underthrusting (subduction) and overthrusting (obduction or piggyback riding), the Indian plate's assault on the Asian plate also resulted in a thickening of the earth's crust, from the usual 35 kilometres to 70. Nowhere else in the world is the crust anything like as thick. Himālaya's highest peaks are thus projected skywards from a Tibetan launch pad that's already 4,500 metres asl. Tibetans, like Greenlanders, would have to adapt to the physical and climatic challenges of living atop a giant carapace.

As well as compression and elevation, a head-on collision of plates can have a lateral effect. Such sideways slippage of the Indian plate's leading edge is now thought to account for what, on first glancing at a map, appears the Himalayas' most unusual feature – the symmetry of their extremities. At either end, the great curtain of snows draped across the Indo-Tibetan borderlands looks to be secured by a gigantic massif: that of Nanga Parbat (8,126 metres) in the west and of Namcha Barwa (7,782 metres) in the east. Each of these colossi is isolated by deep gorges – of the Indus round Nanga Parbat and of the Yarlung–Tsangpo–Brahmaputra round Namcha Barwa. And each presides over a radical change in the direction of the whole mountain range. In what geologists call a syntaxis, the structural grain of the range twists through 90 degrees to swing southward. On the map, the resulting appendages look like distinct ranges and are indeed so named – as the Hindu Kush of Pakistan and Afghanistan in the west, and as the mountain spine of northern Burma and South-east Asia in the east. Geology, however, knows better. The composition of these ranges reveals that all were once part of the Indian plate and that their unusual disposition is the result of the same India–Asia collision.

Because of its extreme inaccessibility in the far eastern corner of the Himalayas, not a lot is known about Namcha Barwa and its geology.

A lethal combination of high precipitation, Tibetan xenophobia, ravenous leeches and Sino-Indian border sensitivities has meant that even sightings of the great peak are rare. The mystery of the Yarlung–Tsangpo's reincarnation as the Brahmaputra would not be conclusively resolved until 1912, and Namcha Barwa itself remained unclimbed until 1992.

Nanga Parbat was a different matter. Despite and because of its killer reputation, the Naked Mountain has attracted considerable attention and provided plenty of surprises. Oxford's Mike Searle acknowledges the work of Darashaw Nosherwan Wadia of India's Geological Survey. It was Wadia who correctly identified the mountain's gneiss and 'granitoid' composition when he criss-crossed the whole syntaxis area in the 1920s and 1930s to compile a geological map. But as to the age of this gneiss, Wadia could not have been more wrong. Like others denied the certainty that came with radiometric dating, he thought the granites and gneisses 'pre-Cambrian', so half a billion years old and nearly off the scale of geology's deep time. But in fact, according to Searle, 'the Nanga Parbat gneisses are [now] understood to show the youngest metamorphic event ... anywhere in the world'.[26] In terms of John McPhee's one-year calendar of deep time, instead of dating from, say, Easter, their metamorphism didn't begin till early on New Year's Eve morning. By then dinosaurs had long been and gone and the hills were already alive with the stomp of massive ruminants. Indeed Nanga Parbat's gneiss may still have been metamorphosing when the first hominins came padding across the passes.

Pausing beside a rock outcrop on the Srinagar–Gilgit trail – which could well be that which was exposed by the 1841 Indus landslide – Searle muses that the visible strata here 'summarise the incredible geology of the Himālaya'. Rock that was formed within the earth's crust billions of years ago sits happily atop gravel deposits of less than a million years old. And this is not the latest instalment in the mountain's story. 'Even more incredible was the history of the Nanga Parbat gneisses,' says Searle. Having settled at the base of the earth's crust for 500 million years, these were subject to another metamorphism when India and Asia collided to trigger the Himalayan orogeny. 'Finally, during the recent and active uplift of Nanga Parbat, the once deeply buried and hot gneisses of the lower crust were thrust up and out onto the recently deposited gravels of the Indus River.'[27] It is all part of 'the great crustal recycling scheme'.

As 'the greatest continental collision the Earth has seen',[28] India's head-on encounter with Asia has done more than juggle with the strata. The southern edge of the zone of contact between the advancing India and the resisting Tibet is marked by a still active fault running the length of the outer Himalayas. This Main Boundary Thrust is where all the worst earthquakes occur. But further north another massive fault, the Main Central Thrust running along the base of the Great Himālaya range, seems to contest this reading; here the strata indicate a 'south-vergent' thrust that redirects material from the leading edge of the Indian plate back the way it has come. Is India being repelled, then? Is Tibet holding its own? Or, as the twisted strata of Zanskar, the liquorice-like extrusions and upended slabs suggest, is there more going on here than has yet been revealed by the geologist's hammer?

As Searle might cheerfully admit, nothing in geology is written in stone. 'Every model that has been proposed for the Himālaya, Karakoram and Tibet is almost certainly wrong; some may be slightly useful, many are wildly inaccurate.'[29] There are still scholars who, like Wegener, change their minds, and there will surely be more technologies, like radiometry, which will change them anyway. The rocks have further revelations, none more intriguing than the story of life in Himālaya. In 2018, when the first chapters of this book were being written, it was generally supposed that the earliest evidence of a human presence in Himālaya dated from about 40,000 years ago. Since 2020, reanalysis of an existing find has pushed this back to 160,000 years ago and come up with an intriguing explanation of how the first Tibetans managed to adapt to life in such a harsh environment.

Earlier still, Himālaya's notably weird fauna had flourished thanks not so much to adaptation as to a more benign climate. Just as the fossilised molluscs noted by Hooker and Godwin-Austen had encouraged Suess to propose the existence of the Tethys Sea, so it may have been the faunal remains unearthed in the outermost Himalayas that emboldened Dèzes to identify Zanskar as African. For the odd thing about the rediscovered zoology of the region would be that it too seemed more typical of Africa than Asia.

3

A Domain of Animals

The snowy domain to the north [Tibet] is presently a domain of animals, so even the word 'human being' does not exist there – it is a vast darkness. And all who die there turn not upwards but, like snowflakes falling on a lake, drop into the world of evil destinies.

The Great Chronicle of the *Mani Kabum*[1]

The news broke on the evening of 4 November 1834. As he rose to address the eminent membership of the Asiatic Society of Bengal, James Prinsep, Calcutta's outstanding scholar and the Society's secretary, could scarcely contain his delight. Recognition had been a long time coming, but at last India's natural history had as good a claim on the world's attention as its man-made history. And the key to both, it seemed, lay buried in Himālaya's periphery. At the time Prinsep and his antiquarian colleagues were busily unravelling the Indian origins of Buddhism by studying coin finds and mastering the rock-cut scripts in which India's earliest history was recorded. Now almost simultaneously India's prehistory, and along with it that of all humankind, was being prised from the hills and reconstructed on the evidence of fossilised bones. In both cases the aim was a better understanding of antiquity; the deductive process was similar; and the materials to be studied were comparable. As one leading palaeontologist put it, 'the remains of extinct animals are our coins, our bas reliefs, our indecipherable annals'.[2]

For centuries there had been rumours of fossilised vertebrates embedded in a low outlying range of the western Himalayas known as the Siwaliks. Back in 1360 CE massive bones had been excavated thereabouts during the digging of a canal from the upper Jumna (Yamuna) to Delhi by the then sultan, Feroz Shah Tughluk. One bone, thought to be that of a forearm, had measured nearly three metres. The chronicler responsible for recording the find noted that such monstrous limbs were supposed those of an extinct race of giants. It was an identical explanation to that given by St Augustine when, in about the year 400 CE, what was almost certainly an elephant bone had washed up on the coast of North Africa. Augustine, himself an African and bishop of the nicely named Hippo (in what is now Algeria), recorded the incident in *The City of God*, his great theological apologia for Christianity.

In the nineteenth century, Indians still called such relics 'hero bones' and usually attributed them to the larger-than-life protagonists of the *Mahabharata*. But not all the finds were gargantuan; most were just fragments and splinters. And not all were from the sub-Himalayan Siwaliks; the first British reports of fossilised bones cited a collection made on the other side of the Himalayas in Tibet. In the wake of the 1814–16 Anglo-Gurkha (Anglo-Nepalese) War, the British had annexed the western hill districts lately acquired by the Gorkha (Gurkha) kingdom. These comprised a cross-section of the Himalayas extending from the plains and the sun-baked Siwaliks north to well-wooded straths or 'duns', like Dehra Dun, then to the steep inclines and alpine pastures of the outer ranges, and on up to the snow-choked passes of the Great Himālaya. Surveying their new acquisitions in the 1820s, British administrators and road-builders couldn't resist pushing up and over the more accessible of these passes. They were not reassured by what they saw. From atop the Niti Pass on the lip of the Tibetan plateau, the only notable feature was the dazzling cone of sacred Mount Kailas standing among some 'not very peaky snowy heights' in the far north-east. Otherwise there was nothing. 'Right in front there stretched a dreary plain, shrubless, treeless, and houseless, terminated ... by a low range of rounded brown hills utterly without shrub or tree or jutting rock ... Had there been heather instead of stone and brown grass, it would have resembled a Highland moor.'[3]

The pass did, though, yield a plentiful supply of fossilised ammonites, while from the shrubless tundra beyond, willing accomplices retrieved the first fossilised bones. These were fragments of fragments; only a geologist would have recognised them as organic and only a palaeontologist would have strained credulity by claiming them as the mortal remains of a rhinoceros. It was as improbable as attributing them to a Tibetan whale. Even supposing a rhino could survive at 5,000 metres above sea level, which was unlikely, even if it had acquired a woolly coat like the mammoth, plus the ability to shin up cliffs like a chamois and bulldoze its way through the ten months of snowdrifts like a yak, there was nothing for a rootling rhino to eat.

Yet rumours of even less likely finds would keep on coming. In the early 1830s Rajah Fateh Prakash of Sirmoor (Sirmour), a Siwaliks principality west of the Jumna in what is now the Indian state of Himachal Pradesh, seems to have made a habit of presenting visiting dignitaries with pocketable fossil bones said to have been found in his hills. He called them *bijli ki har*, 'lightning bones', and he rated them highly as talismans, like the fossilised ammonites or saligrams offered for sale in Nepal and with which, as in Tibet, they were sometimes found.

Among visitors favoured by the Sirmoor rajah with such keepsakes was a party of military engineers working on a new canal project that would incorporate what remained of Sultan Feroz Shah's waterworks. Led by the thirty-year-old Captain Proby Cautley, in 1832–3 these engineers heeded the rajah's fossil clues whenever occasion offered, 'but', says the father of one of the engineers, 'their labour was scantily rewarded'.[4] Meanwhile the young Dr Hugh Falconer, he who ten years later would submit the first report of the great Indus flood, had arrived to take up his post at the East India Company's botanical outstation in nearby Saharanpur. Hugh Falconer and Proby Cautley (Cautley's unusual first name being his mother's maiden name) immediately hit it off. The engineer became a frequent house guest at Saharanpur, botany and canal-digging were brushed aside in favour of fossilised bones, and between the Scots doctor from Moray and the English officer from Suffolk was forged one of the most productive and harmonious partnerships in the history of palaeontology.

Yet during his first two years Falconer enjoyed no more success in locating animal fossils than the engineers. He confirmed that the gravels and sandstones of the Siwaliks were a Tertiary formation (65 million to 1.64 million years old) that might well contain fossil beds and also lignite (or coal), the discovery of which would put a more favourable gloss on the fossil-collectors' diggings. Indeed

Falconer had found both bones and coal during a short tour into the hills in 1831. As he told Dr Forbes Royle, the Saharanpur garden's superintendent whom he was about to replace: 'I returned loaded, not only with lignite but with noble fossils of the monsters of the deep! Bones of crocodiles, fragments of the skulls of large turtles and a fragment of a bivalve shell as large as an oyster.' Better still, he was confident of further finds; 'remains of mastodon and other large extinct mammalia' were surely awaiting discovery. But, for all his enthusiasm, the spoils of Falconer's first foray into the hills were disappointing. His lignite would prove of poor quality and his fossils, however 'noble', were not that big and belonged exclusively to aquatic species; as Prinsep at the Asiatic Society would recall, 'confirmatory evidence of animal remains was wanting'.

*Botanist and palaeontologist Hugh Falconer
led the hunt for Himālaya's fossilised fauna*

In April 1834 Falconer was back in the field and recovered part of the shell of a tortoise. This time it did indicate a creature of unusual proportions and justified his calling on Cautley for help. 'With characteristic energy' the Captain obliged, but perhaps a little too energetically when 'by means of blasting' he revealed more remains 'including Miocene mammalian genera'.[5] The Miocene is the fourth of the Tertiary epochs and covers the period 23 million to 5 million years ago, but to which mammalian genera these remains belonged, let alone which species, was uncertain; presumably the evidence had been blown to smithereens.

Though far from being the end of Falconer's tortoise discoveries, it was the end of that particular dig. The season was by now too advanced. Falconer headed south to Saharanpur to sit out the monsoon, and Cautley retired north to his field camp at Dadupur, there to meet up with his two subordinates. Lieutenants William Baker and Henry Durand, both in their early twenties and keen as mustard to make a name for themselves, were directed to prepare for one more fossil foray as soon as the rains ceased.

October 1834 thus found the engineers again 'hammering away at rocks' and now in the vicinity of Nahan. Nahan's fort being the home of the Sirmoor rajah, they soon had a visitor. The rajah wondered what they were up to. He was shown the fruits of their labours, namely 'a few crocodile teeth'. He was unimpressed. India boasted at least two still extant crocodile species; the teeth could have come from one of the Jumna's lately deceased denizens. Stifling a smile of condescension His Highness sent for his prized 'tooth of a Deo [a god]' and graciously bestowed it on Lieutenant Baker. Weighing all of 12 pounds (5.4 kilos) it was the most gigantic fang any of the engineers had ever seen. Falconer would later identify it as the molar of a mastodon, an elephant-like genus often confused with the mammoth and at the time thought to be peculiar to North America. More to the point, the rajah obligingly directed the engineers to where the tooth had been found. They struck camp forthwith and were not to be disappointed.

On reaching the site, Cautley could hardly believe his eyes: 'The hills were covered with fossils like all the others. (How they could have escaped observation before, must remain a source of wonder.) Mastodon and hippotamus's remains looking one in the face at every step!'[6] One deposit contained the bones of small mammals including porcupine and rat, all 'too perfect to admit of any doubt [and] each

consisting of the palate, with the two lines of molars!!!' Other beds
yielded specimens of what the more anatomy-confident Dr Falconer
confirmed as mastodons and hippos. In a matter of days the fossil-
hunters' bag was one of the most varied ever collected:

> The remains at present dug out consist of portions of an
> anoplotherium or palaeotherium [a large tapir-like creature],
> rhinoceros, hog, horse, ruminants of the most giganic proportions,
> with those of the smaller classes, carnivora, hyaena, canis [jackal,
> wild dog], tiger (or lion) etc ... Many of the bones of these animals
> are coming out perfect, and some have been found with connected
> joints ...
>   ... You will join with me in an exclamation that has been upon
> my lips, day after day, since the discovery of the first fragment of
> bone – 'What shall we have next?'[7]

Such was the gist of Captain Cautley's report as delivered to the Asiatic
Society by Prinsep on that November evening in 1834. It was followed
almost immediately by a letter from Falconer:

> You have heard from Capt. Cautley and Lieut. Baker about the late
> fossil discoveries up here ... I got a hint of where they came from,
> and on going to the ground, reaped a splendid harvest. Conceive
> only [of?] my good fortune: within six hours I got upwards of 300
> specimens of fossil bones.[8]

To this haul, Lieutenants Durand and Baker added their own finds in
1835. Baker reported on the remains of a (presumably wild) camel and
on the discovery of bone and antler fragments belonging to an elk,
the first ever recorded in India; Durand found further bones of the
hippo, which was also unknown in India and was supposedly unique to
equatorial Africa. In a synopsis of their combined finds dated November
1835, Cautley and Falconer listed twenty-four genera and twenty-nine
species, twelve of them *Pachydermata* ('thick-skinned' mammals like
elephants, rhinos and hippos) and five of them *Ruminantia* ('cud-
chewers', like the ox, camel and deer). In time these genera would
generate additional species and be joined by other genera, including
giraffes, antelopes, buffalo, bison, bear and pig. As Prinsep warned his

Calcutta audience, 'the Asiatic Society is now in a fair way of possessing a splendid museum of this newly discovered, or rediscovered, tract of country'. In an age when fossil finds were largely restricted to the cold or temperate regions of Eurasia and America, thanks principally to Cautley, Falconer, Baker and Durand 'a sub-tropical mammalian fossil fauna was brought to light, unexampled for richness and extent in any other region then known'.[9]

At the risk of stating the obvious, it is worth a reminder that these creatures were not dinosaurs. The term 'dinosaur' as referring to 'monstrous saurians [lizards]' was unknown in 1834. It would be coined only in 1842, and not till much later was it realised that dinosaurs, with the exception of those capable of flight, had long predeceased the Siwalik fauna. Some dinosaurs may even have witnessed the break-up of Wegener's Pangea into Gondwana and Laurasia 250 million years ago; they were certainly around when the Atlantic split Gondwana into the African and American land masses 180 million years ago; and when, a mere 65 million years ago, extinction overtook the dinosaurs, India and Asia had still not collided and Himālaya had yet to be upheaved. In other words, millions of years had elapsed between the disappearance of the dinosaurs and the advent of the Siwaliks' fauna, or indeed of the Siwaliks themselves. Adverting again to John McPhee's one-year calendar of deep time, if the dinosaurs disappeared on Boxing Day, Falconer's hippos were still wallowing in the Jumna on New Year's Eve. Clearly the Siwalik finds were too recent to promise insights into the origins of life on earth. They would, though, provide pointers to the origin of the species most likely to be responsible for bringing life to an end, in other words *Homo sapiens*.

Topping Falconer and Cautley's 1835 synopsis of their Siwalik finds were an elephant species, *Elephas primigenius*, and two mastodon species. Mammals with proboscises, or trunks, are known as *Proboscidea*, and partly on account of their size, partly because of the discovery in the Siberian permafrost of nearly intact mammoths, *Proboscidea* were given precedence over other genera by nineteenth-century palaeontology. They had been heading the stampede of fossil species into the world's museums ever since 1796 when the great Georges Cuvier at the National Museum of Natural History in Paris had demonstrated the anatomical differences between an Indian elephant, an African elephant, a Siberian mammoth and (later) a North American mastodon. These differences warranted each being considered a separate species and got Cuvier's lifetime tally of some forty-nine newly identified and mostly extinct species off to a head start.

The father of palaeontology and the pioneer of comparative bone study (osteology), Cuvier worked from the fossil collection in his Paris museum and from specimens and drawings received from his far-flung correspondents. He was not a fossil-hunting fieldworker like Falconer and Cautley but a brilliant anatomist and bold theoretician whom historical circumstance happened to favour. Thus his skeletons of extant Indian and African elephants, originally from the Dutch colonies in Sri Lanka and South Africa, had reached Paris among the cultural plunder seized by the French after Napoleon's takeover of the Netherlands. Enjoying the patronage of revolutionary France's Directorate and then of the Napoleonic authorities, Cuvier also had preferential access to the hoard of fossil specimens being unearthed in the limestone basin around Paris and especially in the gypsum mines beneath Montmartre.

As the fossil bones were carefully detached from their clayey matrix and deposited in the museum, Cuvier sorted them into the species to which each might belong. After minute scrutiny of especially any teeth or joints, he then attempted a reconstruction of whole skeletons, incorporating prosthetic limbs where necessary. This was a formidable task and it was further complicated – for Cuvier as it would be for Falconer – by the realisation that many of these species were not only extinct but apparently 'intermediate in character between the pachyderms and the ruminants, or more precisely between the camel and the tapir'.[10] The likely explanation for such aberrant forms lay in the evolution of species by natural selection as soon to be championed by Charles Darwin. But Cuvier had had no time for what in the 1830s was still called 'transformism'. Instead he ascribed the absence of such forms among the world's still extant fauna to the selective impact of mass extinctions as a result of climatic or geophysical change. Variant or intermediate animal forms – mammoths, say – must be those of an elephant-like species that had been overtaken by some global disaster at a comparatively early stage in the life of today's surviving elephant species. Species in general, Cuvier believed, were fixed for all time. They were invariable, and the idea of them emerging by natural selection was preposterous. 'Did ducks by dint of diving become pike?' he scoffed, or 'hens striving not to get their thighs wet, succeed so well in elongating their legs that they became herons or storks?'[11] Not a chance. There was, though, evidence for entire species being wiped out by rising sea levels, ice-age glaciations, volcanic eruptions and other natural catastrophes.

Aided by considerations of scale and mechanics, and sometimes by comparison with the skeletons of still extant equivalents of the species in question, Cuvier proceeded to reconstruct (*resusciter* was his word for it) his skeletal models according to what he grandly called 'the principle of correlation'. Hence, for instance, teeth designed to tear flesh rather than grind vegetation should come from a creature with clawed feet, a sturdy jaw, a predator's physique and a gut capable of digesting meat. On to a skeleton thus constructed he could 'reconstitute more or less confidently the musculature and whole body form'. Habitat, diet and disposition could then also be inferred. From careful study of a single tooth and a few bone fragments, long-extinct fauna could be magicked, not back into life, but into a static and often outlandish vision of what life had once been.

Starting with that investigation of elephants and mammoths, Cuvier published his findings in a succession of papers. These were then revised and incorporated into his *Recherches sur les ossemens fossiles de quadrupèdes* (or, in full and English, *Researches on the fossil bones of quadrupeds, wherein have been restored the characters of several animals whose species has been destroyed by global upheavals*). The work appeared as four volumes in 1812 and, frequently augmented, as five in 1825. It served as a compendious update of the great man's latest findings and, thanks to its magnificent engravings of fossilised bones from every conceivable angle, also as a field guide. Hugh Falconer certainly knew it well and followed its classifications to the letter. He must have consulted it at the Geological Society in London where, before sailing to India, he had studied some fossilised mammalia bones collected in Burma (Myanmar); and he had probably found a copy in Calcutta when, as he awaited his appointment to Saharanpur, he had compiled a list of similar Burmese fossils in the Asiatic Society's possession.

How and when he acquired Cuvier's volumes in remote Saharanpur is less clear. He didn't bring them with him because, according to Prinsep in Calcutta, 'from time to time they [Falconer and Cautley] requested that Cuvier's works on osteology might be sent out to them, and expressed their disappointment when, from various accidents, these volumes failed to arrive'. Falconer's published references to Cuvier suggest that they must eventually have arrived, but initially the fossil-hunters had to make do with DIY expedients.

> Thrown entirely on their own resources [recalled Prinsep] they soon found a Museum of Comparative Anatomy in the surrounding plains, hills and jungles, where they slew the wild tigers, buffaloes, antelopes and other Indian quadrupeds, of which they preserved the skeletons ... They were compelled to see and think for themselves, while comparing and discriminating the different recent and fossil bones and reasoning on the laws of comparative osteology, till at length they were prepared to appreciate the lessons which they were taught by the works of Cuvier.[12]

But Cuvier's *Recherches* would not always have been of help. In opposition to Cuvier's single elephant genus with its three species (African, Indian, mastodon), Falconer was struck by the enormous variety of extinct *Pachydermata* – twice as many genera as now exist and five times as many species, he thought. To distinguish them, he paid minute attention to the composition and arrangement of their teeth and so 'laid a new foundation for the classification of elephants ... through the study of dental anatomy ... In breaking the narrow framework of Cuvierian classification, this impressive work actually led to a new vision of the fossil world.' By demonstrating a gradation from 'primitive' to 'evolved' forms that accommodated 'intermediate' forms, Falconer's paper 'implicitly told a new history of living beings and raised the question of the modality of evolution'.[13] And all this in a monograph first published in 1846, thirteen years before Darwin's *Origin of Species*.

Nor, in the case of some massive remains first reported by Cautley, would the corpses of extant quadrupeds bagged in the neighbouring hills have been of much help. In announcing their largest-ever discovery, Falconer and Cautley awarded it the designation of *Sivatherium giganteum* ('Giant Shiva-beast'). Falconer had no doubt that it would live up to its name. It was 'a new acquisition to extinct zoology' and 'one of the most remarkable of the past tenants of the globe hitherto detected'. Cautley simply announced 'one of the most superb fossils, I suppose, ever found'. He regretted, however, that he could give no details of his specimen. 'It was unfortunately discovered and excavated by a party of workpeople employed by a gentleman with whom I was unacquainted.'[14] Who this interloper was, and what became of his superb find, is not known. Evidently fossil beds in the Himalayas could be as hotly contested as those in Europe. But Cautley stuck to his guns. He'd seen the monster *in situ* – femur, tibia, tarsal and metatarsal embedded in sediment – and, though not permitted to take measurements, he'd recognised it as a *Sivatherium* when Falconer obtained the skull of another specimen in 1836.

Falconer's fossilised cranium was largely intact. It was also 'remarkably perfect', it was quite unlike any other known creature, and it was enormous. From the teeth it obviously belonged to a cud-chewing ruminant, and at first Falconer was inclined to class it as a pachyderm related to the rhino. But judging by the dimensions of the skull, plus other bones found in its vicinity, the *Sivatherium's* size was more like that of an elephant and far in excess of any known rhino or buffalo species – or indeed the Himalayan yak, a species known to Falconer only as the shaggy cross-breed used to carry loads over Baltistan's higher passes, not as the much bigger wild yak of the Tibetan Changthang. Like a yak, the *Sivatherium* had had horns; but there seemed to have been four of them and, unless Falconer was very much mistaken, it also had the nasal bones needed for a short trunk. So here was a pachyderm with the jaw of a ruminant and the snout of a *Proboscidea*. It was 'without an analogue in its order', declared Falconer and Cautley; even Cuvier might not have shed much light on 'so singular and grotesque' a monster.[15] And it was indeed colossal. Remains since discovered in Africa and elsewhere in India have led to the conclusion that the *Sivatherium* was the largest of all known ruminants, weighed 1,250 kilos, stood 3 metres tall and was more closely related to the giraffe than the rhino.

The *Sivatherium* may also have been the last of the extinct Siwalik species to have been extinguished. It is known to have survived well into the Pleistocene ('most recent') epoch of the Quaternary (post-Tertiary) period and so was still roaming the floodplains of the Jumna less than a million years ago. It, or a closely related subspecies, may have lasted even longer in central India. But there the evidence is pictorial rather than palaeontological. In 2006 a group of elderly amateur archaeologists stumbled upon a network of caves near Amravati in Madhya Pradesh. Some of the cave walls are described as being painted with a 'breathtaking' variety of wildlife ranging from the aardvark (otherwise unknown in India) to the large, horned and long-necked quadruped depicted in red that has been tentatively identified as a *Sivatherium*. The paintings have been dated to just 8,000 years ago. Moreover they lend credence to a small Sumerian figurine of a similar creature found at Kish in Iraq in 1928. This is presumed to date from a mere 5,000 years ago.

In other words the *Sivatherium* may well have been around long enough to witness the advent on to its wooded pastures of early man. Here, then, was a member of Himālaya's fauna that very probably straddled the frontier between geology and anthropology, that had strayed from the 'bone age' into the Stone Age and from the palaeontological into the palaeolithic.

Dr Falconer would have been delighted. If there was one field of study that intrigued him even more than fossils and flowers, and if there was one contemporary concern as burning as that of the origin of species, it was the vexed question of the antiquity of mankind. 'In the early 1830s the question of human antiquity remained one of the most refractory problems in geohistory,' writes Martin Rudwick of the universities of Cambridge and California (at San Diego). 'In practice the question devolved into that of the contemporaneity – or otherwise – of humans and extinct "antediluvial" mammals.'[16] The *Sivatherium* could be a case in point. But, for the idea of mankind overlapping with the Siwalik fossils to be taken seriously, more such crossover finds were needed.

In the early 1840s, after a decade of bombarding the Asiatic Society with reports of the Siwaliks' ever more outlandish mammalia, Falconer and the canal-building engineers parted company. Cautley moved on to survey, and then construct, the great irrigation scheme for which he is still best remembered. This was the 500-kilometre Ganges Canal, the most ambitious such project anywhere in the world at the time and the forerunner of the great network that still irrigates India's north-western states. Baker meanwhile was sent south to oversee canals in Sind. He later built railways before being co-opted by the India Office in London as its military secretary. And as for Durand, he finally got a taste of the military action after which he'd long hankered and for which the scientific dissection of an occasional elephant was no substitute.

Having secured a posting to the East India Company's Afghanistan-bound Army of the Indus, in 1839 he achieved near-immortality by braving heavy fire to reset the charge that blew up the gates of Ghazni and so opened the road to Kabul. Thirty years later, by when he was lieutenant-governor of the British province of Punjab, Durand would renew his acquaintance with *Proboscidea* and portals. Entering the walled town of Tonk, his mahout misjudged the clearance afforded by a monumental gateway and the lieutenant-governor was sent head first from his howdah to the stone roadway below. He died soon after of injuries sustained. Elephants and gateways had proved his nemesis.

By chance all three of the fossilising engineers reaped identical honours. They all attained the rank of general and were all awarded knighthoods. Falconer, on the other hand, remained plain Dr Falconer. Distinction never entirely deserted him, but neither did provocative discoveries and a hint of contemporary condescension.

In 1842, soon after reporting on the great Indus flood, he returned to Britain on health grounds. He brought with him seventy chests of botanical specimens and forty-eight cases containing 5 tons (about 4.5 tonnes) of fossil bones and geological samples. In the same year there arrived Cautley's much larger consignment of 214 hefty chests. The average weight of the chests was given as 4 hundredweight (200 kilos) and the cost of shipment as a whacking £602. Cautley being an officer on assigment at the time, this expense was generously 'defrayed by the Government of India', which may explain why many fossils that had been jointly discovered were consigned in Cautley's name. The bulk of the combined collections eventually found their way into the British Museum, which was the only repository with enough space for them, and there they remain.

Falconer is next glimpsed in September 1844 delivering an 'extempore discourse' at the annual meeting of the British Association for the Advancement of Science, held that year in the Great Concert Hall in York. According to a reporter for *Chambers' Edinburgh Journal*:

> The members, on entering this evening, were surprised by the picture of a tortoise, displayed on the green screen above the speaker's head, exhibiting an animal the same in form as ordinary land-tortoises, but about twelve feet long. Strange as it may seem, remains of this huge animal, to which the name of *Colossochelys*

atlas has been given, are found in the superficial gravel upon the
Siwalik hills ... showing that the Tertiary species may have lived in
certain districts down to a time nearer to our own era. And this idea
Dr Falconer connected in a very interesting manner with mythic
traditions of India descriptive of enormous tortoises, one of which
was fabled to support the elephant by which the whole world was
supported. It seemed not unlikely that these legends referred to
animals which had been living in the early ages of mankind, but
which have for many centuries been extinct.[17]

The giant tortoise in question had first been named *Megalochelys*, *megalos*
being the Greek for 'big' and *chelys* the Greek appellation for reptiles
of the tortoise, or *Testudo*, genus. But this had quickly been changed
to the still bigger-sounding *Colossochelys* and was further amplified by
the addition of *atlas*, Atlas being the titanic figure who held aloft the
heavens in Greek mythology.

No complete skeleton of *Colossochelys atlas* was ever found; but in
the course of their decade in the Siwaliks, Falconer and Cautley had
encountered enough bones and shell fragments to hazard a reconstruction
of one, and then, by analogy with smaller land tortoises having a similar
bone structure, to calculate its dimensions. These were nothing if not
impressive. *Colossochelys atlas* was heftier than a large rhino and far and
away the most massive tortoise or turtle ever recorded. Its carapace alone
was estimated at 12 feet 3 inches long and 8 feet in diameter which,
allowing for head, neck and tail, suggested an overall length of 22 feet
3 inches (6.7 metres) and a height of 6–7 feet (1.83–2.13 metres). Bullock
carts seldom came any bigger. And quite coincidentally it was this idea
of the tortoise as a vehicle which most excited Falconer.

As an example of the 'rapturous enthusiasm with which the doctor
pursued his researches', the editor of his memoirs quotes verbatim from
what was either a letter or the actual text of his British Association address:

What a glorious privilege it would be [declared Falconer], could
we live back – but for an instant – into those ancient times when
these extinct animals peopled the earth! To see them all congregated
together in one grand natural menagerie – these Mastodons and
Elephants, so numerous in species, toiling their ponderous forms and
trumpeting their march in countless herds through the swamps and

reedy forests: to view the giant Sivatherium, armed in front with four horns, spurning the timidity of his race and, ruminant though he be, proud in his strength and bellowing his sturdy career in defiance of all aggression. And then the graceful Giraffes, flitting their shadowy forms like spectres through the trees, mixed with troops of large as well as pigmy horses, and camels, antelopes and deer. And then last of all, by way of contrast, to contemplate this colossus of the Tortoise race, heaving his unwieldy frame and stamping his toilsome march along the plains that hardly look over strong to sustain him. Assuredly it would be a heart-stirring sight to behold![18]

As a result of what the *Edinburgh Journal*'s reporter called Falconer's 'plain and perspicuous, yet arresting' address to the British Association, the presidents of the Royal Society, Asiatic Society, Geological Society and Geographical Society joined forces to petition successfully both the government of the day and the East India Company for a grant towards the cost of cataloguing and arranging the Cautley–Falconer collection in the British Museum. Falconer was given the job of supervising the work but failed to complete it. It was quite probably beyond the capabilities of one man, especially one with more than enough irons in the fire already.

Most of these other commitments were botanical. Falconer's first contribution to the Asiatic Society of Bengal's journal, a paper on 'the Aptitude of the Himalayan Range for the Culture of the Tea Plant', had resulted in the acquisition of tea seedlings from China and the experimental establishment of tea plantations in Himachal Pradesh and more famously Assam. Costus, the medicinal root of a thistle-like plant found on pastures above 2,500 metres, was another candidate for exploitation. Falconer named it *Aucklandia* after the then governor-general and wrote a paper commending its conservation. Along with the potato, for whose introduction to Himālaya Falconer was also partly responsible, costus is now a principal cash crop in Lahul (south of Ladakh) and in similar areas in Nepal. Kew Gardens were as often favoured with Falconer's discoveries as Calcutta's museum. As a Himalayan botanist he anticipated the work of Darwin's friend Joseph Dalton Hooker and that of later plant-collectors like Frank Kingdon-Ward. Hooker would acknowledge Falconer's pioneering work by naming one of the loveliest of his own discoveries *Rhododendron falconeri*.

Hugh Falconer depicted as an elephant reading his work on the Siwalik fossils while supporting the world from atop a giant tortoise, all as per Hindu mythology. Sketch by Edward Forbes

Although there would be no return to the Siwaliks for Falconer, the palaeontological findings kept on coming. As noted by the anonymous Edinburgh reporter, Falconer was of the opinion that the lifespan of the species he'd called *Colossochelys atlas* had probably overlapped with that of *Homo sapiens*. Since smaller but otherwise almost identical tortoises had been found in the same fossil beds and could still be seen in the flesh in the Saharanpur garden, the *Colossochelys* itself, like the mammoth, could well have survived into post-Tertiary times. To Falconer's mind, it was the only possible explanation for the frequency with which massive tortoises featured in ancient mythology. Even native Americans awarded the tortoise a creational role. Ejected from heaven, it was a tortoise that served as a spacecraft for the mother of the Iroquoi and which, on splashing down in the ocean, had formed 'a small island, which gradually became the earth'. Pythagoras's followers had also subscribed to the notion of tortoise travel. In their India-derived cosmology the earth was placed on the back of an elephant which was itself 'sustained on a huge tortoise'. Obligingly, Edward Forbes of London's Geological Society drew sketches of this balancing act in one of Falconer's field notebooks. Forbes's elephant wears glasses, like the doctor, and rests a tusk on an open book; according to the caption, he is 'deciphering the mysteries of *Fauna Antiqua Sivalensis*', that being volume one of Falconer's never completed *Palaeontological Memoirs and Notes*.

But the classic locus for tortoise lore was of course Hindu mythology. As noted by Falconer, the tortoise was the second of Lord Vishnu's ten avatars. To create the world, Vishnu had churned the ocean using a mountain as a twizzle-stick while himself, in the guise of a tortoise, stabilising the stick's gyrations. Thus was the world formed atop a tortoise; and there, with the addition of an elephant between carapace and globe, it has remained ever since in orthodox opinion. Falconer was happy to make due allowance for Hinduism's 'uncurbed imagination' and its 'maze of exaggeration', but his point was simply that the pairing of an elephant and a tortoise made no sense if the only tortoise known to early man belonged to one of the small, still extant species. If, on the other hand, the authors of Hindu mythology were quite familiar with the *Colossochelys atlas*, it all made perfect sense. The matching scale of elephant and tortoise, or what Falconer called their 'image of congruity', then sustained the myth's 'harmony of representation'. Ergo, concluded Falconer, 'there are fair grounds for entertaining the belief as probable that the *Colossochelys Atlas* may have lived down to an early period of the human epoch and have become extinct since'.[19] Like the giant *Sivatherium*, the giant tortoise, far from being extinct when mankind first made his debut, could have been decidedly extant.

Falconer's dream, like that of all his palaeontological contemporaries, was to discover fossilised human bones among those of the extinct mammoths and mastodons. That would clinch the matter of 'contemporaneity'. But in the Siwaliks, as at comparable sites in Europe, there was a worrying absence of anything pertaining to early man, whether bones or artefacts. Indeed primates of any sort seemed unrepresented in the fossil record. As a result, Cuvier had favoured a late, post-fossil date for the advent of mankind and had thereby afforded biblical fundamentalists some welcome reassurance; for science, like Genesis, seemed to be saying that God's making of man in His own image was a distinct and deliberate act, indeed the climax and conclusion of the divine work of creation, and not a sideshow featuring some gradual random transformation of an existing species.

Or so, at least, it had seemed until 1836. In that year, in the remote sub-Himalayan Siwaliks, Falconer and Cautley, and then Baker and Durand, had augmented their findings with the first ever bones of a fossilised '*Quadrumana*'.

Under an obsolete system of classification used by Cuvier, the difference between monkeys and men as implied by the apparent time lapse between fossilised fauna and any evidence of *Homo sapiens* was bolstered by an important anatomical distinction. Apes were classed as *Quadrumana*, or 'four-handed', while what we now call hominins were classed as *Bimana*, that is 'two-handed'. Both genera had two arms that terminated in hands and fingers, but only the *Quadrumana* had two legs equipped with hands and fingers. *Bimana*, of course, had feet and toes instead. At what point a hand becomes a foot, an arm a leg, or fingers toes may seem an arcane matter, but it could supposedly be determined by osteology. Hence, in the case of a *Quadrumana* like Cautley and Falconer's find, the bones of the feet (or of the 'rear hands') are critical for identification purposes. And, as luck would have it, the bone that first caught Falconer's attention was an astragalus, a wrist/ankle bone of a rear arm/hind leg. It was only 3 centimetres long and, according to Cautley, it had been in Falconer's pocket for some time. The doctor had evidently been so overawed by its significance that he forebore to make an announcement pending further corroboration. Meanwhile he noted its similarity to the astragalus of the Himalayan Grey Langur, a large, rather sombre ape that still frequents Indian hill stations. It was only when, later in 1836, Baker and Durand reported finding the fossilised jawbone of another much bigger *Quadrumana* that Falconer stressed the two-fold importance of the creature: 'The discovery is interesting in itself as supplying a deficient link in the series of the former tenants of the globe, but greatly more [so] in connection with the races with which the fossil was associated.'[20]

So to Falconer's way of thinking the *Quadrumana* was – or *Quadrumana* were, for he proposed several different subspecies – firstly a 'missing link'; it filled what Professor Rudwick calls 'the glaring gap in the fossil record at the very point at which it was closest to the human species'. Indeed it was even closer to the human species than Falconer had at first supposed. On later comparing the tooth of a *Quadrumana* with its equivalents in the jaws of some primate skeletons from Sumatra, he suspected that the *Quadrumana*'s closest relative was not the langur at all but the much more human-looking and endearing orang-utan. The height of one of the Sumatran skeletons was reckoned at 7 feet (2.13 metres), which agreed with the King Kong dimensions proposed for some of the Siwalik *Quadrumana*. Even the doubtful Prinsep would concede there were grounds for believing that 'the extinct Siwalik form had been a large ape allied to *Pithecus satyrus* [the orang-utan]'.[21] In the impending controversy over evolution, Falconer's sympathies would look to be with Darwin. But, more excitingly, the little astragalus was proof that at least one species of primate had leapt along with that 'grand menagerie' of stampeding extinct fauna that so excited the doctor's imagination. And where monkeys had been, it was looking increasingly likely that men would not have been far behind.

Prinsep now grasped the importance of Falconer's *Quadrumana*. It was 'the nearest approach to a human being that has yet been found in a fossil state in company with the extinct monsters of primeval antiquity'.[22] There was, moreover, a certain logic in looking to South Asia for the all-important overlap between fossilised fauna and humankind. In more northerly latitudes an ice age now known as the Last Glacial Maximum had plunged most of Europe and North America into extreme cold storage from roughly 30,000 years ago to 16,000 years ago. Average temperatures dropped to around 6 degrees below those of today and sea levels to around 125 metres lower than today's. Britain became a European peninsula, like Scandinavia; the Arctic ice sheet reached south to Yorkshire. Conditions proved catastrophic for many life forms, especially mammals; each species had either to adapt (like the now woolly rhinoceros), migrate (like Europe's suddenly fossil-unrepresented giraffes) or fail (like the mammoths and the mastodons). But these same harsh conditions did not apply along the southern fringes of Eurasia. There, despite considerable fluctuations in humidity and temperature, there was no white-out. How much of Tibet was subject to year-round glaciation is uncertain, but sub-Himalayan India was spared, as were Iran, Arabia and the Horn of Africa, all of which, as a result of the lower sea levels, were reconnected to one another by land bridges spanning the Gulf and the Red Sea.

Thus the markedly African character of the fauna unearthed in the Siwaliks may be explained by the ease with which quadrupeds and bipeds could, in the comparatively recent Late Pleistocene age (say down to 16,000 years ago), have extended their range from continent to continent around the Arabian Sea and even up Asia's Pacific coast and across what would become the Bering Strait. Some species presumably took their time, like the hippos, and it's not obvious that all were going from west to east. Some, like horses and camels, appear to have migrated from north to south, from glaciated Central Asia to the southern slopes of Himālaya and, in the case of the Siwaliks fauna, from southern Himālaya to central India. As Falconer had rightly concluded, the terrain that best suited the Siwalik fossil finds would have been lusher and swampier than anything in the vicinity of the Siwaliks today. This argued for higher humidity and warmer winters than nowadays, and for extinction or evacuation when these conditions ceased to apply.

From today's standpoint the most obvious feature of the fauna unearthed by Falconer and his colleagues was the great variety of forms identified and the unexpected dimensions attained by some of them. Large mammals as a whole were much more plentiful than they are today and their 'sudden and widespread reduction' is what the *Encyclopaedia of Indian Archaeology* calls 'a most startling event for geologists and biologists'. But it was the multiplicity of species that struck Falconer and Cautley most forcibly. 'Of the nearly 30 species of elephants that were present [in the Siwalik fauna] only one is found living today.'[23] Of *Bovidae* (cattle) there were six new species; and of *Equidae* (horses) there were at least two distinct species, neither of which still exists. It was the same with the hippos, neither of whose Siwalik representatives was identical with the still extant African species. Rhinos came in at least three different forms, antelopes, giraffes and camels in something similar. The carnivores, though fewer, were equally bewildering, with otherwise unknown kinds of tiger and bear 'essentially distinct from existing or extinct species ... and remarkable also for large size'.

In the matter of size, diminutive breeds were as notable as big ones. They added much to the charm of Falconer's imaginary cavalcade of extinct species, though they complicated the work of identification. The pygmy rhino had to be distinguished from immature versions of

Giant wild yak shot by the Sven Hedin expedition of 1907–9

its larger brethren, and the toy hippo from the toddling offspring of the standard hippo. At the other end of the scale the giant *Sivatherium* and the *Colossochelys* remained unchallenged – and unseen. No seeker of hidden Shangri-las, and no explorer of the remotest Tibetan tundra, would ever rate his chances of stumbling on a surviving pocket of either. Nor, so far as is known, has anyone suggested linking Baker and Durand's giant *Quadrumana* with reports of the elusive yeti. But the association of prodigious fauna with lofty locations was noted. It would tempt Himalayan sportsmen into the most inhospitable terrain on earth and sometimes reward them. In the Changthang the wild yak (*Bos grunniens*), the world's largest cattle species, was comparatively plentiful into the twentieth century as was the Marco Polo sheep (*Ovis poli*), the world's largest sheep, in the Pamirs.

Contemporaneity, so desirable in the case of fossils, could be a mixed blessing in terms of one's career. For their Siwalik discoveries, Cautley and Falconer had been awarded the Geological Society's Wollaston gold medal. But Falconer's notions about the antiquity of man would surely have attracted greater attention had they not coincided with Charles Darwin's *On the Origin of Species by Means of Natural Selection*. The two men were of similar age; both had studied medicine before being drawn to natural history; and they were certainly well acquainted. Up to a point their findings complemented one another. By airing the idea that man's history started not 6,000 years ago as deduced from Genesis but the several million years ago implied by extending it back into the fossil past of deep time, Falconer opened up a chronological perspective long enough to accommodate natural selection's extreme gradualism. Darwin would admit as much: 'The high antiquity of man ... is the indispensable base for understanding his origin,' he declared in the 1871 *Descent of Man*; he therefore took this antiquity 'for granted'.[24] But while Darwin's publishings were major intellectual events, Falconer's trickle of monographs scarcely caused a ripple. He was never in the happy position of being able to devote years to a major synoptical work, let alone to fund a passage on a five-year circumnavigation of the globe. When in the 1830s the *Beagle* had been wafting Darwin round the world's choicest natural history sites, Falconer had been tramping the dusty Siwaliks, of which no one had heard, in search of bones, about which no one much cared.

In 1856, back from a second stint in India, this time as superintendent of Calcutta's botanical garden, Falconer transferred his interest in extinct mammals from Asian to European fauna, or more precisely to 'the Species of the Mastodon and Elephant occurring in the fossil state in England'. From a study of known finds he distinguished three species of each genus, and then did much the same for England's rhinoceros fossils. He seemed to have lost interest in the search for early hominins. While not despondent, he grew more cautious in predicting fossil evidence for the antiquity of man. Having failed to find any indication of a human presence in the fossil beds of the Siwaliks, he now had less confidence in Hindu tradition, let alone Iroquois creation myths.

Meanwhile his achievement in placing primates (*Quadrumana*) among the Siwaliks fauna had been confirmed – yet somewhat eclipsed – by similar and almost simultaneous reports from elsewhere. A gibbon's fossilised bone had been found in a Tertiary deposit in south-west France and other primate bones were reported in a cave stuffed with fossils in Amazonia. The comparatively undisturbed nature of the calcified deposits found in such sealed 'fossil caves' and inaccessible 'bone caverns' now looked to offer the best hope of establishing the all-important contemporaneity between extinct fauna and early man. Just as geologists, by assigning sedimentary deposits to a particular era – Tertiary, Pleistocene and so on – afforded clues to the antiquity of the fossil bones embedded in these deposits, so palaeoanthropologists could deduce the antiquity of man from that of the fossil bones among which human artefacts, and perhaps human bones, might be discovered.

In 1858, while revisiting such British sites as he thought promising in this respect, Falconer got wind of a newly discovered cave complex near Brixham in south Devon. Within a week he and a colleague had installed themselves in a Torquay hotel. A brief survey of the site led to its being leased for excavation, and an appeal to the Geological Society produced the necessary funding. Initially Falconer 'did not regard the cave as a test case for determining the age of the human race'.[25] He was still mainly interested in the presence of the fossilised pachyderms familiar to him from the Siwaliks though not usually associated with the south coast of England. But on returning to the Brixham site after its layered excavation had begun, he found grounds for second thoughts. As well as mammalian bones, the excavators had encountered what he recognised as 'human industrial remains'.

These comprised numerous 'flint-knives', later identified as chipped hand axes. Better still, they had been 'exhumed from different parts of the cavern, mixed in the ochreous earth indiscriminately with remains of *Rhinoceros*, *Hyaena* and other extinct forms'.[26] There was nothing to suggest the cave had ever been disturbed. Reindeer antlers found on top of the fossil- and flint-bearing strata were evidently later accretions, implying that the reindeer, prolific during the Last Glacial Maximum but long since extinct in England, had 'continued to be an inhabitant of Britain after the appearance of man in the island'. In other words, the reindeer provided another clue to the antiquity of mankind: man's advent must have preceded the last great Ice Age. For Charles Murchison, editor of Falconer's *Memoirs and Notes*, that clinched the matter: 'The great and sudden revolution in modern opinion, respecting the probable existence at a former period of man and many extinct mammalia, has been universally attributed to the results of the exploration of the Brixham Cave.'[27] Falconer was less sure. Men and extinct mammalia had indeed coincided. But who were

these makers of flint hand axes? Where were *their* fossilised bones?
How far back in deep time could they be dated? And were they really
our ancestors, or were they perhaps representative of some other strain
of hominin? Every discovery just raised more questions.

In 1859, obliged for health reasons 'to seek a warmer climate for
the winter', Falconer extended his enquiries to France and Italy. Sicily
rewarded him with coastal 'bone caverns' containing the remains of
a 'vast number of hippopotami', plus 'siliceous' slivers that 'closely
resembled obsidian knives from Mexico, and flint knives from
Stonehenge, Arabia and elsewhere'. But it was France that afforded the
most rewarding sites. At Abbeville in Picardy he sought out Jacques
Boucher de Perthes, an eccentric septuagenarian who had amassed a
collection of stone artefacts, including hand axes, extracted from the
gravel terraces of the Somme. The meeting went well: Falconer secured
authentication of his Brixham 'flint-knives' from the curator of Europe's
finest assembly of Stone Age implements; and de Perthes secured the
support of the most internationally respected palaeontologist in his
efforts to get the Paris savants to accept his ideas on the Stone Age
antiquity of man.

At Falconer's urging, there followed him to Abbeville the eminent
British geologist Sir Joseph Prestwich and the archaeologist John Evans
(the father of Sir Arthur Evans, excavator of Minoan Crete). Both
agreed with Falconer about the authenticity of the Abbeville hand axes,
so providing de Perthes with an international endorsement which the
scientific establishment in Paris could not but respect. Yet no sooner
were both London and Paris edging towards acceptance of mankind's
remote antiquity than serious doubts began to circulate. It was rumoured
that de Perthes, whose scholarly credentials had always been suspect,
had 're-labeled many of the flint tools in his collection'. By tinkering
with the details of their provenance and altering his own records he
'lessened the contributions of others while inflating the magnitude and
orginality of his own achievement'.[28]

Matters came to a head when in 1863 one of de Perthes' workforce
received the princely sum of 200 francs for uncovering half of a well-
preserved human jaw. A couple of teeth were associated with it, and more
of the chipped flint hand axes were found in its vicinity. Here surely was
incontrovertible proof that tool-making humans had coexisted with
extinct mammals, whose fossils had also been found in the Somme

Valley deposits. 'At last it was thought that the objects so long sought for in vain had been discovered,' recalled Murchison. Prestwich, Evans and Falconer all happened to be back in France at the time. Within ten days of the find being announced they again descended on Abbeville. And having inspected both the site and its yield, the suspicions of all three 'were instantly aroused'.

Prestwich and Evans 'controverted' the authenticity of the jawbone, and Falconer, after a second examination, declared some of the hand axes 'spurious', all of which was of course vigorously contested by their French counterparts. Although the British declined to accuse anyone of deliberate deceit or organised counterfeiting, their denial of any such intent was in itself deeply offensive. On both sides illustrious reputations stood impugned, academic integrity was being questioned, national honour was at stake. Only a competent review of all the evidence by a cross-party panel composed of those most qualified to judge could decide the matter.

Accordingly, in Paris and then Abbeville, for five long days in May 1863, ten of the most eminent members of Europe's scientific community sawed up an old bone and picked over a collection of stones and detritus in what an irreverent wit would later call 'the first battle of the Somme'. The results were inconclusive, the outcome even more acrimonious. In Falconer's words, 'the case throughout maintained a perplexed and contradictory character, not to be surpassed probably by any *cause célèbre* on record'. Professor George Busk, the surgeon and anthropologist on the British team, declared the jawbone not very old and quite consistent with what one might find in any cemetery. As for the hand axes, John Evans had 'not the slightest doubt' that most of them were modern fakes; the artisans responsible for them had perpetrated 'a fraud, and a most ingenious and successful one'. But perhaps the most damning admission came from de Perthes himself. His workforce, he said, had acquired great skill in producing replica hand axes; in fact the results were so good that he had himself sometimes been deceived by them.

Of the British experts, Falconer was the most reticent. No doubt he was painfully aware of the irony that it was he, the champion of the antiquity of man for thirty years, who was now instrumental in discrediting the most eagerly sought proof of that antiquity. He kept on good terms with de Perthes and in a letter of March 1864 applauded

the first discoveries of Stone Age art which were being made in the caves
of Dordogne by Édouard Lartet, one of the French experts from the
Abbeville panel. Falconer's letter appeared in *The Times* in March 1864.
It seems to have been his last. Still in his fifties, Hugh Falconer died the
following January, possibly from pneumonia following a winter trip to
Gibraltar to inspect more caves.

By the time of Falconer's death the case for the high antiquity of
man, bolstered by Darwin's theory on the origin of species and little
discredited by the Abbeville fiasco, was being taken as proven by most
men of science. Prestwich decided man had been around for at least
20,000 years; Sir Charles Lyell, the greatest living geologist, plumped
for at least 100,000 years. Yet still there were very few undisputed
finds of fossilised human bones; moreover it wasn't certain that such
finds actually belonged to *Homo sapiens*. The first Neanderthal skulls
had already been discovered in Germany and Gibraltar, but only in
1864 were they correctly recognised as belonging to a distinct species.
Carbon and then radiometric dating would reveal that man's high
antiquity was higher even than Lyell had supposed. Genomic studies
would revolutionise the process of identification, yet also complicate
it by revealing the extent of human–Neanderthal interbreeding. There
was much still to be learned. And, in the case of Himālaya, much is still
being learned.

4

When Men and Monkeys Meet

Great things are done when men and mountains meet;
This is not done by jostling in the street.

William Blake (1808–11)[1]

It was a conceit among officers of the 1904 Younghusband expedition that the Tibetans were an ill-favoured race. Adjectives like 'doltish', 'simian' and 'stunted' were as freely applied to them as the usual 'dirty' and 'unwashed'. 'The Tibetans carry about on their persons a particularly luxuriant parasite fauna of the familiar type,' noted Perceval Landon of the London *Times* in a brief excursion into Tibetan entomology. The women were no better than the men. Their complexions were actually quite fair, but they too never washed and, on attaining their early twenties, 'they adopt the custom of besmearing the forehead, cheeks and nose with crimson kutch'. This cosmetic pigment blackened as it dried, like blood, giving the wearer an owlish appearance and turning even a pleasing countenance into that of a witch. Some said it was supposed to 'mitigate the glare of the sun', others that it was 'to save the ladies of Tibet from the sin of vanity'. Landon was not convinced by either explanation.[2]

No one could accuse the Tibetans' forebears of vanity. In pre-Buddhist mythology, their earliest ancestors are described as red-faced hominins, 'something between men and monkeys with hairy bodies and curtailed hind parts' – or, just as mystifyingly, 'bottoms like rudders'. Given their mixed parentage, these creatures could never have been other than hideous. The father of the race, the Tibetan Adam, was an actual ape, though a king

Tibetan nuns shave their heads but on festive occasions may don abundant wigs

and bodhisattva (a preincarnate Buddha) to boot; and his Eve was a quite repulsive 'rock-ogress', albeit with a big heart and a beguiling alter ego.

According to a later and suitably Buddhist-adjusted version of the same creational pairing, it had all begun when Amitabha, the Buddha of Boundless Light, became so overwhelmed with compassion for the wretched plight of the living that his head 'burst into a thousand pieces'. Out of this explosion 'there appeared an emanation in the form of a monkey-king by name Halumantha, and Tara [Avalokiteswara's consort] sent forth an emanation in the form of the wrathful Ogress of the Rocks'.[3]

> 'Could the monkey-king see his way to becoming a monkey-hermit?' asked [the Buddha of] Boundless Light. 'And could he meditate in the Land of Snows and thereby lead many beings to the buddha-fields?'
>
> 'I can,' came the reply.

So the monkey-king was ordained a hermit and sent away to meditate in a cave in the Land of Snows. He was not alone for long. 'Just then the Ogress of the Rocks appeared … and made lustful and lascivious gestures.' The hermit was having none of it.

> Eight days later the Ogress reappeared in the form of a very beautiful woman and said, 'You and I must set up house.' Then she showed her breasts and her sides, but the monkey-hermit struck the ground and sat with his back to her.

The Ogress of the Rocks was in despair. Either the monkey-hermit must break his vow of chastity or she would die childless. Nay, worse still, she might 'set up house' with an ogre, she said, and terrorise all creation.

> 'Every morning we shall kill tens of thousands of living creatures. Every evening we shall kill thousands of living creatures. Then innumerable ogre-children will be born and this snowy realm of Tibet will become a realm of ogres and not of men. Moreover these ogres will eat all living beings in the world. So you must have compassion on me.'
> And so saying, she wept with wails of affliction.

By now it was all getting too much for the monkey-king turned hermit. Being responsible for disappointing a suitor was one thing, but being responsible for loss of life on a genocidal scale was quite another, especially for a bodhisattva. After a quick excursion to 'Mount Potala' to receive Boundless Light's blessing on his proposed course of action, the monkey-king returned to the Land of Snows and capitulated to the Ogress of the Rocks. 'So they set up house and lived together.' But that was only half the story.

'Nine or ten months later' a child was born. He was not much like either his father or his mother. 'He stood upright [so presumably his parents were on all fours] and had a flat red face and no tail. He ate red meat and drank warm blood; he was impure in his ways and would never stay still … He was like a wild monster who no one could tame.' Fearing that, as soon as he was big enough, he might devour his mother, his father led him away and 'left him in the wood named "Peacock-Flock"'. It was somewhere 'in the south'.

At this point the two best-known versions of the legend diverge. One says there were six sons, that they did have tails, that hair covered their heads and bodies, 'and that, moreover, they knew how to speak'. It also says that all six of them were deposited in Peacock-Flock Wood and that they there became 'a-hungered', as the English translation has it. The wood was no Eden, but it afforded shelter and was full of monkeys 'for them to live with'. The other version simply says that the solitary ne'er-do-well son was dumped in Peacock-Flock Wood and there 'disported himself' with monkeys.

The outcome was the same anyway. When after only a year the monkey-king returned to the wood to check on his offspring, he was confronted by around 500 grandchildren who 'thereafter increased still more and filled the whole area'. In fact the population explosion was threatening their very existence. Those that had had tails were losing them and those with hair were shedding it. 'They were neither monkeys nor yet men, and they were exposed to the rain and the sun's rays in summer and the wind and the snow in winter, and they had neither food nor raiment.' And so they came to their sire, 'holding out beseechingly their helpless hands, exclaiming "Father, what shall we eat? With what shall we clothe ourselves?"'

The monkey-king knew not what to do; but answer came from him of Boundless Light who was now revealed as none other than the great bodhisattva Avalokiteswara, the embodiment of compassion. Taking up a handful of gold dust Avalokiteswara scattered it over the land of Tibet, 'consecrating it as five kinds of grain that it might be food for those households, namely barley and wheat and rice and sesame and peas'. The monkey-king understood; agriculture was the key to his descendants' survival. The land was theirs to be worked. The seeds must be sown and the rest of the gold likewise so that in time it could be mined. For as Boundless Light explained to the monkey-king, 'these thy children and grandchildren have finally become men'. Some would farm, others would trade, and if Peacock-Flock Wood was in the south-east, it was well sited geographically and climatically for both. All being well, the upright, tailless, talking monkeys who were now men would be honoured with a succession of semi-divine kings who, from a favoured valley in the south-east, would initiate Tibetan history. And in due course, according to one pious recension, these Tibetan humans would qualify for the guidance and redemption that the Enlightened One himself would bring.[4]

As with most myths – and nearly all of Tibet's ancient history – the marvels described in these popular accounts predate the events themselves by hundreds, perhaps thousands, of years. In the course of this long gestation it is reasonable to suppose that material from non-Tibetan sources was freely incorporated. For instance, the monkey-king cum founding father named as Halumantha in the Tibetan versions sounds a lot like Hanuman, India's monkey-deity best known as one of the principal protagonists in the Hindu *Ramayana*. Hanuman also features in Chinese legend. He is the likely inspiration for Sun Wugong (Wukong), the monkey-king whose exploits fill the pages of the sixteenth-century novel known variously as *Journey to the West*, *Monkey-king* or just *Monkey* and which remains China's best-loved work of fiction and fantasy to this day. Tibet's Halumantha could be directly descended from India's Hanuman or indirectly through China's Sun Wugong.

Moreover it's not entirely inconceivable that such myths cohered to some dimly perceived framework of still greater antiquity. The ogres and ogresses could represent an ancestral memory of contact with, say, Neanderthals or some other primitive hominins whose presence may once have challenged that of *Homo sapiens* and resulted in the interbreeding recorded in the myth. As will appear, coincidence need not obscure significance.

Stranger still, though, is the whole idea of monkeys gambolling across the Tibetan plateau. In a land for the most part devoid of trees or shelter, short of fruit and other monkey fare, intensely cold and with an atmosphere so poorly oxygenated as to require a degree of hypoxic adaptation in most mammals, the existence of monkeys seems unlikely. Nineteenth-century listings of Tibet's fauna rarely, if ever, mention them; nor do those sportsmen who with telescope and gun roamed Himālaya's least frequented heights in search of trophies.

There were, though, a couple of exceptions. In 1889 a French expedition led by the experienced Gabriel Bonvalot and the young Prince Henri d'Orléans (great-grandson of Louis-Philippe, the last king of France) made a winter crossing of a corner of the Changthang en route from Lop Nor in Xinjiang to Hanoi in what was then French Indo-China. Mid-January 1890 found the expedition well to the north of Lhasa, about 4,500 metres asl, and in some peril. Bonvalot thought they must be near the source of the Yangtse and was cheerfully conferring a proud Gallic name on every feature that drew his attention – 'Lac Montcalm', 'Dupleix Montagnes' and so on. They were hoping to strike the pan-Buddhist trail connecting Mongolia and

Lhasa but were otherwise lost. The temperature was down to minus 28 degrees Celsius, rice wouldn't cook because boiling water wasn't hot enough, and a wild yak shot by the prince was so hard frozen it had to be chopped and sawn up like firewood, then singed in a flame as the next best thing to braising. It was weeks since they had met a soul. Instead Bonvalot reported what he must at first have thought was a hallucination. 'Today we have seen monkeys, crossing the frozen river and playing on the rocks which form its banks. But we cannot kill one of these animals, which are very short with red hair, small heads and almost imperceptible tails.'[5] None of which was very convincing, especially the failure to explain why no specimen had been procured. No doubt Bonvalot's report occasioned some merriment among those who thought they knew better. If so, the know-alls were wrong. Thirteen years later and in a very different quarter of Tibet, Captain Henry Wood was quite definitely not hallucinating.

> One day near the village of Sang Sang, Wood had been detained
> at survey work, and whilst hurrying down the hill saw, five or six
> hundred yards away, many animals moving amongst the rocks.
> Looking at them through his glasses, he found them to be a troop
> of monkeys. Unfortunately the day was fast drawing to a close and
> he was still far from camp, so felt little inclined to set out in the
> opposite direction on what might have been a fruitless quest. Had
> he been able to obtain a specimen then, all doubts as to their nature
> would have been put at rest.

At the time, November 1904, Wood was one of four British officers returning to India after the Younghusband expedition by way of the Tsangpo–Brahmaputra and western Tibet. He was a highly rated surveyor and had already made a detour to inspect the Tibetan approaches to Mount Everest. When he sighted the monkeys he had been following a route parallel to, and north of, the great river. The terrain was not quite as bleak as the Changthang but neither was it monkey-friendly. 'Now it must be remembered that the country was destitute of all vegetation but grass and that the hills were bare and rocky,' noted Colonel Rawling, the man in charge of Wood's party. Rawling was mildly suspicious. He had not read Bonvalot, he knew of only one other doubtful sighting, and he confidently pronounced that 'it had never been believed that monkeys existed in the uplands of Tibet'. But on asking around he learned that in summer a few apes had been known to cross the high passes from Nepal. Why they did so and where they went was unclear. Rawling didn't believe they could have crossed the Tsangpo without being seen, nor that, having done so, they could have survived a Tibetan winter. Yet it was now November and, according to Wood, monkeys were still around. 'The conclusion to be drawn is, therefore, that they are indigenous to the country.'[6]

Rawling was right, though for the wrong reasons and with no idea of what species they belonged to. If they were visible at 600 yards (say 500 metres) they may have been as big as or bigger than Bonvalot's monkeys. They were perhaps Nepal Grey Langurs, which are common enough throughout the sub-Himalayan region. On the other hand Bonvalot's monkeys, with their stub tails and small heads, sound like a species of macaque, the most widely distributed monkey genus in Asia and well represented in China where one species, the Tibetan or Chinese Stump-tailed macaque, seems the obvious candidate. Bonvalot's impression of red hair may be explained by these macaques having pink to reddish faces, and their unusually thick coats appear to be an adaptation for survival.

Since 'the black snub-nosed monkey [another species of stump-tailed macaque] has the highest elevational range known of any non-human primate', it has attracted scientific interest. 'In search of clues as to how animals survive in these challenging conditions', it was reported in 2016 that a team of Chinese biologists had studied the genomes of thirty-eight individuals from four of the five snub-nosed monkey species and had discovered 'variants of 19 genes that seemed to have been under positive selection'. The findings may be of medical importance. One of the genes, ADAM9, is found 'in some oxygen-starved cancer tumours'. It might be under selection in the monkeys because of 'the thin high-altitude air to which they are exposed'. According to the report of these breakthroughs in *New Scientist*, a further study found that 'ADAM9 also seems to be under selection in another high-altitude species, the Tibetan chicken.'[7]

More obviously, perhaps, it could be under selection in a human ethnicity with a well-known aptitude for high-altitude conditions like the Sherpa. Genomic study of how hominins may have adapted to life on 'the roof of world' looked to hold the key to an understanding of how and possibly why the first Tibetans got there.

Climatic conditions on the Tibetan plateau are known to have varied enormously in the geological past. Recent findings by palaeontologists from the Xishuangbanna Tropical Botanical Garden in Jinghong and the Institute of Vertebrate Palaeontology in Beijing indicate that 47 million years ago (mya) – so in the early stages of the Indo-Asian continental collision and the Himalayan uplift – the central plateau was a mere 1,500 metres asl and enjoyed ample rainfall plus a muggy average temperature of 19 degrees Celsius. From plant fossils 'a thriving sub-tropical forest system' has been deduced; and even 'as recently as 5–10 mya pollen deposits indicate the presence of deciduous trees comparable to those in modern temperate climates'.[8] Assuming this benign environment pertained, say, 15 mya, it would have witnessed the demise of the Last Common Ancestor from whom the family trees of humans and chimpanzees diverged and may well have provided a congenial Tibetan habitat for the first great apes, including the *Quadrumana*. Thereafter temperatures dipped, but they did so for the most part incredibly slowly, though sometimes erratically, as during the Last Glacial Maximum around 30,000 years ago.

The evolution of *Homo sapiens*, like that of life itself or indeed of the rocks and oceans that support life, took a very long time, longer even than Darwin and Falconer had supposed. It was also a notably wasteful process with an infinity of false starts and dead ends. Nature is profligate. Darwin might quite reasonably have called his theory of 'natural selection' one of 'natural rejection'. Today's surviving species are vastly outnumbered by extinct species, of only a few of whom we are even aware. In those evolutionary trees whose topmost branch presumptuously culminates in *Homo sapiens*, dozens of other hominoid species are represented by inferior branches withered to a brittle extinction by nature's caprice.

As recently as the 1990s the *Encyclopaedia of Indian Archaeology* admitted that 'the complete absence so far of pre-*Homo-sapiens* in India is undoubtedly the biggest lacuna in Indian prehistoric archaeology.' This is in marked contrast to the situation in China where, in the 1920s and 1930s, discoveries made in caves just outside the metropolis then known as Peking first primed Chinese interest in palaeoanthropology. In yielding a collection of fossilised human bones and stone tools dateable to around half a million years ago, the site was just the kind of 'bone cavern' that Falconer and his contemporaries had looked for in Europe. The caves' several crania, mandibles and teeth were duly awarded to an erect (*Homo erectus*) but short (56 centimetres) hominin whom the Han Chinese proudly proclaimed their ancestor as 'Peking Man'. The discovery served as proof that the Han – and, by expansion or migration, supposedly related ethnicities in neighbouring Mongolia and Tibet – had occupied the same region of East Asia for 500,000 years. At a time when the great antiquity of hominoid finds in Africa had still to be appreciated, it was supposed that Peking Man with his stone tools and apparent use of fire might even be the first *Homo sapiens*.

The Beijing institute where Peking Man's fossils were studied went on to become China's Institute of Vertebrate Paleontology and Paleoanthropology and to play a leading role in the DNA analysis of other fossilised relics. But meanwhile Peking Man's fossils had themselves gone missing. Deposited in an American-run medical college in China for safekeeping at the beginning of the Second World War, the skull fragments either went down in a vessel carrying them to the USA, or were stolen before they left China, or were hidden close to their find-site by someone who didn't live to reveal the secret – or just chose not to. The twentieth century's greatest find in palaeoanthropology thus became 'the single greatest loss of original data in the history of paleontology'.

Happily, just before the war, plaster casts of the original skull fragments had been made, and a detailed record of their discovery filed. But plaster casts are of no use to genome analysts. Meanwhile the mystery of the whereabouts of the originals continues to be 'a magnet for con artists and amateur investigators, and the subject of several English-language novels', according to Sheila Melvin of the *New York Times*. Hiding the skulls, if they were indeed hidden, may have been almost as random as finding them in the first place, and refinding them looks to be nigh impossible. Amid suspicions of skulduggery, no doubt, in 2005 an official Working Committee to Search for the Lost Skullcaps of Peking Man was announced and a hotline set up for informants; Melvin reported that it had yielded '63 clues' in the first

few weeks. Gao Xing, a member of the working committee, was less
sanguine. The finds 'may well have been destroyed', he conceded, 'but
we have to look'.⁹

What, if anything, Peking Man might have revealed about the origins
of the Tibetans or any of Himālaya's other peoples is a sensitive subject. In
China 'archaeology is seen not only as a research tool, but as a means by
which to create a significant connection between Tibet and the remainder
of China', warns prehistorian and anthropologist Mark Aldenderfer of
the University of California, Merced.¹⁰ Establishing a common ancestry
for the Han and Tibetan peoples bolsters Beijing's case for giving political
expression to the post-1950 unification of Tibet and China.

Palaeoanthropology can't always be counted on to oblige, though.
New finds overturn former certainties and even old finds can hold
a few surprises. In 2010 what might be called 'a tale of two caves'
opened a whole new vista on Asia's earliest inhabitants. Both of the
caves in question had been known for some time. One was south
of Barnaul in the Altay (Altai) Mountains of southern Siberia near
the Mongolian frontier and had first been investigated by Soviet
scientists in the 1970s. They knew the place as Denisova, so named
for one Denis or Dionysus, a schismatic 'Old Believer' who had taken
up residence there in the eighteenth century to evade persecution.
Denis no doubt appreciated the cave's comparatively easy access,
opening into a roomy main chamber with side galleries and with
what irreverent journalists would later describe as 'high ceilings and
a lovely view'.

Dating and identifying fragments of interest deep in the Denisova
cave's Stone Age landfill has aready taken several decades and is still in
progress. New scientific advances, like DNA analysis and a method of
'fingerprinting' protein residues using mass spectrometry, have greatly
extended the range of obtainable information, the time taken to obtain
it – and the challenges all this poses for an uninitiated generalist and
his gentle readers.

Of the fossilised bone fragments found at Denisova, some have been
identified as those of extinct hyenas, mammoths, rhinos and cave lions
that might have been familiar to Falconer and Cautley. No evidence
of monkeys has been reported, though the numerous stone and
bone artefacts already found include items typical of sites in Europe
associated with Neanderthal occupation. This Neanderthal link has

been borne out by DNA analysis of Denisova's yield of hominoid bones. The cave's earliest occupants appear to have been Neanderthals; and Denisova being thousands of kilometres east of any other known Neanderthal site, this is taken as evidence of Neanderthal mobility and of a much wider distribution than previously supposed for one of man's closest cousins. Indeed the Neanderthal DNA obtained from Denisova's bones exhibits more unique sequencing than that from anywhere else and now serves as something of a benchmark in Neanderthal identification.

But the big surprise came as the excavations at Denisova proceeded: the Neanderthals, it seemed, were not alone. By 2008 DNA analysis of the, by now, thousands of bone fragments was revealing the presence of other hominins who were neither *Homo neanderthalensis* nor our own *Homo sapiens*. Known as Denisovans, the new hominins are genetically quite distinct from Neanderthals, though both are descended from a common ancestor (who is also *Homo sapiens*'s ancestor). Whether Neanderthals and Denisovans had ever actually shared the cave seemed unlikely given its estimated 280,000-year record of occupancy. Yet in 2018 it was reported that a toe bone had been identified as that of 'a girl whose mother was a Neanderthal and whose father was a Denisovan'. 'This hybrid child lived between 55,300 and 84,100 years ago,' reported Carl Zimmer in the *New York Times*, after which long interval evidence of Neanderthals declines.[11]

Denisovans and/or their hybrids may have lingered on. Tools, ornaments and beads dating from 45,000 years ago could be their work or that of the cave's first human occupants. Much has still to be learned of the Denisovans, not least where and when their DNA came to be represented in that of *Homo sapiens*. Recent findings indicate that Denisovans, like Neanderthals, travelled far and interbred with other hominins, including China's Han ancestors. They even reached Australasia where Denisovan DNA reportedly constitutes as much as 5 per cent of the genome of the peoples of Papua New Guinea.

The final surprise in this tale of two caves came as this book was being written. Until 2019 the airy Denisova cave in the Altay Mountains remained the world's only known site of Denisovan occupation. But in that year everything suddenly changed; a second cave grabbed the headlines as palaeoanthropology's focus was redirected south and up on to the Tibetan plateau.

Known as Baishiya Karst Cave, this latest bone cavern is located in Xiahe county of China's Gansu (Kansu) province, about 180 kilometres south-west of Lanzhou (Lanchow), the provincial capital, and amid the rocky skyline of the Xiqing Shan, an extension of the Kun Lun. Access being through an aperture in a cliff face, ladders are needed to reach it and it's not as commodious as Denisova. Until recently it was known only as a place of Buddhist retreat and, to a privileged few at the Lanzhou museum, as the find-site of half of a fossilised hominin jaw complete with a couple of large teeth. The jawbone had been discovered by a Buddhist monk, presumably a Tibetan, who was meditating in the cave and who duly passed it to his spiritual mentor, 'a living Buddha', who deposited it in the Lanzhou museum. That was in the 1980s. For the next thirty years the bone seems to have been ignored. Unlike the bone fragments from Denisova the Baishiya jawbone contained no DNA; there was therefore no way of identifying who it belonged to; nor was there any reason to suppose the Baishiya cave was in any way linked to Denisova.

The breakthrough came when in 2019 it was announced that proteins in the dentine of the bone's two molars had been subjected to analysis by the latest in hi-tech diagnostics. The results corresponded with those obtained from proteins found in teeth from Denisova that did have Denisovan DNA. The Baishiya jawbone was therefore allocated to a Denisovan. It was the largest Denisovan relic yet found, and unless it had at some point been removed from Siberia and deposited 1,700 kilometres away in that part of historical Tibet that was now in Gansu, it was the first Denisovan find from anywhere other than Denisova itself.

Without direct DNA support the evidence was still open to doubt, but in late 2020 painstaking analysis of sediment from the Baishiya cave produced both DNA sequences and radiocarbon datings, each of which supported the Denisovan theory. Previously, on the basis of palaeolithic artefacts, the earliest hominin settlement in Tibet had been put at 40,000–30,000 years ago. Now it appeared that Denisovans, and probably Neanderthals, had been there c.120,000 years earlier and may have still been there as recently as 45,000 years ago. Interbreeding with newly arrived members of *Homo sapiens* seemed feasible and, according to Dongju Zhang and others of Baishiya's Sino-Australian palaeontological team, it may have been the Denisovans who 'helped today's Tibetan Plateau dwellers on their evolutionary journey of adapting to high-altitude life'.[12]

At 2,380 metres asl the Baishiya cave is not that high but is still high enough for life there to be much more challenging than at Denisova's 700 metres asl. Quite apart from the drop in temperature, higher aridity and greater exposure to UV rays, the barometric pressure at this altitude reduces the oxygen content of each breath to about a third that at sea level. The snub-nosed macaques developed an adaptive capacity to deal with this. So apparently had the Denisovans. And perhaps the first Tibetans had too. Or maybe the Denisovans obligingly shared their own adaptive genes with them. A 2014 study found that adaptive DNA sequences could indeed have been 'introduced into the Tibetan gene pool from a Denisovan-related population'.[13] Another study of the same year found that two DNA sequences in particular 'can only be explained by the introgression of DNA from Denisovans or Denisovan-related individuals into humans'.[14]

When and from where *Homo sapiens* first ventured on to the Tibetan plateau is still uncertain. No doubt some form of protracted migration to the uplands from the lower-lying surrounding lands played a part. It is supposed unlikely that any such population movements were from the south because of the Himalayan barrier, but they could well have originated from Central Asia, Siberia or East Asia. And there may well have been several migrations. Genomic research suggests 'two periods of upward migration onto the Tibetan plateau, one in the late paleolithic (before the Last Glacial Maximum [aka Last Ice Age] approximately 20,000 years ago) and one in the neolithic period'. The major population sources were 'north-east Asians including Han Chinese and Japanese … with some contributions from Eastern and Central Asia'.[15]

Back in 2010 a comparison of Tibetan and Han Chinese genomes conducted by the University of California, Berkeley had concluded that 'the Tibetans split off from the Han less than 3000 years ago'. The genetic codes for a protein that senses oxygen levels had apparently 'spread from fewer than 10% of the Han Chinese to 90% of all Tibetans'. But if the 'less than 3000 years' was right, it was 'the fastest genetic change ever observed in humans'. And since change in genes, as in species, works by natural selection, it may also have been the most wasteful. 'For such a very strong change, a lot of people would have had to die simply due to the fact that they had the wrong gene.'[16]

The Denisovan discoveries of the next decade put this in a different light. So sudden and so recent a genetic change could now be discounted. Having hosted genetic adaptation to oxygen deficiency over untold millennia, the Denisovans had been in a position to provide unacclimatised newcomers with a ready-made adaptive mechanism for survival at altitude. 'Between 50,000 and 30,000 years ago, some Denisovans and the ancient ancestors of Tibetan and Han Chinese peoples had sex, merged their genomes, shuffled the genes like a pack of cards, and produced children who would grow up to have offspring of their own.'[17] To the Denisovans, then, the first Tibetans, howsoever mythologised as simian or ogre-like, owe their survival in the high heart of Himālaya.

5

Of Flowers and Towers

The mountains are as a rule a world apart from civilizations …
Their history is to have none, to remain always on the fringe of the
great waves of civilization, even the longest and the most persistent,
which may spread over great distances in the horizontal plane but
are powerless to move vertically when faced with an obstacle of a
few hundred metres.

Fernand Braudel (1949)[1], *The Mediterranean and the*
Mediterranean World in the Reign of Philip II, p. 34

Vehicles involved in a head-on crash are as likely to be flung on to the
roadside verge as to be piled on top of one another on the carriageway.
The sudden impact may also overturn them and reverse their
momentum making their original direction of travel hard to discern.
It's much the same with colliding continents. As the Indian plate
rams into the Laurasian plate and begins elevating the Himalayas and
thickening Tibet's continental crust, some crust is deflected sideways.
Professor Searle calls this movement 'lateral extrusion' and sees it as
accounting for the syntaxis, that 90 degree change in the direction of
the mountains, that characterises Himālaya's extremities. Trending east
and then south, 'large tracts of crust slid[e] past one another along
enormous San Andreas-type transform faults – fault lines that extend
from the high plateau of Tibet all the way to Vietnam and the South
China Sea and south to Burma and the Andaman Sea'.[2] This rifting
of the South-east Asian appendage is mirrored at Himālaya's western

extremity by similar fault lines running down through the earthquake-prone Hindu Kush of northern Pakistan and Afghanistan.

As with Antarctica or Amazonia, so with Himālaya: it is the periphery and extremities that have attracted the most attention. Here, from Himālaya's arid western ramparts to its sodden eastern declivities, is where the hills most readily bare their geology, yield up their fossilised fauna and seduce the visitor with dazzling blooms and alpine vistas. Here too came some of Himālaya's earliest settlers, and here too is found archaeological evidence to authenticate their presence. Whether or not Fernand Braudel was right about Europe's mountains having no history, the same could hardly be said of Asia's mountain citadel.

Take, for example, the Pamirs in the far west. To them, and more specifically than to Tibet or the Himalayas, Arabic writers awarded the term *bam-i-dunya*, 'roof of the world'. Barely habitable, little cultivated and politically fragmented, the Spain-sized Pamir region might reasonably be described as a great waste of space. Devoid of conifer-crowned crags and sawtooth horizons, the Pamirs' lumpy upland of humped ranges and gravel-filled troughs disdains the picturesque. Without shade or shelter it must have seemed an unlikely candidate for settlement, let alone archaeological distinction. Of noteworthy towns there are still none, and of roads few, mostly bad. The average altitude is usually given as 13,000 feet (4,000 metres), and so cutting is the wind and so unvarying the terrain that most travellers failed to notice any antiquities at all. In the nineteenth century British and Russian explorer–geographers traced the region's hydrography while decimating its wildlife and festooning their maps with fanciful frontiers, but it was left to an expedition from Denmark in the late 1890s to direct attention to the archaeology of the Pamirs.

Primed on tales of survival in Greenland, which was already a Danish dependency, Lieutenant Ole Olufsen and his colleagues came to Himālaya well prepared. Following a reconnaissance in 1896–7, in the autumn of 1898 they arrived overland from Osh in Russian Ferghana (now in Kyrgyzstan) along with twenty-two local attendants, eighty-three porters, thirty horses and '6000 Danish Pounds' of instrumentation (chronometers, thermometers, theodolites, astronomical circle, cameras, glass plates and so on). They had also equipped themselves with twenty-two 'magazine guns and revolvers', an unspecified number of 'fowling pieces', 'a phonograph with appurtenances' (for linguistic recordings as well as amusing the locals with the sound of their own voices), a five-man collapsible boat with ensign (Olufsen was a naval lieutenant) and a cast-iron cooking range in kit form. The Danish government and the Carlsberg Foundation had largely funded the expedition, and the imperial Russian government was supporting it in return for a share of its findings. The British government of India merely insisted it not cross the frontier into Afghanistan, a regrettable stipulation since the expedition's stated objective was 'the exploration of the South Pamir [which the frontier bisected] up to the Hindu Kush [which was on the other side]'.

For the most part the agreed border followed the fast-flowing River Panj, one of the headwaters of the Oxus. Running east–west through the Wakhan valley, the Panj's south bank had been designated the Wakhan Corridor, a long strip of supposedly neutral Afghan territory interposed as a buffer between British feudatory territories to the south in the Hindu Kush (like Gilgit and Chitral) and Russian-controlled regions to the north in what had been Bukharan territory (and would eventually become Tajikistan).

Thirteen years earlier, in a stand-off known as the Penjdeh Incident, the British and the Russians had nearly come to blows over this northern frontier of Afghanistan. It remained a highly sensitive region, although Olufsen showed himself, unusually for an explorer, to be both tactful and amenable. He kept to the Russian side of the Panj, made no attempt to ford it and agreed to make his winter headquarters well to the north of the contentious corridor. Accordingly, in late October 1898 the expedition repaired to Khorog, 200 kilometres downstream, where they windproofed a crumbling habitation, hoarded five months' food and fuel, hoisted the Danish flag, assembled the cast-iron cooking range and dug in for the long Pamir winter. They were joined there by an obliging *kazi*, an official from the Afghan side of the river who would provide much of the ethnological and geographical data collected by the expedition; winter was not to be time wasted. But having already paid one visit to Wakhan, Olufsen's appetite for the unexpected was whetted. In early March, as soon as the snows permitted, his team set off upriver again, threading the gorges of the Panj where it ricocheted through the mountains till they regained the Wakhan valley and could turn their attention to Wakhan's improbable array of what they were already calling 'castles' or 'fortresses'.

In his *Through the Unknown Pamirs* Olufsen lists seven of these fortress sites on the Russian side of the river and supposes there could be more on the Afghan side. Often some distance from the river itself, they were built entirely of large uncut stone slabs and sun-baked mud bricks, and they blended so well into the background of steepish hillsides that they could be hard to spot at a distance. They were not, though, insignificant. At a place called Zunk towards the top of the valley, the fort of Zengi-bar occupied the entire surface of a stone platform 458 metres in circumference. Its surviving walls still stood as high as 7 metres and connected at least twelve square towers. The fort was 'so large it would hold several hundred people', thought Olufsen. At the lower end of the valley, the walls of Yamchun (Yemchun) hinted at something even bigger. The builders of this fortified enclosure, with its 'considerable circumference of about 12 kilometres', must have commanded resources of a value to warrant its erection as well as fund the actual work. 'It is almost incomprehensible that they were able to build such a mighty construction in defence of their poor mountain province, and still more incomprehensible that they were able to support a garrison large enough for its defence.'[3] Who the 'they' of these gargantuan constructions were is still unclear even today; so are the dates to which the fortifications should be ascribed. On the evidence available to him, Olufsen became convinced that Wakhan had once been overrun by the Siah Posh, the infidel 'Kafirs' of the Afghan Hindu Kush whose last mountain retreat in what is now the Afghan province of Nuristan was being forcibly incorporated into the kingdom of Afghanistan even as the Danish expedition went about its work. 'There is not the slightest doubt that they [the Siah Posh], not so very long ago, possessed [W]akhan,' says Olufsen, and 'after a detailed examination of all the fortresses, and a thorough interrogation of the natives, we made sure that all the fortresses and strongholds were built by the Siah Posh'.[4]

Olufsen was right enough about the Siah Posh having raided into Wakhan and perhaps even ruled there. Right, too, about the Kafirs' warlike reputation and their constructional skills. But his chronology was highly conjectural. Some of his informants insisted their grandparents actually remembered the Siah Posh occupation. The know-all *kazi*, on the other hand, said it was at least 300 years ago and that the fortresses 'may possibly be 500 or 600 years old'. Moreover, though clearly built

Ruined Yamchun, one of several Cyclopean fortresses in the Panj valley of the Pamirs. Their age and function are still uncertain

for defence, no one was sure what, in a wilderness like the Pamirs, they were meant to be defending – or against whom.

Twenty-first-century discoveries have done little to resolve these problems. Stone towers have been identified further up the Panj River on the approach to the none-too-difficult Baroghil Pass over the Hindu Kush to Yasin and Gilgit. And, much further down the Panj, atop a hill in the Tajik province of Darwas, an entire 'lost city' – no less – is reportedly being unearthed at Kalai Khumb. Dated on stylistic grounds to the second millennium BCE, this 'Castle Karon' is credited with 'multi-storey buildings with their staircases intact, broad streets and even an altar niche with remnants of its original paint', not to mention a ziggurat-like Zoroastrian temple and a polo ground with a 10,000-capacity 'stadium'. 'Will this "Machu Picchu of Tajikistan" become one of the most significant archaeological discoveries of our time?' asked a breathless features writer in the *Telegraph* in 2019.[5]

Olufsen's informants, in awarding the Wakhan fortresses to the Kafirs – that is, the unconverted (to Islam) Siah Posh – had slotted these puzzling constructions into an Islamic time frame: the forts were seen as part of an infidel counter-offensive against the inexorable eastward advance of Islam as spearheaded in the Pamirs by the Iranian-speaking Shi'ites whom Olufsen called 'mountain Tajiks'. But a much older, pre-Islamic time frame is suggested by recent research; and this owes everything to another of history's great speculative trajectories, the westward-leading skein of trails nowadays popularly known as the Silk Roads.

From their archaeological studies on the northern side of the Panj River, Soviet and Russian scholars have suggested a millennium-long period for the construction of the forts, running 'from the 3rd–1st century BCE to the 6th–8th centuries CE, with subsequent occupation in later periods'. Although the archaeologists found no material that lent itself to the more precise parameters of carbon-dating, evidence suggestive of a place of Buddhist retreat has led to this rather generous chronology being whittled down to the more manageable c.50 BCE–250 CE, these being the probable dates of a Buddhist-supporting empire that flourished both north and south of the mountains and is known to South Asian history as that of the Kusana or Kushans.

As the Yuezhi, the Kushans were also known to the Chinese. In c.130 BCE the Han dynasty had attempted to establish contact with these Kushan/Yuezhi with a view to enlisting their support in the Celestial Empire's ongoing struggle with the Xiongnu, its insatiable neighbours in Mongolia. To this end in 138 BCE Zhang Qian, an imperial official, had been despatched on an odyssey of exploration that lasted thirteen years and took him across the Pamirs as far as Persia. Explorer Zhang's reports opened Chinese eyes to the existence, beyond the deserts of what is now Xinjiang, of 'great states' with wealthy cities and literate societies. Evidently Han China was not alone in the world; it might not even be its 'middle' kingdom. The effect of these revelations on Chinese imperial policy has been compared to that of the discoveries of Columbus and da Gama on the fortunes of the Iberian kingdoms. Military and commercial missions following in Zhang's footsteps feature frequently during the Later Han period and would climax under the Tang dynasty in the seventh century CE. 'Explorer Zhang deserves recognition as both the pioneer of the "Silk Road" and the first to play "The Great Game".'[6]

The idea that the Pamirs more often served as a conduit than as a barrier has been further developed in research undertaken by John Mock of the American Institute of Afghanistan Studies. Relying on later (seventh–eighth century CE) Chinese and Tibetan sources, plus the abundant archaeological evidence provided by the excavation of sand-buried ruins further east in Xinjiang's Tarim basin, Mock sees the establishment of strategic garrisons and watchtowers as key features in imperial China's 'patron–client relationship' with its frontier neighbours. It was a relationship that partook of the nature of trade to the extent that bulk consignments of silks and silver coins were disbursed to the empire's client satellites in return for their so-called tribute of livestock (especially horses), skins and gems. And such, it seems, may have been the resources at stake in the southern Pamirs as early as the dawn of the Common Era.

The location and design of the Wakhan forts, built with massive walls and towers on high ground above narrow river corridors, comports well with a network of control over trade routes.[7] Whether under Han or Kushan control then, by the first century BCE thanks to an exchange that wasn't quite trade the Wakhan was already serving as a corridor of military, cultural and quasi-commercial significance. By cutting across Himālaya's western extremity this 'Silk Road spur' linked the shifting frontiers of China's western provinces to the world of Buddhist India beyond the mountains and to that of Zoroastrian Persia beyond the Oxus. From all three directions – east, south and west – a lot more traffic would wind its way into the mountains, turning the Himālaya of the second CE millennium into something of a cultural crossroads. Unwittingly Lieutenant Ole Olufsen of His Danish Majesty's Royal Navy had given the lie to that Braudelian suggestion of the mountains being beyond the reach of history and constituting an impediment to the spread of civilisations.

There are, as it were, two Tibets. There is th[e] plateau country including the lake region, called the Chang Tang, and the upper courses of the great Tibetan rivers, where they flow eastwards or south-eastwards in comparatively wide shallow valleys; and there is the little known and far more formidable country comprising the middle courses of these rivers, where, having dug themselves in, they change direction to the south and force the barrier ranges to flow down to the plains of India and China.[8]

As a botanical explorer Frank Kingdon-Ward was not particularly drawn to the first of these Tibets. Its 'plateau country' was as arid and wind-blasted as the Pamirs. To someone trying to make a living out of discovering plants suitable for propagation in the moist conditions of the British Isles, such terrain was of only marginal interest. The other Tibet, however, the country's south-eastern extremity of teetering gradients, dripping gorges and lush vegetation, was a different matter. To a plant-collector it was a veritable Eldorado.

> A country of dim forest and fragrant meadow, of snow-capped mountains and alpine slopes sparkling with flowers, of crawling glaciers and mountain lakes and brawling rivers which crash and roar through mountain gorges; and, where men dwell, of lonely monasteries plastered like swallows' nests against the cliffs, and of frowning forts perched upon rocky steeples, whence they look down on villages clustered in the cultivated valleys at their feet. Such is the other Tibet, a land unknown to the outside world, yet presently to be known by some of the most wonderful flowers ever brought out of the cold heart of Asia.[9]

Though small compared to the rest of Tibet, this other Tibet covered an area equal to that of France and Switzerland combined. It extended right across the Himalayan axis, so confounding several ethnic divides and locating, for instance, the botanically rich Tibetan district of Pemako south of both the Great Himalayan chain and the north–south watershed. The geography had a lot to answer for. 'No country in the world is so deeply rent by rivers, so rugged with great mountain ranges, so bristling with high peaks,' noted Kingdon-Ward. And no country in the world was quite so demanding of the traveller.

> Add ... a scanty population confined to the main valleys; a climate which varies from sub-tropical to arctic, the only thing common to the whole region being perpetual rain; snakes and wild animals, giant stinging nettles and myriads of biting and blood-sucking

ticks, hornets, flies and leeches, and you have some idea of what the traveller has to contend with.[10]

Kingdon-Ward (born without a hyphen, he subsequently adopted one) became accustomed to these inconveniences in the course of some twenty-five botanical forays over fifty years. Nearly all of them took him into Himālaya's eastern syntaxis where in an almost unmappable tangle the modern frontiers of India (Assam and Arunachal) intersect with those of Burma and China (Xizang, Sikang, Sichuan and Yunnan). Between journeys Kingdon-Ward served, without relish, as an Indian army captain in two world wars. At first his travels were funded by a horticultural seed-merchant; later they were supposed to be self-financing. Belying his wiry physique, his trim military demeanour and the bristle of a regulation moustache, Kingdon-Ward enhanced the skills of a natural travel-writer with lyrical observations that captured perfectly the botanist's enthusiasm for his discoveries. The twenty-five journeys yielded nearly as many books, some of them classics, and his carefully packed plants and seeds did indeed 'bring from the cold heart of Asia' some 200 of those blooms best loved to this day by gardeners with a moist 'west coast' climate.

The plants themselves – the poppies and primulas, the magnolias and rhododendrons – get the star billing in Kingdon-Ward's narratives. His other companions feature sporadically, including 'dog', possibly a border terrier, and 'Cawdor', a young Scottish earl who, in between fretting over the endless deviations that plant-collecting necessitated, attended to those expeditionary details – like porterage and provisioning – of which a botanist on the loose in paradise could be neglectful. Cawdor, or John ('Jack') Duncan Campbell, 5th Earl Cawdor, was present throughout the 1924 expedition to the 'other Tibet' which resulted in the immensely popular *Riddle of the Tsangpo Gorges*; and in the course of a trip that promised nothing in the way of antiquities, it may well have been his lordship who first directed attention to a scatter of stumpy towers on the horizon.

At the time Kingdon-Ward was taking time out to indulge in an excursion – or what, as if it was something totally alien to his avowed purpose in life, he called 'a journey of exploration'. It was August, a month in which the botanical collector had to be patient while desirable plants, identified in flower earlier in the year, set the seeds which could be collected later in the year. Hence the expedition was marking time by breaking out of Pemako's sodden forests and misty gorges to cross the Tsangpo–Salween watershed and 'travel northwards as far as possible'. This took them through the district of Pomé to that of Kongbo. Both were part of Tibet though largely autonomous in that, at the time, they acknowledged Lhasa's authority only in religious matters. Judging by the title of the *Riddle*'s relevant chapter, the original object of the excursion was a 'Journey to a Lost Lake', although this was soon elbowed aside.

> All this country to the west of the Tsangpo bend [where the river starts nearly to loop-the-loop round the 24,000-foot (7,300-metre) Namcha Barwa] is very dry and we were rapidly approaching plateau conditions again. There was a certain amount of coniferous forest up the glens, but not a tree in the valley, only bushes and thorn scrub; the further west we went the drier it became … with not a stick of wood anywhere.[11]

Instead, it was at this point, with nothing much in the way of vegetation to distract him, that Kingdon-Ward allowed his attention to be drawn to the towers. They stood mostly on the sites of ruined villages, sometimes in a good state of preservation, sometimes more or less battered, and as much as '40 or 50 feet high [about 15 metres]'. They were also of 'a curious shape … tapering slightly from the base'. They came in groups, three or four together 'for mutual support', and Kingdon-Ward reckoned there were 'several hundred' in total.

> From a distance they look not unlike factory chimneys but they are as a matter of fact – or rather were once – watch-towers, hollow inside, loop-holed for defence. Staging was erected inside at various levels – you can still see the holes for the cross-beams, and the garrison could enfilade the attackers on any side.[12]

There was no evidence of associated fortresses on the scale of those of the Pamirs, although the towers themselves were a distinct advance on those identified as lookouts by Olufsen half a continent away in the Wakhan Corridor. The ones guarding the Baroghil Pass are described by Mock as being of an unhollowed and unloopholed construction and can only have served as watchtowers, perhaps for signalling the approach of danger. With their several storeys, thick walls and apertures, the Kongbo towers were designed to repel the intruder. 'Such technology is not the work of nomads,' says Rolf Stein in his *Tibetan Civilization*. A nine-storeyed tower was 'reported in Kongpo back in the early twelfth century' and others still stand in central Tibet.[13] But who was responsible? Kingdon-Ward was mystified. There were several theories. One held that they were the work of Kongbo's original inhabitants and 'were captured and destroyed by the Tibetans when Kongpo passed under the control of Lhasa'. Chinese historians are said to associate them with pre-Tibetan tribes like the Qiang or even Indo-Europeans like the Tocharians and to account them 'the prototypes of Tibetan architecture in general'.[14] Another school of thought awarded them to the intruders themselves, who were obliged to build them 'as a defence against the wild Pobas [the people of Pomé]; though in that case who pulled them down?'

The stone towers of Kongpo (Kongbo) in eastern Tibet were typical of similar structures scattered through Himālaya

There is no doubt an element of truth in both versions, which are not difficult to reconcile; after all, the only point of real interest is,

who built them? And that is the one thing we don't know. It may
be remarked that such towers are common throughout the Tibetan
marches, from Kongpo to Szechuan [Sichuan]. In the Mantzu
marches (Yunnan and Sichuan) they are usually octagonal in plan;
in Kongpo they are polygonal.[15]

The ruinous state of some, and that of the adjacent villages, could even
have been the result of the 1910 Chinese invasion of Tibet (in the face of
which the thirteenth Dalai Lama had fled Lhasa to India); but Kingdon-
Ward discounted this. The damage was older and the culprits were more
probably Tibetans 'expanding eastwards from the Lhasa region towards
the gorge country'. Such migrations had been going on for centuries, but
it was impossible to track them because, 'in the 5000 years history goes
back', the landscape had changed beyond recognition. A fast-flowing
river like the Tsangpo was forever devouring its surroundings.

> If it digs down only 3 inches a year, it will dig a bed 1250 feet deep in
> that time. If a mountain range is being uplifted only 3 inches a year,
> it will rise 1250 feet during the same period. If a glacier is retreating at
> the rate of 1 foot a year, it will go back nearly a mile, quite far enough
> to expose a pass, for example. At the same time vegetation and
> climate are shifting and changing; there is no stability, no finality.[16]

Doubtless familiar with works like Eduard Suess's *The Face of the
Earth* and Charles Lyell's *Principles of Geology*, Kingdon-Ward had no
illusions about the earth's permanency. In what sounds like an echo of
James Hutton's 'no vestige of a beginning, no prospect of an end', he
seems to hint at the futility not just of historical reconstruction but of
exploration in general. With 'dog' and the faithful Cawdor snapping at
his heels he would return downriver to Pemako, there to forage across
the heights and indulge the leeches while wrestling once more with 'the
riddle of the Tsangpo Gorges'.

Roughly midway between the Tsangpo's exit-right from the Tibetan
plateau and the Oxus's exit-left from the Pamirs, it's the turn of the Satluj, a
tributary of the Indus, to force its way through the Himalayan barrier. The
Satluj rises somewhat mysteriously in what was once a vast lake in Tibet's

south-western corner and soon exits the country heading west-sou'-west. Unlike the Indus, the Satluj maintains this direction of flow throughout its course, wriggling through the outer Himalayan ranges past deodar-clad ridges like that on which the British had built their summer capital of Simla, and bypassing the badlands of the Siwaliks in which Falconer and Cautley had been so richly rewarded with fossil trophies. The consistency of this course had been recognised as offering the most direct line of communication between India (including Delhi and the Punjab) and the supposedly 'forbidden' Tibetan plateau beyond. By the mid-nineteenth century a trail grandly known as the Hindustan–Tibet Road roughly followed the river's course, entering Tibet by the adjacent Shipki Pass. The road was sporadically promoted as a vital trade route down which Tibet's production of the precious shawl wool could be hauled to the looms of Jalandhar and Amritsar for weaving into Kashmir shawls.

It was also the route chosen by Captain Cecil Godfrey Rawling and his three British companions on their return from Lhasa in 1904. The Younghusband mission had brought an end to Tibet's cherished isolation, and the Dalai Lama had made the first of his twentieth-century escapes from Lhasa, this one to Mongolia. Edmund Candler, Perceval Landon and the other war correspondents had headed back to Darjeeling. Meanwhile Younghusband had extracted concessions from the Tibetans that included a British right to station two trade agents in Tibet. One was to be at Gyantse on the main route from India, the other at Gartok, the administrative capital of far-off western Tibet. This Gartok was Rawling's immediate objective.

Rawling's companions in the assignment were two surveyors, Captains Ryder and Wood, the latter the man who had reported that sighting of monkeys, and Lieutenant Frederick 'Eric' Marshman Bailey, the ice-skating subaltern who had commanded a unit of the mounted infantry during the campaign. The most persistent of Britain's Tibetan trespassers, Bailey was known to Rawling as 'Hatter', presumably because Rawling thought him as mad as one. Their main task was to locate Gartok and prepare the ground for what would rank as one of the bleakest outposts in the entire British empire. They were then free to find their way back to India by whichever route best suited them. They chose the Shipki La and the Hindustan–Tibet Road because it was the shortest. And it was thus, in triumphalist mood and with much end-of-term bonhomie, that they had set off across Tibet. Stalking everything that moved and putting to flight everything that didn't, they reached the headwaters of the Tsangpo and then scaled a pass over the watershed into the Satluj basin.

The pass was called the Ayi La. It was now mid-September and at nearly 19,000 feet (5,800 metres) the pass was the highest, coldest and most rarefied they would encounter. Rawling and 'Hatter' Bailey were among the first to the top and, when they looked down into the Satluj basin, Rawling thought he must be hallucinating. It was 'a country', he would recall, 'the like of which I had never seen before'. Instead of glaciers and scree stretching away as far as the eye could see, he was staring into a subterranean hollow of dun-coloured sandstone that had been worked into the most fanciful constructions imaginable. After draining its one-time lake, the river's feeders had apparently been cutting down into the lake's soft alluvial floor for centuries to carve out a maze of 'great clefts and chasms, many hundreds of feet deep, to the bottom of which the rays of the sun never reached'.

> Narrow precipitous ridges, spire-like pinnacles and isolated plateaux rise on all sides. It is a weird and wonderful sight, but also a depressing one, for neither shrubs nor grass can grow, and a death-like silence reigns supreme.[17]

This was just a foretaste of what lay ahead. After a long descent into the 'weird and fantastic' underworld of stalactite-like pillars and frowning precipices, Rawling and Bailey sighted the chortens and prayer flags of Toling, the religious capital of a one-time kingdom called Gugé (pronounced 'Goo-gay'). They were not impressed. 'Tooling's' temples were too 'gaudily painted', thought Rawling; but high above them and some distance away, he spied a fairy-tale skyline of hoarier walls and crumblier towers. Here, they were told, was the orginal Toling. What remained of it teetered on table-land atop 300 metres of sheer cliff. It could be reached only with scaling ladders and ropes and by crawling through caves and up miles of unlit 'funnel-like passages'.

Rawling and Bailey decided against it. They hoped to reach Simla for Christmas. Time was short. They spent just a night at Toling and never so much as sighted Gugé's equally weather-worn political capital of Tsaparang (Chabrang) only a day's march away. For a description of both places they relied on Ram Singh, an indomitable surveyor provided by the Survey of India who actually clambered up through old Toling's 'funnel-like passages'. Thus, though Ram Singh was no archaeologist, it was his account of Gugé's 'lost cities' as related to Rawling that was destined to provoke a flurry of interest in an unlikely quarter of British India.

Rawling duly published an account of his journey in 1905. He published another on his adventures in *The Land of the New Guinea Pygmies* in 1913 and died at Passchendaele when the Great War drowned out his brand of imperial insouciance. But by then Father Henry Hosten, a Belgian Jesuit based in Darjeeling, had been reminded of something. Three hundred years earlier Portuguese Jesuits had actually established a mission in Gugé. Enjoying the patronage of the Mughal emperor Jehangir, in 1624 Father António d'Andrada had twice travelled north from Goa to Agra, Lahore and Tibet in the not unreasonable belief that, in terms of ritual, monasticism and even theology, Tibetan Buddhism and Roman Catholicism already had much in common. Nothing d'Andrada had found in Gugé persuaded him otherwise. He had apparently been well received there, had converted the king, trounced the lamas in debate, stayed for over a decade, built a church in Tsaparang and, before returning south to India, established a mission that would last at least seventeen years.

Whether, three centuries later, d'Andrada's church might still be standing was what interested Father Hosten when he read Rawling's account in Darjeeling. Perhaps there were even a few long-lost Catholics, descendants of the original converts. He dug out relevant correspondence and prepared some additional notes. As a result, in 1912, eight years after Rawling's reconnaisance, a British official sent 'to enquire into the delinquencies' of the latest trade agent at Gartok set off well primed not just on Gartok's commercial prospects but on a largely forgotten episode in Europe's seventeenth-century dealings with – and beyond – the great Mughal empire.

The troubleshooting British official of 1912 was Gerard Mackworth Young of the elite Indian Civil Service. A formidable scholar, impromptu baritone, lanky fell-walker and future director of the British School of Archaeology in Athens, Mackworth-Young (he too eventually embraced a hyphen) was just the man for Gartok and Gugé. Gartok, Tibet's 'western capital', he found to be not an administrative centre but a sorry collection of 'some fifteen squalid huts huddled together on damp ground in the middle of a bleak and isolated plain'

which, at 4,500 metres asl, was 'perpetually swept by cold winds'.
Winters lasted nine months, during which the trade agent and his
staff of two (clerk and medic) were described as living 'the existence
of prehistoric cave-men'.[18] For 200 rupees a month, life in Gartok as
British India's trade agent was not worth living. Mackworth-Young
recommended withdrawing the agency for at least six months a year.
Then, taking the incumbent agent with him, he headed back to Simla
via Gugé.

Like Rawling, on descending from the Ayi La he was taken aback.
He too could hardly believe his eyes. 'On the plateau I had had the
sensation, common in Tibet, of wandering in another world. At Toling
I thought I was dreaming as well.' The crests of the precipices were so
crumbled and castellated they looked more like the work of man than
of nature. There were 'boulders incredibly poised on slender columns
of earth, like the glacier tables of the Alps but ten times as high ...
[and] now and again when in this country, I rubbed my eyes to see a
real castle in ruins, its jagged outline merged indistinguishably in the
ridge from which it sprang.' Even the weather was unexpected. Rain
was rare and the meandering Satluj sulked in its coils as darkly as the
Styx. Where not in perpetual shade, the canyons pulsated with a heat
worthy of Dante's Inferno. 'True, the Satlej valley is notorious as a sun-
trap throughout its course in the hills; but at 12,000 feet [3,657 metres]
above sea level in these latitudes one feels aggrieved at a temperature
rising well over 100 degrees [Fahrenheit; 37.7 degrees Celsius] even in
June.'[19]

Mackworth-Young, like Rawling, was pushed for time. He characterised
his impressions of Toling and Tsaparang as those of 'a weekend-tripper',
and though hopelessly superficial he justified recording them on the
grounds that there hadn't been many other weekend-trippers. Also he
spoke no Tibetan and could see that 'the orthography of Tibetan names
presents some difficulty'. (He was good at understatement.) But he did
manage a tour of both Tsaparang and the two Tolings. With the help
of the Gartok trade agent he had long interviews with Toling's 'abbot'.
And after entering almost all of Tsaparang's surviving houses he was
pretty confident that d'Andrada's church had long since disappeared.
He also tested the market for Buddhist antiquities. The monks wouldn't
sell and he, of course, feigned indifference. Nevertheless, terms for the
clandestine removal of some statuettes and a lot of loose block-printed

pages were agreed. All the items were from a deserted ruin that had once been a monastery and library and was now unrecognisable as either; they would be safer in the care of the Lahore museum, reasoned their new owner.

On the information available to Mackworth-Young, Guge's glory days had already been ancient history in the seventeenth century. The kingdom's recorded past reached back, it seemed, 'to about 1000 AD' when its rule embraced not only most of western Tibet but all those neighbouring districts of Ladakh, Lahul, Spiti and Kulu which by Mackworth-Young's time were part of British India. Guge had once served as a conduit through which the teachings of the Buddha and the scriptures of Vajrayana had entered the rest of Tibet. In the eleventh century it had attained a spiritual and political eminence that eclipsed that of central Tibet. Dynastic rivalries and dogmatic disputes eventually brought invasions from Lhasa and then Ladakh; no doubt, too, the climate had become a lot drier and much cultivable land had had to be abandoned. But even in 1912 Guge's wealth, mainly derived from the wool trade and some high-altitude gold-diggings in the upper Indus and Tsangpo basins, was still evident. Mackworth-Young noted 'the famous gold roof' of Toling's main temple, the gilded statuary and utensils to be found in all the temples, the richly executed frescoes that covered the walls of so many surviving buildings, and the drifts of unbound pages written in gold ink that lay knee-deep in the Tsaparang library.

'The existence of all the riches and civilization implied in the remains of Toling and Tsaparang is at first sight astonishing,' he declared. Here, in the surreal canyons of the upper Satluj, instead of enigmatic watchtowers and unattributable castles, Himālaya cradled cities whose fame might indeed eclipse that of Machu Picchu – if only they were a bit more accessible.

> Enough however is known of the past of Toling and Tsaparang to stimulate further curiosity, which can only [be] satisfied by the despatch of a competent Tibetan expert to both places, with full permission to study their antiquities at leisure, and to hunt for inscriptions ...[20]

'Unfortunately this is not as easy as it seems,' he added. It wasn't easy in 1912, and it isn't now. In the meantime Mao's Red Guards and Beijing's

Han-first policies have added their destructive venom to nature's erosive tendencies.

Yet Mackworth-Young's hopes for informed research would not go entirely unheeded. The account of his visit can have been known to few. It was first aired in a short address to the Punjab Historical Society in 1913 but was not published, and then only in that society's journal, till 1919. Mackworth-Young duly married, rose to dizzy heights in the British raj, inherited a modest fortune and withdrew from government service to study the antiquities of ancient Greece. 'His words remained for a long time without an echo,' recalled Giuseppe Vincenzo Tucci in the 1930s.[21]

Happily, though, not for ever. It would be Tucci himself, the twentieth century's greatest, if somewhat controversial, Tibetologist, who took up Mackworth-Young's none-too-hopeful suggestion.

6

Scholar, Explorer, Writer, Pilgrim

In recent years an uneasy peace on both sides of the Himalayan border has done little to diminish Tibet's extraordinary isolation. Even with the arrival of satellite photography it remains to this day the least known, least explored country on earth, rich in mysteries, still beckoning us with its secrets and still denying us the answers – a vacuum at the centre of the world.

Charles Allen (1982)[1]

Tibet, even when not politically 'forbidden', has often been found unapproachable. The obstacles to understanding its culture seem as formidable as those to accessing its terrain. For a start, there's the challenge of defining the country. To mollify the Chinese, in the aftermath of the Younghusband expedition the British proposed a division, like that of Mongolia, into an Inner Tibet and an Outer Tibet. Inner Tibet ('inner' to China, that is) was to be the Tibetan-speaking regions to the north and east of Lhasa, including Amdo (Qinghai) and much of Kham, which border the Chinese provinces of Gansu, Sichuan and Yunnan; here Beijing, while acknowledging Tibetan 'suzerainty' (a diluted form of sovereignty) and the spiritual authority of the Dalai Lama, could station its own troops and set up its own jurisdiction. Outer Tibet, comprising the rest of what had been Tibetan-speaking 'Greater Tibet', while acknowledging Chinese suzerainty, was to remain 'in the hands of the Lhasa government', immune to Chinese interference and amenable to British overtures.

This Inner–Outer division was never recognised by all the interested parties, but it was effectively revived in the 1960s when Tibet was reconfigured following its forcible reclamation by the People's Republic of China and the 1959 flight of the Dalai Lama to India. The new Tibetan Autonomous Region (TAR), or more correctly Xizang Autonomous Region (XAR), roughly corresponds to what would have been the Lhasa-centred Outer Tibet. Most of what would have been Inner Tibet, on the other hand, was no longer recognised as being Tibetan at all. Instead its component areas were incorporated into the neighbouring Chinese provinces. Tibet of old was being dismantled just as its name was being erased.

Yet what politics discards, academia retrieves. As Tibet shrinks, its cultural reach expands. Just as the concept of Himālaya was once confined to Tibet, so the concept of Tibet may now be seen as entrusted to Himālaya. 'Failure to locate Tibet within a wider regional context has serious implications,' writes Alex McKay; 'it threatens our understanding of Tibet as a political unit in the "real world".'[2] The post-1950s diaspora of Tibetans to lands beyond the Himalayas is one factor in this realignment. India and Nepal now host around 150,000 Tibetan exiles, with over 20,000 more scattered across the Western world. Less obviously, a finer appreciation of the peculiar nature of Tibetan culture has led to some further repositioning of it, plus a greater awareness of the formidable challenges to an understanding of it.

Giuseppe Tucci may have been the man largely responsible for this development. In the 1930s he advanced the idea that no one 'deficient in that intimate comprehension of *realia*, without which it is difficult to understand fully the significance of Tibetan literature and art', could hope to do justice to Tibetan history and archaeology. According to Tucci, one needed first to 'acquire' Buddhist culture and attain 'that profound and direct knowledge of India and its civilization without which it becomes difficult to understand and assess many of the cultural and historical manifestations of Tibet'.[3]

Naturally Professor Tucci himself possessed such qualifications in abundance. A brilliant scholar, he read and spoke an impressive variety of Asian languages including Sanskrit, Tibetan and Chinese. He was an authority on the art, iconography and literature of Mahayana Buddhism before he set foot in Tibet. By scrutinising *realia* (texts and artefacts) – which he did as a collector and connoisseur – and by 'acquiring' Buddhist culture – which he would do as a devotee and, sometimes, a professed convert – he seems to have been asking for a much more indulgent, even initiated, treatment of the relevant source materials than would be expected of scholars working on, say, Italy's state papers. *Realia* and the 'real world', like 'reality' itself, mean different things to different people. For the Buddhist the only reality is spiritual; all else is illusion. For others reality is decidedly corporeal; it's the world of the spirit that's illusory. Unfortunately, acquiring Buddhist culture in order to reach out for the truths of a less easily communicable reality may entail loosening one's grip on the more tangible reality.

The Italian professor Giuseppe Tucci, here with a Tibetan official, combined unrivalled Buddhist scholarship with the appetite of an insatiable collector

Inspired by Mackworth-Young's account of his 1912 visit to Toling and Tsaparang, in 1933 Tucci and companions – a captain in the Italian navy called Ghersi who was to produce a photographic record of the expedition, and a Tibetan mastiff named Chanku who was to chase away the wolves – headed east via the Kulu and Spiti valleys to the canyons of the Satluj. With twenty men and a like number of horses, the Tucci Scientific Expedition to Western Tibet was not particularly large. It was, though, generously funded. Persuading western Tibet's lamas to part with the precious statuary, artworks and texts of what Tucci deemed their ancient but hopelessly doomed civilisation would mean paying handsomely for them. His original proposal to Italy's Royal Academy had been quite candid on the subject: acquiring a stake in Tibetan culture would be expensive. The expedition had nevertheless been approved thanks to 'the enlightened intervention of the Head of the Government'. This was Benito Mussolini. At *il Duce's* prompting, the Tucci Scientific Expedition to Western Tibet had become something of a national undertaking. The Italian navy, the Bank of Italy and the Savings Bank of Milan had all chipped in, followed by a host of other bodies, among them the National Fascist Confederations of Public Works, of Agriculturists and of Italian Industry. The unhappy association between fascism and Tibetan studies that would soon be epitomised by swastika-flying Nazis converging on Lhasa was already taking shape. Tucci himself, small and intense with tousled hair, untidy moustache and sometimes a goatee, sympathised with the rejuvenating ideals of National Socialism and willingly lent his scholarly authority to Mussolini's campaigns.

The expedition's first impressions of western Tibet were not particularly encouraging. High in the hills above Miang the expedition encountered a trail of body parts scattered across the hillside, then paused to witness a grisly committal ceremony. Eight days after the dismemberment of her corpse, the spirit of the deceased, a young girl, was being entrusted to 'certain mystic forces' prior to being born again. A bell was being frantically rung and a *vajra* or *dorje*, a symbolic thunderbolt, shaken. 'The bell indicates the verity of "void", that is the undifferentiated beginning of all things,' explains Tucci, 'and the *vajra* is ... that which serves for the complete realisation of same.'

> The priest places on his head ... a kind of crown made with five pentagons of papier mâché, that is the five supreme Buddhas, who symbolise the five germinal centres of cosmic evolution made visible in the heart of the undifferentiated beginning of existence; the ministrant is in this way identified with the beginning almost of existence, with the powers which govern the evolution of the world, and he can command all their forces.[4]

Neither Tucci nor his English translator (Tucci wrote in Italian) cares to make life easy for the reader; but the concept here taken for granted is what Professor McKay calls 'spiritual transmission'. Transcending the conventional limitations of movement through time and space, spiritual transmission enables Tibetans to make extravagant retrospective claims and confident future predictions by ignoring things like causality, distance and the passing of the years. An obvious example might be the head of their theocratic state. Living embodiments of the Dalai Lama invariably die, but the Dalai Lama's spirit, being immortal, is always transmitted to another embodiment. It can skip across the mountains or jump generations. The continuity is infallible. Many cherish the hope that the same will be true of Tibet itself.

Tucci's first encounter with Tibet began with his witnessing a traditional rite of committal. The bones of the deceased were usually accorded 'a sky burial' courtesy of the vulture

Miang village, 3,500 metres above sea level and poised over a tributary of the Satluj, struck Tucci as very old and perhaps of importance. A ruined castle overlooked it from the spur of a nearby mountain, but there was no sign of any inscriptions. A cursory excavation of a chorten in the village brought to light only 'a dedicatory folio' containing references to Senge Namgyal, the Ladakhi king whose forces had overrun Gugé soon after Father d'Andrada departed. The chorten was therefore post-1630 and not that old. But the castle's construction was interesting. The foundations were of enormous stones and the defensive walling was of equally hefty blocks of sun-dried clay interspersed with courses of brushwood and small stones. Identical constructional features had been noted by Ole Olufsen in the Pamirs. And just as the Wakhan fortresses had invariably been attributed to the incursions of the Siah Posh Kafirs, so this and other ruined castles in Tibet were often credited to the Mon.

As a referent, the term 'Mon' crops up variously. It can indicate indigenous peoples like the Mishmi and Abor tribesmen of the Tsangpo gorge country (and hence it appears in a linguistic connotation as 'Mon-Khmer'). It can refer to a despised caste of musicians in Ladakh and Baltistan. And it can occur as a generic name for Bhutan and Sikkim. Tucci had nothing to say on any of these. He did imply a connection between Mon and *bon*, but for the time being he left it at that; *bon*, a basket term for all manner of un-Buddhist beliefs, was a big subject.

The nearer the expedition got to the twin sites of Toling and Tsaparang, the more Tucci bemoaned their imminent demise. 'One can understand why this people lives so intensely in its religion, almost entranced and lost in visions of its own. The landscape is the natural background for Lamaism, for its rites, for its demons; all is gigantic and mysterious, infinite and sad.'[5] Toling had had 500 monks in d'Andrada's day; now there were just thirty-eight. Tsaparang, once home to thousands, now had a population of ten. Desertification had already killed off cultivation, the river was carrying away the ground on which the remaining temples stood, and the melting snows were dissolving their famous frescoes. By making off with whatever documents and artefacts he deemed of value, Tucci was not pillaging but conserving.

> The road runs through piles of ruins, alongside *mani* [walls built of flat stones engraved with a sacred mantra], with very ancient *chorten* on the right. There is nothing here that does not give the impression of decay and death. The cave dwellings look like empty eye sockets which watch malignantly the desecrated recesses. The monk who accompanies us is more of a beast than a man.[6]

Tucci had seen enough; and perhaps the monks of Toling had seen enough of Tucci. Fosco Maraini, who would succeed Captain Ghersi as

Tucci's photographer and amanuensis on a later expedition, found the great man's company decidedly taxing. After months together, Tucci still deployed his scholarship to snub his countryman and insisted on being addressed as 'Your Excellency'. 'There is something of the night about Tucci,' wrote Maraini, 'something feline, tantric, not to say sinister.' It was as if 'a living computer [had been] grafted onto a petty, shifting personality, capable of shabby spiteful behaviour, inordinately clever at trimming his sails to every puff of wind'.[7] The lamas of Toling were simply overwhelmed. Tucci's understanding of Buddhist iconography far exceeded their own; he was indeed 'the great foreign lama'. Backed by a bottomless exchequer, the Italian's combination of scholarly credentials and wheedling ways made a convincing case for handing over whatever statuary, relics and textual materials took his fancy.

Soon overladen with one of the greatest ever hauls of Buddhist texts and devotional materials, the expedition headed back to Simla via the Shipki La and Bashahr-Kunavar (Bushahr-Kinnaur) in what is now the Indian state of Himachal Pradesh. At Poo in Kinnaur they were in a hybrid world where, though both Tibetan and Hindi were spoken, the locals used among themselves a language related to the pre-Sanskritic Munda of central India. Other pockets of Munda-related speakers found in the skirts of the mountains from Pakistan to Burma would lead ethnolinguists to propose for Himālaya a pre-Aryan matrix of Munda-type languages.

The religious allegiance of Kinnaur was equally archaic. 'The most ancient aboriginal cults are grafted on the ruling Lamaism,' noted Tucci; 'non-Buddhistic divinities are even today venerated as tutelary patrons.' Indeed a host of practices and liturgies highly offensive to Buddhism were here 'vigorous up to recent times'; human sacrifice was 'common enough' and so were fertility rites involving veritable 'holocausts of victims'. The beliefs responsible, 'the origin of which is lost in the night of time', were dismissed by evangelising Buddhists as *bon*, the religion which Tucci believed to be 'that which preceded theirs in Tibet' and about which he was notably ambivalent.[8]

Bon (pronounced 'po' or 'p'on') and *bonpo*, its adjectival form (used as a noun too to mean a *bonpo* adherent), Tucci often takes in this sense of the pre-Buddhist beliefs and practices of the Tibetan people. Thus a *bonpo* sorcerer whom he had failed to locate in Spiti was 'one of the superstitious followers of the indigenous religion of Tibet'. The *bon* religion's 'doctrinal systematization' he understood to be the work of Shenrap, 'a man of Žaṅ Žuṅ', of whom – and of where – Tucci has nothing further to say in the context of his 1933 expedition. But as himself a Buddhist, indeed an authority on all things Buddhist and 'the great foreign lama', he could not but be of the opinion that *bon* was anathema and Shenrap no better than an Anti-Christ.

Times, however, have changed. Ninety years later many scholars believe that the doctrines of *bon*, the exploits of its semi-divine hero Shenrap, and his realm of 'Žaṅ Žuṅ' (Zhang Zhung, Shangshung and so on) are central to an understanding not just of prehistoric Tibet but of pre-Buddhist Himālaya. The breakthrough may be traced to 1956. In that year David Llewellyn Snellgrove, an English disciple and former pupil of Tucci's who taught Tibetan at London's School of Oriental and African Studies, trekked into Dolpo near the Tibetan frontier in remote western Nepal. He had heard tell of its 'renowned *bonpo* establishment' of Samling, and his men realised they must be getting near when 'we came upon a prayer wall and noticed that the invocations upon the stones were unfamiliar ones'. Instead of the usual formula of 'Om mani padme hum' they read the inscriptions as 'Om matri muye sa le du'. This was the *bonpo* version of the standard Buddhist invocation. It sounded like a corruption of it, and to Snellgrove's porters it served as a reminder that the *bonpo* liked to do things differently. Whereas passing travellers must always keep Buddhist prayer walls and monuments on their right-hand side, *bonpo* structures must be kept on one's left. Likewise *bonpo* prayer wheels had to be spun in what orthodox Buddhists regarded as a backwards rotation.

Snellgrove stayed in Samling for a month collecting 'interesting manuscripts'. On return visits in 1960 and 1961, by when Tibet itself was firmly under Chinese control, he found Samling acting as a repository for *bonpo* texts smuggled out of Tibet and as a focus for 'the few knowledgeable *bonpo* monks now living as exiles in India'.[9] With help from the Rockefeller Foundation – and not without some opposition from Tibetan monks of the more orthodox monastic schools – Snellgrove brought a group of these *bonpo* exponents to Britain. In Berkhamsted near London, he and two of them then set about distilling, from the sixty-one chapters of a twelve-volume compendium of *bonpo* teachings, an English translation of manageable proportions and semi-intelligible substance. Entitled *The Nine Ways of Bon*, it was published in 1967.

Thanks to the Tibetan diaspora, one of the more esoteric manifestations of Himālaya's culture began to enjoy a new currency.

Snellgrove's *Nine Ways* nicely illustrates the complexities of *bon*. The work, with alternating pages in Tibetan and English, requires copious notes, a twenty-page glossary and then a further explanatory note for the glossary. Asked to define the world of *bon* in two words, the French Tibetologist Marcelle Lalou once rather harshly suggested 'blood and poison'.[10] Sacrificial rites tend to be bloody, and the third of the Nine Ways certainly enumerates and describes the rites to be observed when disposing of opponents. Also featured are tantric practices and prescriptive rituals appropriate to divination, exorcism, meditation, propitiation, death and rebirth, and of course the fully enlightened 'Great Perfection'. Less predictable are the punishments awaiting any ascetic who incurs divine displeasure by admitting onions into his diet:

> Defiled by the sinful onion, one suffers in the mud-pit of the Hell of Putrefaction.
> Defiled by the harmful onion, one suffers in the Lake of Pus and Blood.
> Defiled by the debilitating onion, wretched disease appears upon one's body.

And so on. Failing to purify oneself from a whiff of onion means being 'born with the stench of body-odour for 500 births'.[11]

To its practitioners the term *bon* means not just the original religion of Tibet but 'the true religion of Tibet', says Snellgrove. Its symbol is the swastika, signifying auspiciousness and infinity, but with that perversity typical of *bon* the swastika's 'arms' are reversed so that they bend left, or anti-clockwise, as opposed to the clock-wise Buddhist (and Nazi) versions (which *bonpo* regard as heretical). How adherents choose to perceive their own beliefs is of course important, but how *bon* is defined in terms acceptable to Tibetan Buddhists is much trickier. The Norwegian Tibetologist Per Kvaerne offers three 'significations' for *bon*. It can indeed indicate (1) the pre-Buddhist (so pre-seventh-century CE) religion of Tibet which was gradually suppressed by the first wave of Buddhist promulgation in the eighth and ninth centuries. It can also refer to (2) a more formalised religion that appeared in Tibet in the tenth and eleventh centuries and that mirrored (even parodied) the new wave of evangelising

Buddhism which was then being introduced from India. This is the *bon* that is still very much alive today, as witness the *bonpo* monastic establishments from which Snellgrove's Berkhamsted monks came. And finally the term *bon* may be applied to (3) the cluster of popular beliefs, including divination, local deities and conceptions of the soul, which Stein labels 'an unorganized, churchless, doctrineless, priestless and almost nameless whole' and which Tucci encountered in Kinnaur.[12]

Tucci used the *bon* word to describe all three of these but without distinguishing them. Snellgrove treads more carefully yet cannot disguise where his sympathies lie.

> Accepting everything, refusing nothing through the centuries, [*bon*] is the one all-embracing form of Tibetan religion … Normally a *dGe-lugs-pa* (a 'Yellow Hat' [like the Dalai Lama]) scholar would be ashamed at reading a work of any other Tibetan Buddhist order, let alone a *bonpo* work. Yet educated *bonpo* monks clearly have no such inhibitions. They will learn wherever they can, and given time they will reabsorb and readapt what they have learned.[13]

But therein lies a problem. *Bon* in its current signification has been a bit too accommodating. It has adopted and absorbed wholesale, or plagiarised and perverted, nearly all the doctrines, narratives, liturgies, devotional practices and organisational features of Tibetan Buddhism.

> [*Bonpo*] may well be adjudged the world's greatest plagiarists, for they have rewritten the whole Tibetan Canon (*Kanjur*), giving it a supposed *p'on po* [*bonpo*] setting instead of a Buddhist one and attributing it to their founder-teacher Shenrap. There is a very early tradition that the orginal *p'on po* teachings came from the land of sh'ang-sh'ung in Western Tibet.[14]

But the traffic may not all have been one way. A concept like spiritual transmission positively encourages backdating, sectarian encroachment and chronological inversion. Buddhism may have borrowed from *bon* almost as much as *bon* has undoubtedly borrowed from Buddhism. Orthodoxy is in the eye of the adherent.

Where *bon* differs most markedly from Buddhism is in its sacred history and geography. Instead of acknowledging an Indian provenance, as Tibetan Buddhism does, *bon* insists on a western origin. Shenrap's realm and the scene of most of his exploits are located in a land called 'Ta-zik', which is usually read as 'Tajik' and may have embraced not just modern Tajikistan but also northern Afghanistan and Iran. Also featured is 'Drusha', which may be Gilgit and/or Hunza (whose language is still the vaguely same-sounding Burushaski). And then there is 'Zhang Zhung' (Shang-Shung, Zan-Zun and so on), which may have extended to some of these other lands and certainly embraced much of western Tibet including the sacred Kailas–Manasarowar region and the Satluj catchment which would become the kingdom of Gugé.

In effect the land of Zhang Zhung was to *bon* much as India was to Buddhism; it was the *bonpo's* 'promised land', the home of its founder, the fount of its efficacy and the 'paradise lost' of its adherents. 'There is no doubt as to the historical reality of Zhangzhung,' writes Kvaerne, 'although its exact extent and ethnic and cultural identity are far from clear.' There is no doubt either about its conquest by the expanding Tibetan empire in the seventh century.[15] Yet almost all of what is known about Zhang Zhung comes from compilations of much later date (post-1300 CE) as recalled and reinterpreted by the *bonpo* of today. These works conceal more than they reveal. The ethnicity, or ethnicities, of Zhang Zhung's pre-Buddhist inhabitants remains a matter of speculation, their language has yet to be identified and anything like a script, if there was one, has still to be found. Extracting possibly historical data from later accounts of the mind-boggling exploits of Shenrap might be likened to trying to piece together the course of the Trojan War from a badly remembered performance of *Troilus and Cressida*. The history is impossible to distinguish from myth and the geography hard to disentangle from fantasy; dating the one proves as difficult as mapping the other.

None of the surviving buildings at Toling and Tsaparang predates Zhang Zhung/Gugé's seventh-century conquest by southern Tibet's first Buddhist rulers; but scattered about the Tibetan hinterland there are other structures, ruins and stone arrangements for which pre-Buddhist origins are now confidently claimed. In the mid-1980s as part of Deng Xiaoping's liberalisation programme, and for about twenty-five years thereafter, restrictions on foreigners travelling to and within Tibet were significantly eased. Remote regions like Gugé suddenly saw an influx of sunburnt trekkers, scholars and would-be explorers. The damage to the natural and built environments wrought by the People's Liberation Army in the 1960s, by the Red Guards of the Cultural Revolution in the 1970s and by subsequent development and resettlement projects was duly recorded, and many less disturbed sites of interest were logged.

Typical of such initiatives was a series of official-sounding expeditions conducted annually between 1999 and 2008. Thus the Changthang Circuit Expedition (1999) was followed by the Upper Tibet Circumnavigation Expedition (2000), the Upper Tibet Antiquities Expedition (2001), the High Tibet Circle Expedition (2002), the High Tibet Antiquities Expedition (2003), the High Tibet Welfare Expedition (2004) and so on. The Wild Yak Lands Expedition (2007) seems to have been the last. A variety of mainly US foundations contributed to the funding of these projects and most involved a local Tibetan or Chinese partner. All the expeditions relied on motor transport and benefited greatly from the rapidly expanding infrastructure of road, rail and air links; all were subsequently documented in articles, monographs and online; and all were the brainchild of a single affable Tibetanist, John Vincent Bellezza.

Bellezza, since 2002 a research fellow at the University of Virginia's Tibet Center, is reticent about his scholarly credentials. Exploration had evidently been a passion since his childhood, which he describes as having been 'spent roving extensively around in the woods [presumably somewhere in the US] ... to the detriment of schooling and other organised activities'. When as a backpacker in 1984–6 Bellezza first betook himself to Tibet and became interested in its archaeology, the challenges of unravelling the mysteries of *bon* and acquiring some competence in the Tibetan language were matched by that of mastering the conventions of academic discourse. He read up on the classic accounts of nineteenth-century travel, noted the latest discoveries of the few Chinese archaeologists already working in Tibet and, fancying his chances as a Tibetan Indiana Jones, made several excursions north from Lhasa. These took him to Nam Tso, the largest of inner Tibet's great lakes where, in the 1870s, Kishen Singh, one of numerous intrepid 'pundits' recruited by the British to conduct clandestine geographical reconnaissance missions north of the Himalayas, had reported an unusual phenomenon. A 'gigantic pyramid' of ice stood on the lake's shore and, from a hidden chamber within it, an ancient lama could ascend to heaven at will. Bellezza was intrigued. 'Surely, I fancied, one of the great archaeological discoveries of the decade or maybe even of the twentieth century must be waiting.'[16]

It wasn't. The 'pyramid', like a lot of other supposed Tibetan monuments, was found to be just a freak rock formation. Luckily Bellezza's enthusiasm was proof against such run-of-the-mill disappointments. 'I came to hold this lake very dear. Each visit added to my knowledge of its culture and geography, and I made many friends among the shepherds.' Tibet was becoming an obsession. How he came to be smitten he explains in the fulsome tones of a TV voice-over.

> Remote and self-sustaining, Upper Tibet possesses towering mountains, luminous lakes, unbounded plains and profoundly deep gorges. It is a land of transcendental qualities, a supernal sphere in which the mystic remains culturally supreme. To this day, the topographical features of Upper Tibet are [im]bued with an innate divinity, a projection of the nobility nestled in the hearts of its inhabitants.[17]

Refocusing his attentions further north on the wide-open expanses of the Changthang, and wiser now to the ways of academic funding, by the early 2000s Bellezza was clocking up the miles and piling on the expeditions as he scoured the plateau in search of plausible antiquities. All the while he describes himself as grappling with the mysteries of *bon* and 'little by little closing in on Zhang Zhung'. The aim was nothing less than to 'assemble a comprehensive inventory of … every visible ancient remain in Upper Tibet'. As the mileage count soared, the inventory grew to around 'seven hundred monumental and rock art sites, containing many thousands of individual structures and pictorial compositions'. Since many were still under threat from developers and treasure-hunters, the priority was simply to record them. No digging was undertaken; Bellezza logged and measured only what could be seen above ground. The excavation and interpretation of his finds would be left to others; yet the widespread distribution of the find-sites could not but prompt a modicum of speculation.

'Some three thousand years ago, in a country of limpid lakes and radiant blue skies, a civilisation was born,' writes Bellezza. It was 'a realm of ruined castles, wizards, and spirits … Austere yet colorful, remote yet human'.[18] In short, it was living testimony to 'the epic of Zhang Zhung'. He admits that other candidates capable of providing an alternative rationale for his researches were available. There was Sumpa, for instance, an equally enigmatic entity that once comprised much of northern Tibet; and there were the ubiquitous Mon, whose presence may once have been felt over much of the same region. But Zhang Zhung was better suited to Bellezza's purposes. Its existence was amply attested in reliable, if late, sources. Its *bon* culture, however impenetrable, was central to its identity. Its physical and temporal dimensions were obligingly elastic. Its unfathomable antiquity seemed capacious enough to embrace everything in the way of pre-Buddhist structures, stone arrangements, artefacts and etchings that Tibet might possess. And its geographical reach was sufficiently indeterminate to cover all of western and northern Tibet plus any adjacent regions of possible relevance. It remained only to select sites and features that could be regarded as typical markers of Zhang Zhung culture.

Though surprisingly plentiful, these discoveries may disappoint all but the ardent Tibetanist. Anyone familiar with the dramatic skylines of Toling and Tsaparang, the neat towers of Kongbo or the sprawling fortifications of the Wakhan (let alone Lhasa's Potala or Gyantse's great *dzong*) is unlikely to be wowed by the sites featured in Bellezza's *The Dawn of Tibet: The Ancient Civilization on the Roof of the World*. Unworthy suspicions may even be aroused. The book's grainy black-and-white photos signally fail to substantiate either Zhang Zhung's 'audacious vision of architectural spaces' or its 'web of citadels, temples and burial grounds', let alone its 'great variety of acropolises, palaces, temples, hermitages and villages'. A stone-strewn embankment identified as the 'long rampart' of the 'Horned Eagle Valley Silver Castle' (which textual evidence implies was the citadel of the Zhang Zhung kings) looks more like the remains of a glacial moraine. Similarly a bit of wall in 'an extremely remote mountain vale' in the western Changthang may well be a 'one-of-a-kind mortuary temple'; its single recess may even have been designed to 'provide ventilation for the desiccation of human remains and for the funeral priests who officiated'. Or it could just be a bit of wall with a hole in it. In a land largely devoid of timber but well supplied with fractured rock, the corbelled construction of lintels and roofs using overlapping stone slabs is indeed typical but is scarcely proof of antiquity. A pile of such stones which may once have been a structure with some cultural function could equally well have been assembled to serve some latter-day purpose – shelter from the storm perhaps, or a sanctuary for the outcast, a stable for the newborn, retreat for the ascetic, beacon for the benighted.

Bellezza does concede a notable absence of the spectacular. Portals were mostly so small that inmates would have been obliged to enter on all fours. Few structures seem to have been high enough to stand up in and, being windowless and chimneyless, all must have been dark and smoke-filled. The rooms – in those structures where there was more than one – might originally have been caves and were linked by 'warren-like' passages. Either the people were dwarfs or they lived in fear of attack from any who weren't dwarfs. It is of course possible that archaeological excavation will one day uncover beneath these ruins a lost city of ornate chambers and stately banqueting halls. But as things stand it's hard to reconcile the extravagant lifestyle and heroic hyperbole found in the *bonpo* scriptures with this troglodyte architecture of dolls'-house dimensions.

More credible yet even more enigmatic are the sites that Bellezza identifies as funerary or burial grounds. These, he reports, 'have taken the scholarly community by surprise, for virtually nothing was known about them before I undertook to survey highland Tibet'.[19] Some are in the form of single standing stones set within what were once walled enclosures; he claims to have identified 'around one hundred' such sites. Others consist of assemblages of small stone slabs vertically embedded in open terrain. At a distance they look like a gathering of sociable garden gnomes, though on closer acquaintance they could just about seem to 'loom over the grounds of the dead like gatekeepers to the other world'. A single array can consist of anything from a dozen to several thousand stones. Bellezza sometimes calls them necropolises; by 2005 he had found twenty-nine such sites.

All these 'overt symbols of Zhang Zhung cultural glory' are remotely sited, some on islands in the central lakes, others in the emptiness of the bleakest uplands. Bellezza suggests there may be many more such sites awaiting discovery. The stones are uncut and, needless to say, rarely carry anything as revealing as an inscription or pictograph. Their unfrequented locations and, in some cases, their proximity to what may be burial tumuli, certainly argue for a funerary function. On the other hand, whether the Tibetans who created these sites subscribed to the culture Bellezza so confidently calls Zhang Zhung, whether they were *bon* practitioners and whether the sites themselves can shed any light on mortuary practice in pre-Buddhist Tibet is unclear. Carbon-dateable materials, like wood, are rare, and of the few tombs that have so far been excavated, the majority are in the east of the country and post-600 CE, so outside Zhang Zhung's generally accepted parameters.

Most scholars would agree that, whatever the age of these sites, they were not of exclusively Tibetan inspiration. Similar standing stones and stone arrays, often in association with burial mounds, have been found from Siberia and Mongolia to Kyrgyzstan and Kashmir. Across Inner Asia the last millennium BCE and the first CE witnessed frequent population movements as Scythic, Indo-Aryan and Turkic peoples took advantage of the mobility afforded by their horsemanship, equestrian ironmongery, wheeled conveyances and an economy based on pastoralism and transhumance. In a notably balanced analysis of mortuary customs in specific areas of ancient Tibet, one scholar promises to 'make a concerted effort to avoid unreasonable speculation about how observed patterns in variation about mortuary practice can be associated with *bon*, Shang Shung or some other social, religious or political formation'. The variations, says Mark Aldenderfer, could just as well be 'attributed to migration [and] the diffusion of ideas from other regions'.[20]

Much the same goes for 'Upper Tibetan rock art'. Here again Bellezza claims to have made a major contribution, with his own discoveries 'almost tripling the number of [rock art] sites documented'. Included under this heading are numerous painted compositions featuring wild animals and hunting scenes (many of 'a comely appearance that more modern artistic mediums do not capture'), plus a few low-relief etchings and pecked outlines on stone, and a variety of small metallic objects. None can be readily dated and most of them Bellezza ascribes simply to 'Upper Tibet', not Zhang Zhung. 'The numerous thematic and design parallels between this rock art of Upper Tibet and that of other Inner Asian regions including Qinghai, Mongolia and Siberia' prove conclusive. Thus the hunt, 'the headline act of Inner Asian art' and 'the main arena of activity', was treated in equally spirited fashion by peoples with no known affiliation to Zhang Zhung, *bon* or Bellezza's 'Upper Tibet'.[21]

Having forsworn excavation, Bellezza has little to say on the subject of grave goods. In the case of imperial China, the social history of the period coinciding with Zhang Zhung's heyday depends almost entirely on sepulchral finds. Beneath a man-made mountain near Xian the Qin First Emperor (246–210 BCE) was famously interred with a replica of his country's topography along with models of his chariots, palaces, courtiers and of course those much travelled terracotta facsimiles of his entire army. Under his Han successors (210 BCE–220 CE), the mausolea of notable dignitaries continued to be crammed with the wherewithal for a luxurious and prestigious afterlife – bronze vessels, gold utensils, silk furnishings, suits of sewn jade platelets ('jade-mail'), exquisite lacquerware, ornaments, documents, the choicest of victuals and so on.

Sadly the chances of finding anything similar in Upper Tibet are remote. As Bellezza reports, many mortuary sites have long since been ransacked by treasure-hunters. Aldenderfer reports more favourably of recently excavated sites further afield. In Qinghai and Gansu the neolithic peoples of the Qijia culture evidently interred their dead with bronze mirrors, copper items, jade objects, stone tools and fine terracotta jars. These date from about 2000 BCE. Summarising the findings from much later forms of burial in Gugé, Mustang (Nepal), Kinnaur (Himachal) and Malari (Uttarkhund), Aldenderfer lists ceramics, a few textiles, utensils of bamboo, bronze, copper and iron, glass beads, bangles and enough animal bones to suggest that people of substance were buried along with token representatives of their flocks and even their herds.

More sensational has been the discovery of several entombed masks elaborately fashioned in gold. Aldenderfer notes examples of these masks found outside Tibet and Bellezza records their occurrence in Tibet. But both authorities concede that they are also found much further afield – 'the Balkans, Greece and Black Sea region' – and date back to a period lasting over a thousand years. They are not peculiar to Zhang Zhung, though Bellezza believes they 'spread organically, reaching far-flung peoples through a web of subtle interactions as well as through the conventional vehicles of trade, diplomacy, war, and ideological exchanges'.[22]

In 1935 Tucci had complained that 'the prehistoric archaeology of Tibet has not yet begun'.[23] Eighty years later, Bellezza's findings seem to indicate that it is still in its infancy. With the exception of Kashmir and Nepal, the same goes for the rest of Himālaya. In a region defined by its altitude and its natural features, geographical discoveries would always command more attention than archaeological conundrums.

7

Pilgrims' Progress

For all who journey here as pilgrims, may their vision be purified and all obstacles cleared away. May their lives be filled with joy, and may they gain the highest spiritual attainments.

Guide to Pemako: The Vajrasattva Palace which
Liberates upon Seeing[1]

'What is to be done today?' First light in the Tibetan town of Gyantse finds its British trade agent racking his brains as he burrows beneath the covers and waits for bed-tea to arrive.

Four thousand metres above sea level and as dusty all summer as it's frigid all winter, Gyantse in 1910 has little to recommend it beyond its fort (a battered reminder of the 1904 Younghusband expedition), a fine stupa and its halfway-house location between British India's Himalayan frontier and Lhasa. The trade is disappointing, the passing traffic negligible, all of it on foot. Seldom is there anything to do in Gyantse; another quiet day beckons. In fact the agent will make that the title of an article he's planning for *Blackwood's Magazine*: 'From the Outposts: A Quiet Day in Tibet'.

Only to the fittest of fitness fanatics and the exceptionally resourceful can a posting to Gyantse ever have appealed. The current agent is both. With steely physique, short black hair plastered to a bullet-like skull and the toothbrush-style moustache which Charlie Chaplin and Hitler will favour, he looks a caricature of the military man that he is. For Gyantse's current trade agent is Captain 'Eric' (to his friends) or 'Hatter' (to those who can get away with it) Frederick Marshman

Bailey, he of the Younghusband expedition who befriended Candler of the *Mail*, led a mounted infantry unit and then accompanied Cecil Rawling across Tibet to Toling and Simla. Now thirty, Bailey has since acquired a certain notoriety as a do-or-die explorer. Rarely troubled by self-doubt and with a stare of disconcerting belligerence, he's also a crack shot and an avid butterfly-collector. But in Gyantse he's grounded; this being October it's a quiet time even for butterflies. Nor does the walled enclosure which the agent calls his garden boast the blooms that might attract them; the ground is so heavily trampled it looks more like an overworked paddock.

What to do then on this quietest of days in Tibet? 'The obvious answer to the question has just presented itself,' he muses: 'Let us kill something.' Then it dawns on him. It is in fact 'a polo day'. Hooray for polo. The relief is palpable. 'My mind being thus at ease, I doze until I hear my servant arrive.' The tea tray is preceded by a mighty scuffle as an Australian greyhound and three semi-tame bharal (Himalayan 'blue sheep') charge into the bedroom. A ewe shoves its wet nose in his face, a ram shatters a windowpane with its fine span of horns. Eight more dogs appear, including two spaniels and a litter of puppies. All get biscuits, although not William, Clarence and Percy, the threesome of barely manageable wolves which live outside, nor the tame panther which is in disgrace for administering 'a pat' to the greyhound's rear.

Between gulps of tea Bailey rises, robes and goes off to feed the snow leopard cubs. The cubs' day begins with 'raw meat-juice'. Orphaned by his marksmanship, they are really too young to be hand-reared. But he couldn't just leave them. He defends his action with a rough calculation of the Tibetan livestock their mother would have taken had he not shot her. The cubs, he reveals, despite the meat-juice and his best attentions, 'survive only a few days'.

Killing things is a big part of the agent's life, as indeed it is for most other foreigners in Tibet. Sometimes it's justified on grounds of necessity; it might provide the only meat on offer. Sometimes it can be explained away in the name of science; skins and heads are detached as natural history specimens. Sometimes they are destined for display as trophies. And sometimes it is enough for the sportsman simply to proclaim his love of the great outdoors and his God-given right to supremacy over the rest of creation. What's the point of an empire, Bailey might have asked, if you can't hunt in it?

Ironically Buddhism's embargo on the taking of life serves only to encourage all this killing. The arrival of the matchlock, the percussion cap and now the express rifle have made little difference. Himālaya's fauna remains plentiful and much of it is quite tame. Indeed protecting it from hunters is one of the main reasons for the Tibetan authorities' reluctance to admit trigger-happy foreigners; Tibet is 'forbidden', it is explained, as a conservation measure. Yet it is this same protection that makes the country such a sportsman's paradise. Bailey actually extends a grudging approval to the embargo while flouting it whenever opportunity offers. He will probably kill a greater variety of creatures, ranging from a shrew which still bears his name to a wild yak that was bent on trampling him to death, than any Himalayan contemporary. Feather or fin, nothing is safe from 'Hatter' Bailey. Entering Lhasa with the Younghusband expedition in 1904, his diary entry for the big day had been typically clipped: 'Marched to Lhasa morning. (fly) 34 fish.'[2] The journals of even the most celebrated explorers often read like game-books.

Having breakfasted and issued a couple of leave permits to members of his staff, Gyantse's trade agent proceeds to the polo ground. He is followed, like Noah to the ark, by a four-legged entourage composed, he says, of 'the previously described menagerie' plus two young kiang ('the wild ass of the Himalayas'), a gazelle and a fully grown and thoroughly disagreeable serow (a species of goat-antelope). The polo immediately gets under way. The opposing teams are made up mostly of Indian troopers and Tibetan braves. They ride whatever ponies are available. All goes without a hitch until one of the riderless kiangs tries to join in the fun. Then it's time for a picnic lunch on the touchline. Tea is taken back at the Agency where a small monkey rattles the parlour door and is admitted to 'sit on the mantelpiece, making contented noises as he is fed'. Visitors are sent on their way and any other business cursorily despatched. There is still just time to kill something.

As twilight descends, the agent grabs a shotgun and heads off to a nearby pond dear to migrating waterfowl. 'If we are fortunate we may bag a dozen geese and half as many ducks in a few minutes at this time of the evening.' He bags only three incoming geese. It is indeed a quiet day in Tibet. 'We return in time for drinks.'[3]

Bailey's devotion to bloodsports should not be seen as eclipsing his passion for natural history. If few colonial officers can have killed quite so many creatures, few can have befriended quite so many. To

Gyantse Fort from Chang Lo

his passion for butterflies he has added an unrivalled knowledge of birds, especially the ornate Himalayan pheasants, and a good working acquaintance with beetles. To bugs that bite he shows no mercy, but to dogs that yap he extends the comfort of his tent and spontaneous affection. In retirement it will be Bailey and his wife who oversee the introduction to British shores, and to the breeds register of the British Kennel Club, of the Lhasa Apso terrier.

More surprisingly he may be the first non-Tibetan to have attempted to expiate a flagrant act of superfluous slaughter – an act of which even he seems to have been ashamed – by undertaking a pilgrimage round one of Tibet's holiest mountains. First-hand accounts by participants in any of Tibet's sacred pilgrimages may be suspect, especially if coming from colonial pens. It could, too, be argued that Bailey would approach the whole thing in such a questionable spirit that he forfeited whatever absolution was on offer and failed to acquire any insights into the mystical powers of the Pure Crystal Mountain. Nor did the experience cure him of the killing habit; far from it. It couldn't fail, however, in reinforcing his respect for the stamina and devotion of the Tibetan faithful as they performed acts of worship and garnered trophies of unfathomable spiritual value 'in the most challenging and dangerous of all Tibet's numerous sacred places'.[4]

Mountain pilgrimage is often cited as characteristically Tibetan. Elsewhere Buddhist worship is not usually associated with lofty locations. In India it is the Hindus whose gods reside above the snow line and it is the Jains who festoon their parched hills with gleaming white temples. In China Buddhism did espouse elevation, but perhaps as a result of Tibetan prompting. There is evidence of the revered peaks of Wutaishan in China's Shanxi province being visited by Tibetans as early as the eighth century CE, and in 824 the Tibetan authorities addressed a formal request to the Tang emperor for a pilgrim's map of that sacred massif. It looks as if the idea of siting places of worship on high ground owed more to the highlanders of Himālaya than to the plainsmen of India or China.

As a notable form of ritual practice in both *bon* and Tibetan Buddhism, pilgrimage also served a social purpose in uniting all ranks of society. Clergy and laity, orthodox and heterodox, pastoralists and agriculturists, traders and mendicants, rich and poor, men and women – all participated in the great annual (and the even greater twelve-yearly) pilgrimages. In a land with little in the way of urban settlement the experience of group travel engendered an awareness of a shared culture across vast distances, while the hardships and common endeavour involved in clambering round sacred mountains instilled precious memories and bonds of community.

Nonetheless, there is said to be no textual evidence for organised pilgrimage prior to the second wave of Buddhist acculturation reaching Tibet in the ninth to tenth centuries CE. Mention of the mountains in connection with asceticism and meditation comes even later – not until Indian exponents of tantric yoga took up residence in mountain retreats in the twelfth century. But it would seem that in pre-Buddhist times the *bon* masters of Zhang Zhung (Shang Shung and so on) were already aware of a supernatural presence in the most sublime peaks. 'Mountains have, without doubt, been the most venerated and culturally significant feature of the Tibetan landscape ... for the entire period of recorded history,' writes Toni Huber, a German scholar of Tibetan pilgrimage.[5] The *bonpo* acknowledged the majesty of the mountains in more personal terms than Bellezza's 'supernal sphere ... [im]bued with an innate divinity'. A mountain was not, and still isn't, just an 'abode of deities' (*né-ri*); it is itself a super-deity. As such it exercises an awesome agency in keeping with its louring presence, while enjoying a companionable social life with other mountains. Mountains may flirt and fight with one another, quarrel and make up, cooperate and compete; each has its own distinctive mythology and its own localised cults. They are also notably possessive and can be dangerously vindictive. Moreover the life-changing power of their sacred energy radiates downwards and outwards, imparting a spiritual efficacy to all that pertains to them. Their rocks are saturated with this empowering force as is the ground from which they rise, its vegetation, water, wildlife and glaciers.

By 'harvesting' these divinely empowered substances – principally those with medicinal or narcotic properties like minerals, herbs, mushrooms and mosses – pilgrims expect to be cleansed from the twin pollutions of sin (*dik*) and disease (*drib*). More generally they may seek the karmic merit that will improve their eventual chances of enlightenment. Or perhaps they are simply counting on the discreet intervention that will ensure fulfilment of more mundane ambitions.

Not to imperil the sacred ecology of its surroundings, all cultivation is forbidden in the vicinity of the Pure Crystal Mountain. Additionally it goes without saying that the sentient creatures whom intruders like Bailey consider fair 'game' are on no account to be 'harvested'. On the contrary, every life form is here emphatically sacrosanct. Harming even the smallest of a sacred mountain's creepy-crawlies incurs the very *dik–drib* pollution which pilgrimage is meant to cleanse. Pilgrims thus enter the hallowed precincts of Tibet's second most sacred *né-ri* in a state of profound awe tinged with anxiety. They would shudder to imagine a more provocative sacrilege than the wanton slaughter of an antlered 'monarch of the glacier' while it was enjoying the divine mountain's protection.

'Pure Crystal Mountain' is a translation of Dakpa Sheri (Takpo Shiri), the mountain's Tibetan name. It's also known as Mount Tsari, the Tsari Chu and its feeders being the torrents that drain the mountain's northern flank and in part define its hallowed terrain. At 5,735 metres asl, it's by no means the highest of Tibet's peaks though it may be one of the shyest. Rising from a nigh impenetrable knot of gorges, glaciers and rainforest in a remote district of south-eastern Tibet, it's subject to the heavy winter snows and the unbroken summer deluge common to all of the easternmost Himalayas. Adrip in cloud for months at a time, its summit is seldom to be seen in all its icy glory and, so far as is known, has never been climbed.

Any such assault would constitute another sacrilege and would probably provoke an international incident. For, as if its natural setting were not sufficient guarantee of its inaccessibility, Pure Crystal Mountain bestrides two great fault lines in Himālaya's human geography. One is the ethnic divide between the Mongoloid people of the Tibetan uplands and the smaller forest-dwelling tribal peoples of low-lying eastern Assam; the other is the political divide between the territory now claimed by the Republic of India and that claimed by the People's Republic of China. Like the racing waters of the Tsari Chu, the mountain's recognised pilgrimage circuit actually crosses and recrosses this hotly contested international frontier. And that it does so, indeed that there is an international frontier here at all, is a legacy of 'Hatter' Bailey's obsessive devotion to the chase.

In 1913, two years after his menagerie-keeping sojourn in Gyantse, Bailey had returned from home leave with his heart set on more Tibetan exploration. In particular he was interested in tackling what he called the last great Himalayan 'mystery'. A whole decade before the botanist Kingdon-Ward would take up the same challenge, Bailey was bent on discovering what became of Tibet's Tsangpo River when, after majestically traversing the country from west to east for 1,500 kilometres, it suddenly buckled, swung north then south and broke into frothing rapids before being lost to sight in the twists and turns of a deep canyon. Presumably it thus broke through the Great Himālaya and, in a steep flight of cascades or possibly even Niagara-type falls, dropped some 3,000 metres in 150 kilometres to emerge, brown and contrite, from the Assamese forest as the Dihang River. The Dihang being the main feeder of the Brahmaputra, that meant all three rivers – Tsangpo, Dihang and Brahmaputra – were one and the same. (With the addition of 'Yarlung', the name of a tributary with which Bailey was soon to become acquainted, the whole river is now the Yarlung–Tsangpo to the Chinese and the Yarlung–Tsangpo–Brahmaputra to Indians. In Bangladesh it's known under at least three more aliases before its spreading distributaries debouch into the Bay of Bengal.)

So by 1913 there was really no great mystery about it. Geographers had been reasonably confident that the Tsangpo became the Brahmaputra since at least 1882. In that year Kishen Singh, the greatest of the Survey of India's pioneering pundits, had completed the first north–south route survey of Tibet. While skirting the gorge country in search of a way back to India, Kishen Singh (aka 'AK') had deduced that the Tsangpo could have no connection with other nearby contenders like the Mekong, the Salween or the Yangtse and must therefore join the Brahmaputra; it was the region's only other river with a big enough flow to accommodate it. Further probes in the 1880s had confirmed this. 'The identity of the Tsangpo with the Dihang was no longer open to question.'[6] The mystery was not what became of the river but how its discharge managed to cut its way so inconspicuously through the mightiest mountain barrier on earth.

Bailey himself had tried to follow its course down from Tibet to Assam twice before. The first time was in 1904 when, along with the surveyor Ryder and the journalist Candler, he had proposed returning to India from the Younghusband expedition by way of the river. Seven years later, tracing the river's passage through the mountains was to have been the climax of an overland journey he had made from China to India in 1911. On that occasion he had been travelling as a private individual with no official standing, let alone a protective escort, and had been obliged to turn back by an 'outbreak of fighting between the Chinese and the Tibetans'. The fighting was a consequence of the 1910 Chinese invasion of Tibet, itself intended to contest Britain's newly acquired ascendancy in Lhasa as a result of the Younghusband invasion. Bailey was halted at Pomé (Bomé, Poyul), 'the crucial point of my journey', he says. The mountains were already in sight and from there he had planned to strike out down the Tsangpo itself. 'I came so near to success that the journey, so far from discouraging me, merely whetted my appetite.'[7] Refreshed and undaunted, he would now try again, but this time working upriver from India and ideally with the status of an accredited agent of the British government of India.

No one was better qualified. By now Bailey spoke fluent Tibetan, knew the Dalai Lama, had read almost everything that had been written on Himālaya and had established a rapport with the Tibetans that impressed the savants of London's royal societies as much as the mandarins of British India's government. Luck being not the least of a successful explorer's attributes, he had plenty of that too. To ascend the headwaters of the Dihang–Brahmaputra in upper Assam it was first necessary to negotiate a safe passage through a chaos of jungle-choked foothills beset by the interminable squabbles of their Mishmi and Abor tribes. Not for the first time, in 1912 the Abors rose against their foreign overlords, assassinating the nearest British political officer and massacring thirty-eight members of his staff. Bailey thought it all the result of a misunderstanding; yet punitive action was inevitable. An Assam expeditionary force plus survey parties was assembled. Bailey got himself appointed to it as intelligence officer. And hoping to use the expedition's advance into the Assamese hills as a springboard for his own foray into Tibet, in early 1913 he cut his way up through the rainforest, evading the Abors, cultivating the Mishmis (who thoroughly approved of any action designed to cow their more belligerent neighbours) and massacring the wildlife.

More intriguingly, and long before reaching the ethnic frontier between the trouserless tribals and the better-clad Tibetans, Bailey stumbled across a Tibetan settlement. It was at a place called Mipi. The settlers had moved here from Bhutan and Tibet over a considerable period of years in response to a myth portraying the region as a Shangri-la to which they could repair in times of strife. Bailey was doubtful. What with the bloodsucking flies, the ticks, the leeches and malarial fevers, not to mention the Abors' ambushes and the Mishmis' poisoned arrows, the tract fell well short even of a Tibetan's idea of a 'promised land'. In fact most of the settlers had soon given up in disgust and returned whence they came. But a few remained and Bailey listened to their tale of woe with interest. If the Abor war was providing the perfect protective smokescreen for his advance into the hills, the existence of this Tibetan enclave in what might be regarded as British territory promised an introduction to the real Tibet beyond.

For where Tibetans could come and go, so could a Tibetanist. Here was a trans-Himalayan trail, an unknown and unguarded back door into the 'Forbidden Land' that was so remote even Lhasa seemed unaware of it. Clearly British India's north-eastern frontier was soft to the point of incoherence; it was more a zone than a boundary and every bit as undefined as the blank on the map indicated. With a view to rectifying this with a demarcation anchored on natural features and ethnic realities, the surveyors of the Abor expedition were carefully mapping the foothills as they advanced. Bailey needed to do the same as he headed north over the mountains. If he were not to be recalled in disgrace and reprimanded for an unauthorised Tibetan excursion, he needed to interest the authorities in the promise of a survey of the Tibetan side of the frontier that would complement the work of the Abor surveys on the Indian side. Detailed maps and topographical information had a pleasing habit of turning official froideur into warm endorsement. But to generate such maps it was necessary to scale nearby peaks, then with theodolite and plane table to wait indefinitely for a break in the cloud. Obtaining a single fix might take days and involve several such ascents. Bailey alone couldn't possibly combine this work with reconnoitring campsites, managing porters and supplies, quizzing the locals, appeasing their officials and of course decimating the wildlife. He needed his own surveyor.

Henry Morshead with bandaged hand (left), George Mallory (next to him) and others of the 1922 Everest expedition. Bailey thought carelessness would kill Morshead but he survived their Tsangpo journey and Everest

Once again he was in luck. Henry Morshead, a quiet and likeable captain in the Survey of India of about his own age, leapt at the opportunity. Of slighter build than Bailey, Morshead displayed a stamina and stoicism that was, if anything, even greater than Bailey's. So were his recuperative powers and his indifference to discomfort. Within days of joining 'Hatter' Bailey, Morshead is described as standing to his plane table 'covered with leeches and with blood oozing out of his boots as oblivious as a child whose face is smeared with jam'. 'It worried me,' added Bailey, 'because I felt I had to be responsible for Morshead's tropical hygiene as well as my own.'[8] This disdain for basic precautions may have contributed to Morshead losing three fingers to frostbite when, in 1922, he was recruited for the first British assault on Mount Everest. After reaching the unprecedented height of 7,600 metres, he had fallen and been rescued 'half dead' by the legendary George Mallory. 'I suppose [Morshead] took no more notice of the cold than of the jungle leeches,' wrote Bailey; 'I was always afraid that sort of carelessness would kill him one day.'[9] Two years later the missing fingers would disqualify Morshead as a climber on the second Everest expedition which famously claimed Mallory's life. Morshead himself fared only slightly better. In 1931 (not 1923 as Bailey unaccountably has it) he would be shot at point-blank range in jungle near Maymyo in Burma. Though this was evidently murder, neither culprit nor motive was ever identified.

As they waited for the Abor expedition's surveyors to complete their work and withdraw from the hills, Bailey and Morshead speculated on the chances of slipping across the mountains before their own plans became known. Much depended on how one read the Delphic injunctions coming from Sir Henry McMahon, the then foreign secretary of the Indian government and the man supposed most likely to disapprove of their Tibetan escapade. Bailey knew McMahon well and respected his efforts to avoid expensive frontier confrontations by restraining unauthorised adventurers. But he also knew that at heart McMahon's ideas about securing British influence in and beyond the Himalayas were as 'forward' (that is, expansionist) as his own and included making the most of whatever advantage had accrued in Tibet as a result of Younghusband's invasion. Indeed Bailey implied it was McMahon who had arranged his attachment to the Abor expedition in the first place and it was probably thanks to McMahon that he now had Morshead at his side. Certainly the foreign secretary did nothing to countermand these arrangements. In his orders for the Abor expedition McMahon had incorporated mention of the government's desire that Bailey be 'allowed as much scope as possible for the exercise of his talents as regards exploration'. Bailey reckoned this a clause 'of tremendous importance'. Despite directives about awaiting further orders and on no account crossing into Tibet, he interpreted it as 'an implicit permission to enter Tibet'.[10] No doubt wishful thinking was in part to blame. So too, perhaps, was literary licence; in a narrative that he would entitle *No Passport to Tibet*, an injection of dramatic tension and derring-do would not be out of place.

April 1913 was spent in and around the Tibetan enclave at Mipi. It was as good a place as any in which to shelter from the endless downpours and await the melting of the snows that would render the passes ahead passable. To be doubly sure that any recall orders never reached them, Morshead and his theodolite were posted on a distant hill as a lookout, and the only usable bridge in their rear (it was actually a fallen tree) was sabotaged. The wait was spent cultivating Mipi's demoralised Tibetans. It was now, says Bailey, that 'the realisation flashed through my mind that here was my chance of getting through to Tibet from the Assamese side'. As on subsequent encounters with the Tibetans, Bailey dug out his photos of the sacred sights of Gyantse, Shigatse (Xigaze) and Lhasa, and others of himself with prominent figures like the Tashi (Panchen) Lama and the Dalai Lama. These, plus his fluent Tibetan, served to establish his bona fides as a friend of Tibet and were rewarded with letters of introduction from Mipi's headman to his fellow countrymen across the mountains. The letters were not exactly passports but would prove the next best thing.

Equally gratifying was the headman's information about the course of the Tsangpo and its possible falls/cascades. The details forthcoming at Mipi tallied well with the earlier reports from the 1880s and persuaded Bailey to upgrade the river to being his declared objective. It was 'an excellent justification for the exploration that I proposed to undertake', he confided. It would resolve one of geography's great 'mysteries' and, because only about 50 kilometres of river were still unaccounted for, 'it all seemed so matter-of-factual'. Neither McMahon nor the Tibetans need fret over possible political complications.

With three guides provided by the Mipi headman, iron rations scrounged from the Abor expedition and a medley of Tibetan, Mishmi and Gurkha porters to carry them, the twenty-strong expedition slipped out of Mipi in mid-May. Ahead lay two passes across the spine of the Great Himālaya which, though not the highest, would prove, by reason of it being so early in the season, the most difficult of the whole expedition. Climbing steadily from Mipi's 1,000 metres asl to the 4,000 metres of the first pass, they emerged from the ticks and leeches on to the snowfields. Morshead collapsed with a high fever but, dosed with whisky, staged one of his dramatic recoveries. The rain abated not at all, turning the surface snow to slush atop its underlayer of compacted ice. Bailey shot pheasants, some so rare that even he was unsure how to classify them. In retrieving them he was assisted by Roarer, a mongrel puppy acquired from the Mishmi. But on the passes both man and mutt were restrained by the Tibetans, who thought the carnage might 'anger the local spirits'. Bailey dismissed this as 'superstition', yet still desisted lest 'the reverberation of a rifle shot might start an avalanche'. The melting snow guaranteed avalanches enough. Two porters were swept away, 'and later another coolie and I were carried down about twenty feet, but no one was hurt'. Over the first pass, the rain turned to snow. Bailey bagged a brace of ducks as well as more pheasants. Adding game to a diet that was otherwise just *tsampa*, a dough-like substance made from roasted barley flour, proved popular.

The next pass, the Pungpung, took them to over 4,400 metres asl. The going got tougher as they sank to their waists. The soft snow was over 4 metres deep, that being the length of their longest probe. One of the guides was swept down a near vertical slope and spared the drop from its terminal precipice only by being snagged on a bush. Bailey himself had a dangerous fall 'which I thought was going to be my last'. 'I saved myself with the handle of my butterfly net,' he says, presumably by using it as an ice axe. But that was it; the descent began, the sun came out and the worst was over. Next day they were down to a clammy 2,000 metres asl and back with the leeches. Morshead promptly headed into the mountains again to continue his triangulation. Bailey, though this was not the Tibet he was used to, was in no doubt that it was indeed Tibet. The local governor plied them with buttered tea and *tsampa*, and their arrival coincided with a gathering of monks for the consecration of a multicoloured mandala. Suspicions were bound to be aroused by the sudden appearance of well-armed strangers in such an out-of-the-way place, but Bailey sought to allay them with his Tibetan photos and his Mipi letters of introduction.

It was not quite as simple as that. Because of the recent fighting with the Chinese, eastern Tibet was awash with rumours and in a state of high alert. Here and throughout the rest of their journey Bailey and Morshead were repeatedly taken for Chinese agents scouting future invasion routes while cunningly masquerading as British hunters. They carried no British documentation that said otherwise, and few in remotest Pomé knew Their Holinesses of Lhasa and Shigatse well enough to recognise them in Bailey's photos.

Pomé was the district where Bailey had been turned back in 1911. It looked as if there was to be a repeat performance. Again he was intercepted soon after the Tsangpo hove in sight. Desperately short of supplies and told in no uncertain terms to 'regard ourselves as prisoners', he and Morshead were escorted away from the river to Showa, the war-ravaged Pomé capital. There they waited as their fate was decided. In what may have been his finest hour, Bailey describes himself as patiently rebutting the idea they had anything to do with the Chinese, then threatening to appeal against their detention to the Dalai Lama in Lhasa. But in the end it seems to have been a frank admission of the truth – that they were actually unauthorised British travellers hoping to mollify their government by preparing a map – that secured their release. Far from taking umbrage, the Tibetans welcomed the prospect of an agreed frontier and of closer relations with their British neighbour, especially in so far as these might make the Chinese think twice about another invasion. So, within three days, Bailey and Morshead were on their way again, heading for the unknown section of the Tsangpo and proudly possessed of an official *ula* (permit) entitling them to porterage and provisions wherever fancy took them. Morshead jogged ahead with theodolite and plane table; Bailey, 'glad in my heart that we had at last won recognition in Pomé', shot and killed a 'red goral'. It was another species of goat-antelope, like the serow, and one that still bears his name – *Naemorhedus baileyi*.

More lasting fame in the annals of eponymous taxonomy came a week later. His diary entry for 10 July 1913 simply notes that, while crossing a meadow, 'among the flowers were blue poppies that I had not seen before and purple iris and primulas'. This first recorded sighting of what would become the most sought-after of all the

Himalayan plants to find its way into English gardens had almost slipped his memory. Rereading his diary many years later he couldn't believe he could have dealt so brusquely with his first encounter with that most delicately blue of poppies now renowned as *Meconopsis baileyi*: 'What a pedestrian way to record one's assignation with the immortality of a seedsman's catalogue! Not even a sentence to itself.'[11] Nor did he do anything to popularise the plant. It would be Frank Kingdon-Ward who ten years later rediscovered 'Bailey's blue Himalayan poppy' and brought home the first seed. Had Kingdon-Ward not done so, noted Bailey, 'the *Meconopsis Baileyi* would have remained as obscure as the butterflies and other specimens which we brought back'.

The lie of the mountains and the lack of bridges meant that to regain the Tsangpo from Showa they had first to make a long detour to the north, so outflanking the river's great loop where it turned south before burrowing into mountains. This was where the region known as Pemako started, where the river's unexplored reaches began, where it dropped a thousand metres in a crow-fly distance of no more than 40 kilometres and where any falls must be located. But they would now be tracking the river downstream, hopefully until it emerged at the point to which they had tracked it upstream before being hustled off to Showa.

The downstream approach was that supposedly taken by their most dogged and enigmatic precursor. Originally a tailor from Sikkim, this Kintup (he had only one name) had been adopted by the Survey of India as a possible pundit despite his not being able to read or write. Kintup did, though, speak Tibetan and had therefore been assigned to Pemako and the unmapped section of the Tsangpo. There and thereabouts in the early 1880s he had spent several years as the servant of fellow pundits, then as a slave, a fugitive, a monastic dogsbody and finally a pretend pilgrim. It was Kintup's saga that first raised the possibility of a Niagara or Victoria Falls on the Tsangpo; and it was his account that first alerted Bailey to the potential of pilgrimage as a plausible motive for probing Tibet's uncertain borderlands. The two men may even have met. Although Bailey makes no mention of it, Kintup had been part of the Younghusband expedition. He had accompanied Dr Waddell of the Indian Medical Service as a servant and informant and had in fact been Waddell's 'faithful companion during many years' travel among the eastern Himalayas'.[12]

But Kintup being a model of discretion as well as illiterate, his testimony in respect of the Tsangpo was neither well known nor universally accepted. Like Marco Polo, he had been obliged to entrust his story to others. His movements were known only from hearsay – from what he had been able to recall when, nearly five years after his journey, he had been debriefed by a literate colleague, and from what survived of that transcript when translated from Tibetan into English by a third party unfamiliar with the terrain. The result was not necessarily to be trusted. Nor was Kintup's saga inherently credible.

To cut a long and almost unbearable story short, Kintup had braved the Tsangpo gorges not once but four times. In 1878 he had reached the portals of the river's canyon at Gyala while serving in the capacity of servant-porter to a supposedly more competent Sikkimese pundit. The colleague proved more inventive than competent and baulked at the sight of the gorges. Two years later Kintup was back, this time in the service of a Mongolian lama recruited by the Survey of India. The lama favoured a prior visit to Lhasa, where both men adopted the guise of pilgrims to explain their yen for the mountains of the far south-east. The lama also favoured a style of travel more leisurely than the rugged Kintup was accustomed to; en route to Gyala he devoted four months to a dalliance with a Tibetan housewife and only resumed the road when Kintup paid off the lady's husband. They did eventually get past Gyala, reaching a little monastery at Pemakochung, about 25 kilometres into the gorge. Then they had turned back, although not before Kintup had made a mental note of a nearby waterfall. It 'falls over a cliff', he would recall, '… from a height of 150 feet [46 metres]'. He hadn't claimed the fall was on the main river and he would later correct the 150 feet of the transcript to the 30 feet he'd actually said.

It was now March 1881. Kintup and the Mongolian had withdrawn to a fortified village north of the river's great bend. From there, without warning, the Mongolian lama absconded. He was never seen again and only several weeks later did the long-suffering Kintup realise that, as a parting present, the Mongolian had sold him to a local chief as a slave. A whole year then passed as Kintup endured this enforced servitude and debated his escape.

Escape was complicated by his growing appetite for pilgrimage. 'Kintup was increasingly attentive to the sacred landscape around him,' says a later commentator; '… he reports more on the monasteries and pilgrimage routes than on the waterfalls and the course of the Tsangpo'.[13] He was also burdened by a quite crippling sense of duty. Should the pundits fail to follow the river through the gorges, they had been instructed to launch into the stream on a prearranged date 500 specially marked logs. In what amounted to an advanced version of Poohsticks, these logs were to be intercepted downstream by watchers on the banks of the Dihang–Brahmaputra and thus the case for the Dihang being the link between the Tsangpo and the Brahmaputra would be proven. But the whole scheme, while invariably hailed as 'ingenious' by later writers, was also wildly optimistic. Logs tossed into a white-water flood already choked with rocks and timbers, subject to frequent spates and extending to over a hundred kilometres could end up anywhere. Added to which the date for launching the logs had passed, there was no means of alerting the watchers downstream to a new schedule and Kintup had lost the augur with which he was supposed to make holes in the logs for identity tags. After a silence of over two years, the Survey had written off its Pemako pundits – and probably forgotten about Kintup altogether.

Kintup had persevered regardless. Pursued by his former slave master, he struck south, bypassing the gorges to find sanctuary downriver at the monastery of Marpung. There he served the senior monk as a devout dogsbody. He remained for about a year, while nearby he secretly prepared his logs, hid them in a cave and then, resuming the guise of a pilgrim, set off back to Lhasa to get word to India. With masterly understatement his letter-writer informed the Survey how his journey thus far 'had proved a bad one'. He'd been robbed, deserted and sold into slavery; but the logs were ready; he would launch them 'from the 5th to the 15th of the tenth Tibetan month of the year called Chhuluk'.

And this, after tramping the 600 kilometres back to Marpung, he duly did. But his letter, though it reached India, was ignored; its intended recipient was dying and had anyway left for Europe. No watchers were waiting by the riverside in Assam in the tenth Tibetan month of the year called Chhuluk. The logs, if they ever got there, passed unseen down the braided and silt-laden Brahmaputra and out into the choppy waters of the Bay of Bengal.

Kintup, the Survey of India's Sikkimese
pilgrim/pundit in whose footsteps Bailey
and Morshead penetrated the Tsangpo
gorges

When, thirty years later, it was Bailey and Morshead's turn to dodge
the rock slides and edge their way round the lolling glaciers of the
Pemako gorges, they found good reason to revise their opinion of
Kintup. His dates and distances were often awry and his bearings
mere guesswork (his compass had evidently been stolen), but his
itineraries were sound and his observations authentic. His place names

matched those still in use, his grasp of the region's sacred lore could not be faulted and his waterfall at Gyala, though on a sidestream and nothing like 150 feet high, was certainly noteworthy since behind its watery curtain the visage of a deity could sometimes be glimpsed in the wall of rock.

Squeezing between the river's twin portals of the 7,782-metre asl Namcha Barwa and the 7,150-metre Gyala Peri, Bailey and Morshead entered a 'fantastic country'. Like subsequent visitors, they thought it 'astonishing' that, in the 20-kilometre gap between these two magnificent peaks, the great river had carved out a rocky passage all of 5,000 metres deep. 'It was an example of the power of water as startling as that of the Colorado River in the Grand Canyon,' thought Bailey. A century later China's indomitable engineers are inclined to agree. With a series of dams, one of them linking the two mountains, they expect to harness a generating capacity in excess of that produced by the Yangtse's Three Gorges Dam. It will be the world's largest hydroelectric installation and bodes ill for what is surely one of the world's most uniquely sensitive environments.

In summer 1913 the gorge's only access trail was partially submerged by the river's flood. This meant climbing the retaining cliffs, shinning up notched tree trunks that served as ladders or crawling on all fours through water-worn tunnels in the rock; 'the timing had to be correct', noted Bailey, 'and a good speed maintained because every now and then a large wave from the river rushed in and made [the tunnels] impassable'. Persistence was rewarded near Pemakochung by the discovery of a new waterfall. Bailey spotted its cloud of mist from the densely forested cliffs above.

> I first went out onto a rock about 100 feet above the water, which was rushing through a narrow chasm about fifty yards wide. To my left were violently swirling rapids, below and to the right the river plunged over a ledge and dropping about thirty feet sent up clouds of spray ... Morshead, who came to it later in the day, saw a rainbow in it and since the Tibetans had no name for it we nicknamed it the Rainbow Fall.[14]

Having passed the furthest point reached by Kintup, they pushed on, now dangling from roots that served as ropes, now boulder-hopping in the riverbed, now clambering up the gorge's retaining walls to hack their way through dense rhododendron. 'We were exploring country of which nothing was known,' notes Bailey, yet they were making little progress; more ground was being covered laterally and vertically than downstream. Five days out from Pemakochung, with the *tsampa* running low, any game unrecoverable, boots expiring and both men in a state of collapse, Bailey's diary conceded defeat.

> I don't think any of us can cut [through the rhododendron] any more. We saw traces of blood pheasant, takin [a large herbivore said to resemble 'a bee-stung moose'] and musk deer ... We have been living rather cheaply lately, as we can get nothing down here. I was out twelve hours without water.[15]

To eke out their rations, Bailey set off alone in a final bid. A surprise encounter with some locals from downriver promised much; perhaps there was a way through after all. But the strangers rebuffed his company and gave him the slip. Bailey concluded they must be up to no good. He returned to Pemakochung, rejoined Morshead and together they turned their backs on the gorges to follow the river upstream towards Lhasa and India. 'We felt that we had not been entirely vanquished ... the amount that we had achieved was worth it.' In reality their fortnight in and around the gorges had added only a few kilometres to the map and had proved nothing. Kintup's nearly five years thereabouts, plus Kishen Singh's observations, had long since dispelled the river's 'mystery'.

There was, though, a postscript to the Kintup odyssey. After finally releasing his logs Kintup too had headed for home. But being by now an inveterate pilgrim he had first made a detour to Tsari and its Pure Crystal Mountain. No pilgrimage in south-east Tibet was complete without a circuit of the region's main devotional attraction, and no pilgrim had a better claim to the mountain's divinely empowered bounty than Kintup.

Bailey took note. Kintup had served them well thus far; they would keep faith with him. Possibly because they'd followed in his footsteps for so long, or more probably because they'd identified Tsari's Pure Crystal Mountain as a pivotal feature in any future border alignment, Bailey and

Morshead followed suit by opting for an end-of-journey pilgrimage to Takpo Shiri. Nothing in Bailey's *No Passport to Tibet* hints at his having undergone a sudden conversion to Tibetan Buddhism. He was intrigued by Buddhist ritual but not minded to participate in it. He had formed no intention of making a sacred circuit of the mountain until he got there. Even then it seems to have been a spur-of-the-moment decision.

From the Pemako gorges, they had followed the Tsangpo upstream, with Morshead 'fixing' the peaks of the Great Himālaya which marched with the river along the southern horizon. Thanks to the authority of their *ula*, they now rode ponies and covered the 500 kilometres to the sizeable town of Tsetsang (Chetang) in twenty-eight days. The dusty terrain was here more typically Tibetan, the cloudless skies more congenial to triangulation. In Tsetsang a small community of Kashmiri merchants welcomed them as fellow brethren of the raj and brought them up to date about world events. What with war in the Balkans, Russians in Mongolia and other Russians being expelled from Afghanistan, 'it was clear that the world situation was normal,' concluded Bailey. More to the point, Tibetan and Chinese envoys were reportedly converging on Darjeeling for a frontier conference to be convened by McMahon. It was just as Bailey had hoped. Armed with Morshead's maps, and as the only political officer with first-hand knowledge of the region, he was confident of an appreciative homecoming and seemed less likely to be censured.

From Tsetsang they followed the Yarlung tributary of the Tsangpo. It was the route taken by pilgrims to Tsari and led them back into the skirts of the Great Himālaya. Bailey was now calling the mountains 'the natural frontier', though they here trended away from the Tsangpo towards Bhutan and Tawang. The latter was a large monastery that commanded a shortcut down to Assam. With an urgency prompted by news of the Darjeeling conference, the explorers zigzagged south heading for Tsari, Tawang and India. Their six months' perambulation of south-eastern Tibet was nearly over.

Fond thoughts of home comforts were abruptly banished when the chest containing their money was stolen. It was never recovered. The exchequer would be replenished by a loan from their Kashmiri friends in Tsetsang, but the chest had also held most of their ammunition. Bailey would now have to make every cartridge count. He claims to have ceased shooting for science and seldom to have killed for the pot. But, inexplicably, the closer they got to Pure Crystal Mountain's hallowed purlieus, the less he was inclined to respect this resolution.

Midst hail and sleet on the 4,795-metre Kamba La he took off after a herd of bharal. 'I got the biggest (24 inches [61 centimetres])' – this being the spread of its horns – plus another of 19½ inches. They were short of meat but it was the heads that he coveted. In heavy snow on the next pass he investigated another herd of about twenty bharal. 'None had good heads. They were very tame and let me walk within 200 yards. But I did not shoot, as we were now in the holy district of Tsari.' He knew full well that hunting was here forbidden; nor were they still desperate for protein. 'Then on the other side of the river I saw a herd of twelve stag ...'

> It was a good stalk and after about an hour I had come within 150 yards of the only horned male ... I could not make up my mind to shoot because the stag was a small head in velvet [antlers still growing, uncalcified and so worthless as a trophy], but in the end when they got up, I decided that I better get a specimen and shot him. He measured 46 inches at the shoulder.
>
> The people were shocked ... I was indeed rather shocked myself, not merely because Tsari was a sacred place, but also because the stag was in velvet. My reason was that I did not know what sort of stag it was and it might be new to science.[16]

It wasn't new to science; and seldom can so lame an apologia have been offered for a blatant sacrilege. It was in fact a Sikkim stag, a close relative of the nowadays endangered Kashmir stag. Though already rare in Sikkim, Sikkim stags had been sighted in Bhutan and possibly along the Sino-Tibetan frontier. 'It is satisfying', continues Bailey, 'that even if exterminated elsewhere, they still survive in the holy sanctuary of Tsari.' Tragically, their survival would be brief. By mid-century the Sikkim stag had been consigned to listings of the world's extinct species. Bailey's wretched specimen may have been the last on record.

Two days later, and restored to buoyant mood, Bailey declined to dismount when asked to show respect for a roadside shrine. They were now at Chikchar, the starting point for the pilgrims' circuit of the Pure Crystal Mountain. Bailey claimed not to have realised he had committed another 'breach of ritual', but he was still uneasy about the stag. Its skin, hung up to dry, was a grisly reminder. His men were more troubled by the divinely ordained misfortunes that would surely befall them for having been party to such impious conduct. No doubt it was the men who urged him to undertake an expiatory circuit of the mountain. And no doubt he complied because a reconnaissance of the mountain could be a useful addition to his survey.

In a paper delivered to the Royal Geographical Society in 1914 he would scarcely mention the survey work. He simply claimed that 'the sin [of killing the stag] was eventually made an excuse for my going round the pilgrimage'.[17] In other words, the stag was killed to give him a pretext for the pilgrimage, and neither stag nor circuit had anything to do with spying out the frontier. The frontier work was so sensitive that not until the publication of *No Passport to Tibet* all of forty-four years after the travels it describes would the essentially political nature of the expedition become apparent. By then, 1957, the frontier was more sensitive than ever; India and China were about to go to war over it. But India had long ceased to be a British responsibility, Tibet was again overrun by Chinese troops, Bailey was in retirement and the borderland was none of his business. At last he could tell it as it was. The stag had obliged him to make the sacred circuit; the circuit afforded vital insights into the frontier.

Professor Huber's Tibetan informants recall that there were in fact three major, and several minor, pilgrimage routes round Pure Crystal Mountain. Bailey heard of only two, a long one taking three to six weeks and a short one of three to five days. He chose the latter, completing it in four days. Meanwhile Morshead ventured further south to explore the longer route. The important thing was that all the routes must encircle the mountain, and the circuit must be performed within specified dates. Everything was carefully regulated. There were way stations providing firewood and hot water at the overnight halts, and a few pedlars selling basic provisions. With as many as 100,000 pilgrims expected for the great twelve-yearly pilgrimage, troops from Pomé were sent ahead 'to protect the pilgrims from attack by the Lopa tribe of Daflas through whose territory the pilgrims had to travel'. This was of particular interest to Bailey as evidence of an ethnic divide like that between the Tibetans and the Mishmis/Abors further east. But here, as there, it was complicated by unregulated migration, cross-border trade and transit, extra-territorial taxes and subsidies, and intense mutual distrust. Pinpointing the ethnic watershed was as tricky as tracing the actual watershed.

The short circuit as tackled by Bailey took the pilgrim closest to the mountain. 'All day it was up one valleyside, over a pass and down the other side, never crossing a watershed but just moving round the lower slopes of Takpo Shiri.' It either rained or snowed almost continuously; he never saw the actual summit. Negotiating its radial spurs and glaciers entailed at least half a dozen passes of 4,900–5,000 metres. Poor visibility, slippery rocks, steep-sided moraines and glacial crevasses added to the hazards. So did the 'maggots'. 'About as long as a short match', these caterpillars hatched in the late summer in such abundance that it became impossible not to tread on them. Maggot lives being at risk, the pilgrimage season had then to be brought to a close. Such at least was Bailey's understanding as he tiptoed past the first of 1913's hatchings. They were already dead – but drowned in the puddles rather than crushed underfoot.

Another explanation for the mountain having a closed season for pilgrimage was given by Huber's Tibetan informants. Certain esoteric practices associated with the sacred mountain evidently required strict privacy and hence the exclusion of all except the initiated. Huber puts this in the context of what he calls the 'mandalisation' or 'buddhicisation' of pilgrimage sites. As successive waves of Buddhist acculturation swept across Tibet in the ninth–twelfth centuries CE, existing holy sites were gradually accorded Buddhist associations and incorporated into orthodox Buddhist ritual practice. Mountains were reimagined as conforming to the symmetry and concentricity of the mandala; local cults and mythologies were conflated with those of mainstream Buddhist teaching; and the arrival of tantric Buddhism encouraged the practice of meditational and ascetic forms of yoga such as the cold-defying discipline known as *tum-mo*. In Takpo Shiri's sub-zero temperatures, *tum-mo* enabled groups of practitioners to sit stark naked in the snow throughout the night and then rise at dawn none the worse for the experience. Rigid training, strict abstinence and complete seclusion, especially from women, were essential to this exercise. It was possible that the closure of the pilgrimage routes for *tum-mo* just happened to coincide with the annual maggot-hatch.

As a naturalist, Bailey naturally preferred the maggot explanation. He makes no mention of ascetics at Tsari and seems unaware that, with practice, he and Morshead might have managed without their cumbersome bedrolls. It was now September–October. The frosts were getting harder and ahead lay the highest leg of the journey as they returned through the Great Himalāya. 'Extremely cold', 'intensely cold', 'cold intense', noted Bailey as they tackled passes of 5,000 metres asl. Instead of ponies or porters their baggage was now entrusted to yaks. The game on offer included the *Ovis ammon* or argali, a donkey-sized sheep of the high pastures that was as prized as the Marco Polo sheep of the Pamirs. Morshead shot a couple – it was his turn with their only rifle and the remaining rounds – while they investigated what is now the contentious 'Tawang tract'.

Defying all frontier logic Tawang's great monastery dominated a tongue of indisputably Tibetan territory that bounded Bhutan to the west and extended south across the mountains to the border with Indian Assam. Bailey rightly awarded it to Tibet. The British would come to believe that the Tibetan government had subsequently ceded it to them, though they would never actually administer it. China's nationalist and communist governments would reject this British claim and so would the Tibetans. In 1950 New Delhi, rejecting the rejections, incorporated the tract into its North-East Frontier Agency (NEFA, later reconstituted as the Indian state of Arunachal Pradesh). Twelve years later during the 1962 China–India War, Mao's People's Liberation Army (PLA) occupied Tawang, then relinquished it while still maintaining it was unquestionably part of Tibet and so China. This remains Beijing's position. Chinese maps show not only Tawang as Chinese but most of Arunachal. There is no sign of the thick McMahon Line which on Indian maps divides Indian from Chinese territory and whose wriggling course through the eastern Himalayas matches so exactly Bailey and Morshead's 1913 peregrinations between Tsari and Tawang.

Delighted to hear of your safe return hope you are well I would like you to come up to Simla as quickly as possible. McMahon foreign.

The telegram addressed to Bailey 'c/o Thomas Cook and Son, Calcutta' reached him within hours of his disembarking from the Assam Mail at the city's Sealdah station and just as he emerged, bathed, barbered and suited, from Chowringee's Grand Hotel. After much prevarication the Chinese and Tibetan envoys to McMahon's frontier conference had arrived, but since it was now late November the venue had been moved from a possibly snowbound Darjeeling to a possibly sunnier Simla. Leaving Morshead behind to work up the maps, Bailey headed immediately for Howrah station and the Simla Mail.

The 1914 Simla Conference dragged on until the following April. Its so-called convention was never ratified and its 'McMahon Line', the proposed frontier in the eastern Himalayas between Tibet and British India, remained a dead letter even to the British until 1935. Simla was unfinished business not just because the principals failed to reach agreement but because of what the government was calling 'the present time of grave preoccupation'. Minds were elsewhere. An archduke had been shot in the Balkans.

The First World War broke out within weeks. McMahon was promptly whisked off to Egypt. As high commissioner in Cairo, he would be responsible for the notorious exchange of letters with *sharif* Husayn of Mecca that resulted in the Arab Revolt against Ottoman rule. Instead of the gung-ho Bailey, here McMahon relied on the intelligence gained by the gung-ho T. E. Lawrence ('of Arabia'). In predetermining what is still the political geography of the Arab world, McMahon and his Arab Bureau colleagues roughed out frontier lines in the desert that would prove even more contentious than the one that bears his name in the eastern Himalayas.

Bailey himself went off to fight on the western front and at Gallipoli; he returned wounded from both. He was rewarded with intelligence-gathering assignments in the khanates of what had been tsarist Russia and now became Bolshevik Russia. His account of his undercover activities in the Uzbek capital in 1918–19, *Mission to Tashkent* (1946), would win him fame as the consummate 'secret agent'.

Himālaya reclaimed him when in 1921–8 he returned to Tibetan affairs as the decidedly interventionist political officer in Sikkim. 'Did Bailey plan to overthrow the Dalai Lama [in 1924]?' asks Professor McKay; '… the weight of circumstantial evidence definitely points to a coup having been planned under Bailey's direction.'[18] If so, it was without his government's knowledge, and the coup came to nothing. But it was all quite in character with his hands-on approach to managing India's borderlands.

Bailey lived till 1967, long enough to learn of the McMahon Line being overrun by the PLA in 1962 but not long enough to be troubled by news of Pure Crystal Mountain's fate. Only in 2018 did Indian satellite pictures reveal sturdy Chinese bridges and a new all-weather road snaking up the valley of the Tsari Chu. The road reportedly serves 'at least 10 barracks' and advances at least 5 kilometres into what should be Indian territory, so effectively bisecting the region within the 70-kilometre-long pilgrimage route. 'Underground construction' is also suspected, along with further 'PLA posts'. 'China's age-old policy of salami-slicing [incremental encroachment], which has paid it great dividends in South China Sea, is being adopted by the PLA in the Himalayan region too,' says an official Indian posting.[19] The writer regrets 'the paucity of good temporal resolution in the satellite images'; no doubt Mount Tsari is still wrapped in cloud. Though satellite imagery has revolutionised cartography and surveillance, in a region as unstable as Himālaya more data and sharper images are no guarantee of political resolution. Controversy and conflict still abound.

8

The Karakoram Anomaly

But the glory of the [Basha] valley is the magnificent glacier at the
end of it ... the natives say it is slowly but perceptibly advancing. It
occupies the entire valley as far as the eye can see, and a place that
looks more like the extremity of the world does not exist in nature.

Godfrey Thomas Vigne (1842)[1]

Mountains don't change much. Their heights may be recalculated
using the latest technologies but the values obtained seldom differ
dramatically from those arrived at by trigonometrical survey in
the nineteenth century. Orogeny (mountain-making) grinds on,
yet Everest's tectonic upthrust of a few centimetres a year is quite
undetectable to the human eye. Photos of Kangchenjunga as seen from
Darjeeling in the late nineteenth century show the world's third-highest
mountain looking just as improbably ethereal as it does today. And
no doubt Tsari's precipitous spurs, invisible beneath the cloud cover,
still guard the cleansing elixirs of Pure Crystal Mountain. Himalayan
profiles prove reassuringly constant. However young and restive, the
mountains themselves don't appear to be on the move.

Glaciation is a different matter. Glaciers do move. They flow. They
dribble from the flanks of the great peaks like wax down a candle.
From season to season, year to year, century to century, they advance
and retreat in response to flunctuations in temperature and climate,
encroaching in some places and withdrawing in others. Movement –
mostly gradual, sometimes cataclysmic – is a defining characteristic
of glaciers. So is the prodigious force they exert in brushing aside

any topographical obstacles and rearranging them as moraines. No less distinctive are their terminal snouts, from whose icicle-toothed caverns their ice-melt spills to swell the headwaters of half a continent's rivers.

Himālaya's glaciers release most of their melt in the late summer at a time when rainfall anywhere beyond the reach of the Indian monsoon is negligible. It's as if their function in the great scheme of things is to act as obliging reservoirs. The scientific community uses a similar metaphor for the whole mountain region: Himālaya, we are told, is 'Asia's great water tower'. Frigidity and gravity here combine to meet the water needs of some 240 million hill people while, downstream on the floodplains of South and South-east Asia, another 1.65 billion count on the same run-off to irrigate crops and deliver the nutrient-rich sediment in which they grow. In addition, the rivers responsible – from the Indus and the Oxus to the Ganges, Tsangpo–Brahmaputra and Mekong – have an estimated hydroelectric potential of around 500 gigabytes. As yet only a fraction of this is being realised. But every year sees fresh scars gouged into the landscape as new dams come online and still larger ones get the go-ahead.

A lot of livelihoods are riding on those glaciers. Loss of their timely ice melt could mean famine, mass migration, economic breakdown and political turmoil. The world is right to keep an eye on them, monitor their movements and pronounce on their future. It's just unfortunate that glacial masses are so tricky to measure. For one thing they're often hard to reach and for another they're far from constant. Depth and density vary wildly; the sudden surges to which some are prone remain unexplained; and a phenomenon known as 'elevation-dependent warming' (EDW) is surely relevant but has been insufficiently studied. Their blizzard-raked upper reaches being typically inaccessible to all but mountaineers, few have been observed in detail or for long. Yet what goes for one end of a 50-kilometre ice floe may not be true of the other end. 'Despite the large area of ice cover in [Himālaya], there are no long-term measurements available from anywhere in the region,' declares a 2019 assessment. 'In total, only about 30 glaciers [out of several thousand], covering an area of less than 120 km² out of a total glacierized area of 80,000–100,000 km², have had direct glaciological measurements made for one or more years.'[2] Whether based on physical measurements or on data obtained from remote sensing, ground-penetrating radar or satellite imagery, extrapolations and projections in respect of glacier behaviour often need correction.

Sometimes the corrections reach the public arena. In 2007 the scientific world blanched to learn of a truly egregious howler, the repercussions of which would be felt for a decade. The trouble seems to have started with a humble typo. In the fourth of its five-yearly Assessment Reports, the United Nations Intergovernmental Panel on Climate Change (IPCC), 'the world's most widely cited source of climate analyses', had aired a grim and little known fact.[3]

Glaciers in the Himalaya are receding faster than in any other part of the world [declared the report] … and, if the present rate continues, the likelihood of them disappearing by the year 2035 and perhaps sooner is very high if the Earth keeps warming at the current rate. Its total area [presumably, that is, of the entire earth's glaciation] will likely shrink from the present 500,000 to 100,000 km² by the year 2035.[4]

The figures given had been lifted from a report compiled by, or for, the World Wildlife Fund (WWF), which had in turn relied on findings used by the International Commission for Snow and Ice (ICSI) and UNESCO's International Hydrological Programme. But somewhere in this chain of citation an error had crept in. Year Zero for Himālaya's glaciers had been advanced by more than three centuries. Instead of a remote extinction in 2350 as originally proposed, the glaciers were now destined to disappear by the scarily imminent 2035. Someone had scrambled the end-date's digits.

The mistake was soon corrected amid abject apologies. These things happen. But what, it was asked, did this say about a 938-page report acclaimed as 'the largest and most detailed summary of the climate change situation ever undertaken'? How was it that not one of its eminent cast of 'thousands of authors, editors and reviewers from dozens of countries' had spotted the mistake? And what did all this indicate in respect of the IPCC's verification procedures and the validity of the 'over 6,000 peer-reviewed scientific studies' the report proudly claimed to have consulted?

Notwithstanding the document's inordinate length, it was now assured of an avid readership. Close scrutiny was called for – and duly followed. Heads might have rolled – but didn't. For of the several other errors and inconsistencies that came to light, none was deemed to have discredited the report's main findings: namely that glaciers were shrinking faster than expected because greenhouse gas emissions (mainly carbon dioxide and methane) were increasing exponentially, so pushing up temperatures and accelerating meltdown. Naturally, scientists convinced of this Doomsday scenario stood their ground; so did sceptics convinced it was all humbug. Both sides were indebted to that humble typo. The scientific jeremiahs welcomed the further opportunity to ram home their message, while the Doubting Thomases congratulated themselves on having made a point. Future reports would need more robust data and much more robust oversight.

At the time, 2007, 'climate change' was just bobbing to the top of the environmental agenda. Concern in the 1970s over depletion of the ozone layer had been succeeded by the 'global warming' fears of the 1980s, and then by Vice-President Al Gore's 1990s highlighting of the perils (especially in terms of higher sea levels) inherent in glacial shrinkage. 'By the early twenty-first century glaciers had reached celebrity status,' reports Mark Carey, an American historian of science, 'with almost all popular writers who discussed global warming making glacier retreat a key part of their story.'[5]

The endangered glacier was not just a barometer of current climate change; it contained a vital archive of climate-change-through-history. Standing on top of the Greenland ice cap not far from where Alfred Wegener's mortal remains had been consigned to the ice in 1930, science writer Elizabeth Kolbert watched the drilling of ice cores and allocated to each frozen layer beneath her boots a historical milestone.

> A hundred and thirty-eight feet down, there is snow that fell during
> the time of the American Civil War; 2,500 feet down, snow from
> the time of the Peloponnesian wars, and 5,350 feet down, snow
> from the days when the cave painters of Lascaux were slaughtering
> bison. At the very bottom, 10,000 feet down, there is snow that fell
> on central Greenland before the start of the last ice age, more than
> 100,000 years ago.[6]

While analysis of such cores affords insights into the deep history of world climate change, more recent glacial contraction can be tracked using satellite imagery dating from the Cold War and belatedly declassified. Doubts still remain; but they are now less about the science, more about the impartiality of people whose careers may depend on either downplaying the likelihood of catastrophe or amplifying it. The case for recognising 'climate change crisis' as a clear and present danger is generally accepted.

At the time of writing, the most persuasive contribution to the debate is probably *The Hindu Kush Himalaya Assessment: Mountains, Climate Change, Sustainability and People* published in 2019 by ICIMOD. Otherwise the International Centre for Integrated Mountain Development, ICIMOD is a 'regional knowledge hub' based in Katmandu to serve eight member countries within what it calls 'Hindu Kush-Himalaya' (HKH): Afghanistan, Bangladesh, Bhutan, China, India, Myanmar (Burma), Nepal and Pakistan. Its remit, in other words, closely resembles that here proposed for Himālaya. In fact, for the purposes of ICIMOD's 2019 *Assessment*, the Centre's geographical net was spread even wider to embrace an 'extended HKH' including Tajikistan and the Pamirs, so making the match with Himālaya even closer.

At a modest 600 pages, the ICIMOD *Assessment* is not without repetition and copious citations. But it takes a more cautious line than had the IPCC's 2007 document. It focuses less on climate change as a global challenge and more on how Himālaya's society and economy may be affected by it. It concedes the extraordinary variety of the region's habitats and the complexity of its climate zones ('the extended HKH is characterized by extreme topographic and climatic heterogeneity', it says). And it therefore acknowledges the existence of some important exceptions to the 'disappearing glaciers' narrative. Of the few glaciers that have been monitored, not all are contracting at the same alarming rate and some are not contracting at all. The eastern Himalayas (for example, Tsari and Pemako) and the eastern Tibetan plateau (for example, Pomé and Kham) are among the worst-affected regions, the western Himalayas (for example, Gugé and Ladakh) less so, and 'the glaciers in the eastern Pamir, Karakoram, and western Kun Lun ... have since at least 2000 had balanced mass budgets or even slight mass gains'.[7]

This welcome news from Himālaya's most extensively glaciated region is sufficiently remarkable to warrant its own epithet: it's known to climatologists as the 'Karakoram Anomaly'. Here, north of the Indus in what the Pakistan government currently calls Gilgit–Baltistan, more than a dozen peaks of over 7,500 metres asl cluster round a network of colossal glaciers that make up the largest glaciated area not just in Himālaya but anywhere outside Antarctica and Greenland. And happily, it's not as yet shrinking; it may actually be growing. Men like Henry Haversham Godwin-Austen for whom the Karakoram epitomised Himālaya would no doubt approve. The Karakoram have always been an anomaly. To the 'K2 man', treating them as a subsidiary Himalayan range had seemed an injustice.

Though long since stripped of the honour of having a Karakoram peak named after him, Godwin-Austen still has his glacier, and it too may be growing. The Godwin-Austen glacier comes tumbling from the south face of K2 to meet the Baltoro glacier at a point where that 67-kilometre river of ice is joined from the opposite direction by yet another glacial tributary. This is the Vigne glacier, so named for an elusive London barrister called Godfrey Thomas Vigne (pronounced 'vine'). Vigne is the only European known to have preceded Godwin-Austen into Karakoram's 'wintry citadel of the gods'; and, needless to say, the Vigne glacier, like Godwin-Austen's, was so called by a fellow Briton, Sir William Martin Conway. As the man responsible for leading the first assault on the Karakoram peaks and so initiating Himalayan mountaineering, Conway felt entitled to spatter the map with English proper names. He could therefore hardly object to others following suit. In less than a generation the Anglophone names would be matched by other exotic toponyms, many of them Italian.

Arriving in 1835, Vigne had been not just the first foreigner to reach the Karakoram but also the second and third, for he made return visits in 1837 and 1839. During the last he ran into Hugh Falconer. The doctor was devoting his holiday from fossil-hunting in the Siwaliks to exploring the flora of Baltistan and the upper Indus valley. Vigne too was collecting botanical specimens. The travellers shared alfresco meals, discussed itineraries and no doubt swapped seeds. Anticipating 'Hatter' Bailey, Vigne also carried a gun for whatever sport or science might commend to his attention. More surprisingly he was no less adept with bat and brush; he had played cricket for Hampshire and would exhibit at the Royal Academy. There wasn't much that Godfrey Thomas Vigne couldn't do. He was well informed, well connected and well off, the family having acquired their estate in still rural Walthamstow on the proceeds of supplying the East India Company with gunpowder.

Self-portrait of Godfrey Thomas Vigne, sportsman and artist, in Baltistan in the late 1830s

What drew the thirty-two-year-old Godfrey Thomas to Baltistan – or what drew him anywhere – is less clear. In 1830 he had no sooner been called to the bar than he took off for North America. Two years later he was heading for the Middle East, and two years after that, having exhausted the distractions on offer in Persia, he took ship for Bombay. There, he says, 'I determined, not having had the slightest intention of the kind when I left England, to run up at once to the cool air of the Himalayas.' Like most globe-trotting wayfarers he was easily sidetracked. With contacts in the East India Company keen to entertain him wherever he went in India, it took him a year to reach the little-known valley of Kashmir and another before he disappeared over the Himalayan horizon into the utterly unknown Baltistan. Baltistan, he was told, was formerly part of Tibet. The monk-ridden Tibetan plateau ruled from Lhasa was Upper Tibet, the equally Buddhist Ladakh (capital Leh) was Middle Tibet, and this Baltistan kingdom of Shiah Muslims down the Indus was Little Tibet (capital Skardu).

As if to anounce its Tibetan credentials, a 'stupendous peak' greeted Vigne as he left Kashmir and breasted the first of the trail's two passes. The mountain in question was 60 kilometres away but appeared closer, 'rising far above every other around it and entirely cased in snow excepting where its scarps were too precipitous for it to remain upon them'. With its summit poking from a halo of cloud into a peerless sky, Vigne supposed it looked higher than it was. He ventured a conservative '18,000 or 19,000 feet [roughly 5,500–5,800 metres]'.[8] Other Himalayan travellers invariably exaggerated distances and altitudes; Vigne must be unique in consistently underestimating them. The locals called the great peak Diarmul and the Kashmiris called it Nanga Parbat; at 26,600 feet (8,126 metres) asl it is in fact the world's ninth highest summit. It marks the western extremity of the Great Himālaya and, as the 'Killer Peak' of the twentieth century's climbing fraternity, it would prove one of the most unforgiving in all of Himālaya.

Riding where he could and still heading north, Vigne pressed on into the mountains. He became the first non-Asian to cross the Great Himālaya on to the Deosai plains, a lumpy plateau 4,000 metres asl and now a uniquely bleak national park. For company he had his Kashmiri and Balti porters, plus a bear-hunting terrier, a mastiff and a manservant. The last was Mitchell, a lout with a fondness for liquor who, when strung about with his master's indispensable travel items – telescope, kettle, easel, sextant, thermometer, microscope, geological hammer and umbrella – must have looked like a laboratory on legs. If Mitchell didn't complain it was only because Vigne was that rarest of creatures, a companionable explorer who treated his employees with uncommon leniency.

A final pass and then a steep descent into Baltistan brought the little cavalcade in sight of the Indus.

> ... through a long sloping vista formed of barren peaks, of savage shapes and various colours ... I, the first European who had beheld them, (so I believe,) gazed downwards from a height of six or seven thousand feet, upon the sandy plains and green orchards of the

valley of the Indus at Iskardo [Skardu], with a sensation of mingled pride and pleasure, of which no one but a traveller can form a just conception.[9]

The view would provide the backdrop for the only known portrait of Vigne, a self-portrait in which the subject, in a floppy local turban and the knitted gloves and stockings given him by Baltistan's ruler, looks far too foppish for mountain travel. The river itself is hidden from sight, but '[to] the north, and wherever the eye could rove, arose, with surpassing grandeur, a vast assemblage of the enormous summits that compose the Tibetian [sic] Himalaya'. Also known as the Mustagh, this towered obstruction is the Karakoram, a relatively short range at 600 kilometres but surely the world's most spectacular. Geologically speaking it presides over the Karakoram Fault, a 'strike fault' where the advancing Indian plate elbows aside the Asian plate to displace and thicken the Tibetan crust. Hydrologically, and much more obviously, it is also the Asian watershed; rivers to the south, including the Indus, all drain into the Indian ocean, while those to the north make for the oases of Eastern Turkestan (now Chinese Xinjiang) before expiring in the shifting sands of the Takla Makan desert.

Vigne, of course, had nothing remotely useful in the way of maps. But he knew of Turkestan and seems to have set his sights on either Kashgar, the largest of Xinjiang's oasis cities, or Khokand, a Central Asian khanate much further west in what is now Uzbekistan. Both places had featured in the plans of William Moorcroft, the irrepressible pioneer of exploration in Inner Asia who had passed through Ladakh and Kashmir a decade earlier. Denied a right of way through the mountains, Moorcroft had then disappeared into Afghanistan, never to be seen again. To avoid a similar fate, Vigne was now giving Afghanistan a wide berth by opting for Baltistan and the chance of finding a trail through or round the great Karakoram.

There were, he learned, three ways to tackle the mountain barrier: west by way of the 49-kilometre Hispar glacier into the jealously independent mountain statelets of Nagar and Hunza; east by way of Ladakh's Nubra Valley and the 5,500-metre Karakoram Pass; or head-on by way of the giant Baltoro glacier and the disused Mustagh Pass. The first and last called for ropes and ice axes; they were formidable undertakings for a well-equipped expedition, never mind a lone and distinctly casual wayfarer. Vigne nevertheless investigated all three. Hunza–Nagar was ruled out on the grounds that Ahmed Shah, his Skardu host and Baltistan's ruler, opposed it. The Ladakh route was more promising; it was in year-round use, but there Vigne, like Moorcroft, found his plans frustrated by the xenophobia of Ladakh's current rulers. That left the terrifying and probably impossible Baltoro route, its great glacier being, as Ahmed Shah put it, 'interminable'.

A good look at the nearby Basha glacier was enough to decide the matter. Many of the Karakoram glaciers were so strewn with rocks, detritus and dirt that only their yawning crevasses betrayed their icy composition. But the Basha ended in as fine a wall of ice as one could wish for. Vigne reckoned it quarter of a mile (360 metres) wide, 100 feet (30 metres) high, 'clear and green as an emerald', and broken only by a 'lofty, gloomy, and Avernus-like' aperture.

> I have never seen any spectacle of the same nature so truly grand as the debouchure of the waters from beneath this glacier ... The stream that emerges ... is no incipient brook, but a large and ready-formed river ... whose violence and velocity betoken a very long descent and whose force is best explained by saying that it rolls along with it enormous masses of ice, that are whirled against the rocks in its bed with ... a sound resembling that of a distant cannon ...[10]

Local opinion insisted the glacier was advancing, obscuring old trails and severing ancient trade links. In this respect, as in its warlike sound-effects, the glacier's inexorable progress echoed that of the expanding Sikh kingdom of Lahore in the Punjab. Contained by the British frontier advancing from Delhi, Lahore's forces had lately been deflected into the mountains, overrunning Kashmir and Ladakh and even conducting an exploratory foray into Baltistan. They were expected to return at any minute. If Vigne abandoned his Karakoram plans rather readily, it seems to have had as much to do with saving Ahmed Shah and his kingdom from the Sikhs as with conceding defeat to the Karakoram and its glaciers.

Rightly, Ahmed Shah had hailed Vigne's unexpected appearance in Baltistan as a God-given guarantee against Sikh aggression. He treated the Englishman as his honoured guest and pressed him to return as often as possible. While Skardu was hosting a representative of what was by now the subcontinent's premier power, the Sikh forces would not dare to reinvade. Vigne, while disclaiming any official status, played along with this. He lauded Ahmed Shah's rule, talked up his kingdom's strategic attractions and would represent its plight to any British officials who would listen.

Sadly the timing could not have been worse. In 1839, when Vigne was on his final visit to Baltistan, Maharajah Ranjit Singh, the one-eyed founder and undisputed mastermind of the Sikhs' Lahore kingdom, died. The succession was promptly contested; Lahore's army, the most effective in independent India, mutinied; and Lahore itself was rent by civil war as five maharajahs came and went (mostly assassinated) in as many years. Prominent among the powers behind the throne was one Gulab Singh, the Dogra (and so Hindu, not Sikh, and unrelated to the late maharajah) ruler of Jammu. It was this Gulab Singh whose Dogra troops had been largely responsible for the Sikh conquest of Ladakh and who had lately invaded western Tibet and were now eyeing up Baltistan.

At the time Gulab Singh's sights may have been set on Lahore itself. But it so happened that the Sikh kingdom's implosion coincided exactly with the catastrophic British invasion of Afghanistan known as the First Anglo-Afghan War. Britain's 1841 humiliation in Kabul was partially redeemed by a face-saving reinvasion in 1842, followed by another withdrawal. But a surer way of erasing the shameful defeat in Kabul lay in engineering a clear-cut triumph in Lahore. In 1845 the mutinous Sikh army obligingly crossed the lower Satluj into British territory, so triggering the First Anglo-Sikh War.

By 1846 it was all over. Lahore had a British garrison, its seven-year-old maharajah was in British tutelage, and the Dogra rajah Gulab Singh had been rewarded for his wartime neutrality (and a hefty contribution to the ceasefire indemnity) by British recognition of his protected status as maharajah of 'Jammu and Kashmir'. Thus was created the composite state which would become such a bone of contention in the twentieth century. As well as Jammu, it included the famous Kashmir valley and a chunk of the Pir Panjal or outermost Himalayas, plus all of Ladakh, Ahmed Shah's Baltistan (which had been overrun as soon as Vigne departed) and Gilgit and the satellite statelets of Hunza and Nagar (which would soon share the same fate).

Hence twenty years later, when in 1861 the young Henry Haversham Godwin-Austen was posted to Skardu, Baltistan had become an outlying district of one of British India's princely states. Its Karakoram backdrop therefore constituted a section of British India's sensitive northern border, and as such its peaks and passes, including its skein of massive glaciers, cried out to be mapped. To that end, in the 1850s the Survey of India had been extended to Kashmir. And from there, from a peak on the edge of the Deosai Plains, in the summer of 1856, during a lucky break in the clouds, Captain T. G. Montgomerie, the officer superintending the Kashmir survey, caught sight of a cluster of snowy peaks all of 200 kilometres away. Montgomerie sketched a profile of these distant giants and, in the absence of names, allocated to each a number – K1, K2, K3 and so on up to K32. The shy pyramid that was K2 already looked to be the loftiest with a provisional height of 28,250 feet (8,610 metres). At a time when the heights of the Great Himālaya's summits along the Nepal–Tibet frontier were still being calculated, this made K2 a serious contender for planet earth's closest approach to the stars.

As a surveyor, Godwin-Austen was more interested in establishing exactly where K2 was. Did it and its Karakoram neighbours stand south of the watershed and so within British territory? Or north of it and so within Eastern Turkestan/Xinjiang? In terms of man's ongoing encounter with the mountains, the 1860s belonged to what has been called the 'age of passes'; the 'age of peaks' would come later. Mountains were still seen as obstacles to circumvent rather than as pinnacles to conquer. Peaks led nowhere; passes had purpose. And in the high Karakoram, passes invariably meant glaciers.

All of which suited the detached and troubled twenty-seven-year-old who was Lieutenant H. H. Godwin-Austen. He was not interested in scaling the great peaks and he scorned the climbing hardware favoured by professional alpinists. According to one who knew him well, 'he had no knowledge of technical mountaineering: he carried no rope except as an aid to crossing streams and had no ice axe'. On the other hand, 'slight and extremely hardy, with little weight to carry, he had immense endurance and great enthusiasm'.[11] In July 1861, when he first ventured on to the Karakoram glaciers, Godwin-Austen may also have been nursing a tormented soul. Perversely combining the travails of a grieving widower with those of an anxious newlywed, he looks to have turned to the glaciers for solace.

His first marriage had been short and not unhappy, just unconventional. Loving relationships between British men and Indian women had once been commonplace. William Moorcroft, for instance, had kept more than one Indian wife and fathered several Anglo-Indian children. But by mid-century, sharing the white man's burden with an Indian *bibi* meant jeopardising one's career; evangelising churchmen disapproved and house-proud memsahibs closed their doors; promotion would be stalled and invitations withheld.

Godwin-Austen had gone ahead regardless. In 1857 his little Kudikji had given birth to their child and a year later, after declaring his conversion to the Islamic faith of her family, he had been legally married to her. She was the daughter of a Sodhan farmer, the Sodhan being an Afghan clan long settled in the Poonch district of Kashmir and lately subjected to severe mistreatment by both Sikhs and Dogras. The young lovers had evidently met before Godwin-Austen was posted to the Kashmir survey. Perhaps, like Vigne with Ahmed Shah, his attachment afforded Kudikji's family some immunity from marauding Dogra troopers; or perhaps it was Kudikji's company that attracted Godwin-Austen to the Kashmir survey in the first place.

With or without their little son, Kudikji had certainly accompanied her husband on a prior reconnaisance to Baltistan in 1860. This was most irregular and may be the sole mid-century instance of a survey officer enjoying the company of his child-bride while on active service. But thereafter Kudikji fades into oblivion. Either she 'disappeared' or 'died in childbirth', says Catherine Moorehead, Godwin-Austen's indefatigable biographer. Kudikji was not sharing his tent when in 1861 he embarked on what Moorehead rates as 'one of the most remarkable expeditions in British mountaineering history'.[12] Nor perhaps would she have been welcome – for by then her husband had married again, this time at very short notice to the seventeen-year-old and much more socially acceptable Pauline Chichele-Plowden. His family and no doubt his service were relieved. It was not, however, a happier marriage. The new Mrs Godwin-Austen was already living well beyond the means of a survey lieutenant. By 1870, when she died in Calcutta, she would leave Godwin-Austen saddled with crippling debts.

Amid these domestic strains, the Karakoram and its glaciers came less as a challenge, more as a welcome distraction. Godwin-Austen wrote no relevant book on the region, just several hundred scholarly and none too appetising papers. Typical was that delivered to London's Royal Geographical Society on 'The Glaciers of the Mustakh Range'. Here the certainties of science, be they mathematical, geological, topographic or biological, serve as balm; the hardships of extreme altitude scarcely trouble him and, if he is moved at all, it is by the sheer magnitude of the mountains and the undisguised brutality of the extreme terrain.

> For the first three miles the crevasses were broad and deep in places only, and we could avoid them by making detours. They soon became more numerous, and were ugly things to look into, much more so to cross – going down into darkness between walls garnished with magnificent green icicles from 6 to 20 feet long, and of proportionate thickness, looking like rows of great teeth ready to devour one. I tried with our ropes to sound the depths of these fissures, but all of the ropes tied together only made up 162 feet, which was not long enough.[13]

This was just west of K2 as he struck up Chiring Glacier in search of the elusive Mustagh Pass. Like Vigne he failed to reach the pass itself, but he could confirm its existence. 'Four Baltis who had emigrated to Yarkand [in Xinjiang] some years back and had now come over to see their friends on this side' had crossed it just before stumbling into his camp. They had 'suffered a good deal' but this was because they were travelling at night 'on account of the robber tribes'.

Next day Godwin-Austen tried again, reaching 17,300 feet (5,273 metres) asl and getting to within a mile of the saddle.

> The pass was distant about 500 feet above our turning point; we had to beat a hasty retreat down towards Chiring, the snow falling fast. The glacier was making most disagreeable noises – crunching, splitting and groaning to an awful extent – caused by the vast body of ice, two miles across, here forcing itself through a channel only a quarter of a mile broad and with an increased slope ... I had gone as far as was necessary towards the parting ridge; still I would have liked to have crossed the pass ... it was in a disappointed mood that I left Chiring for Punmah [Panmah].[14]

The Panmah glacier, another colossus, brought the survey party to the even bigger glacier that Godwin-Austen calls the Biaho but which was in fact the Baltoro. He was now advancing up the nave of the Karakoram's great cathedral of peaks. Dead ahead, one of Gasherbrum's (K3's) pointy finials reared up to 8,080 metres asl, while to the south the stately profile of Masherbrum (K1) sailed into sight. They were nearer K2 than anyone had ever ventured, and still it was hidden by intervening ridges. Scaling a spur of Masherbrum offered the only chance of a sighting.

> Eagerly I had looked, whenever we stopped to take breath, along
> the line of snowy mountains to the north in search of the Great
> Peak; and now fixing the position on the plane-table, I showed
> those with me the mountain behind which lay the peak I had
> toiled up so far to see. Following up the ridge another 1000 feet of
> elevation was gained, when a distant bit of rock and snow could be
> seen just peering above the nearer snow-line. After another sharp
> push up to a point where it was impossible to mount further, there
> no longer remained a doubt about it. There, with not a particle of
> cloud to hide it, stood the great Peak K2, on the watershed of Asia![15]

So, like the peak on the Nepal–Tibet border just named 'Mont [sic] Everest' after the irascible George Everest whose survey had enabled its measurement, K2 actually bestrode the frontier. It still does. In the wording of the 1963 Sino-Pakistan Agreement, 'the [China-Pakistan] boundary line follows the Karakoram Range main watershed ... passing through the East Mustagh Pass (Muztagh Pass), the top of the Chogri Peak (K2), the top of the Broad Peak, the top of the Gasherbrum mountain (8068 [metres]) ... and reaches its southeastern extremity at the Karakoram Pass'.[16] Yet even political geography has its 'Karakoram anomaly'. The government of India steadfastly refuses to recognise this frontier. On the grounds that Kashmir and its borderlands remain an integral part of the Republic of India, New Delhi insists that neither the frontier alignment nor the Sino-Pakistan Agreement has any international validity.

From the Baltoro glacier, Godwin-Austen turned his attention to the Hispar glacier and from the Mustagh Pass to the Nushik Pass.

Historically the latter had offered a passage through the western Karakoram to the ledges of cultivation that comprise Nagar and Hunza. Thence a precarious trade route led over the watershed to link Gilgit to Turkestan's trans-Asian 'silk roads'. But in the 1860s the Nushik trail was little used. Kashmir's Dogra rulers had as yet only a toehold in Gilgit, while Hunza and Nagar were still pariah principalities jealously guarding their independence. Their mutual hostilities and the robber bands that infested the watershed had brought the trickle of trade and transit to a standstill. It would have been asking for trouble for a British officer engaged on a Kashmir survey to have ventured into either.

As it was, Godwin-Austen was again halted by nature. Within sight of the Nushik Pass his men began succumbing to the effects of altitude. A later traveller put the height at 5,275 metres and confirmed Godwin-Austen's misgivings about the descent on the other side. A 1,000-metre drop down a sheer rock and ice face, it would have meant dangling a man on a rope to cut steps before they could think of groping their way down. The view, though, was magnificent. 'No glacier scene in the whole of the Himalayas can exceed this in the magnitude of its features,' thought Godwin-Austen.

In reflective mood he headed back to Skardu, then Kashmir and India. If not 'one of the most remarkable expeditions in British mountaineering history', his 1861 survey had been one of the most revealing in respect of glaciation. 'Whilst engaged in the work described ...', he wrote, 'I have often been struck by the indications of considerable amounts of change of temperature within what we may call our own times.' Signs of glacial retreat were evident throughout the Himalayas. The entire upper Indus valley had once been under ice. Enormous terminal moraines abutted the river in places from which no glacier was now within sight. But these dated from the last Ice Age. Proof of more recent changes of temperature came from glaciation that had advanced within living memory.

> Many passes which were used even in the time of Rajah Ahmed Shah of Skardo are now [1861] closed. The road to Yarkand over the Baltoro, which before his time was known as the Mustakh, has by the increase of the ice near the pass become quite impractical ... Again, the Jusserpo [Tusserpo] La can now be crossed only on foot whereas in former time ponies could be taken across it ... Certain large glaciers have advanced, such as that at Arandu, of

which the old men assured me that in their young days the terminal cliff was 1½ miles distant from the village. Mr Vigne says 'it was a considerable distance', but it is now only about 400 yards.[17]

Conducting a survey across terrain composed of such restless features must have seemed a thankless task; it was like trying to map the clouds. But it could have been worse. Had Godwin-Austen succeeded in crossing the Nushik La into Hunza, he might have despaired altogether. For in Hunza glaciers were considered a vital amenity, indeed were deemed amenable. Past masters at managing their own hydrography, the Hunzakuts claimed, and still claim, to be able to conjure up glaciers at will.

Neither Vigne nor Godwin-Austen says anything about this practice. Nor does it much feature in subsequent travelogues or in the reports of the IPCC and ICIMOD. While a noteworthy response to the challenge of farming in the Karakoram, glaci-culture needs to be treated with caution.

So do most of the other unlikely claims advanced on behalf of Hunza and its inhabitants: with centenarians supposedly commonplace among Hunzakuts, life expectancy is here among the highest in the world (false); the valley has provided the inspiration for the Himalayan utopia of Shangri-la (unproven); excluding eye problems, disease is practically unknown (false); female education is among the highest in South Asia (possibly true); with its heavy dependence on apricots, the local diet is one of the world's healthiest (an exaggeration); crime is almost unknown (false); and the happiness ratio is off the scale (as it would be were such a benign world for real).

Somehow their life, seemingly hard and austere, has endowed these people with a happiness I forbear to overstate [asserted the nutritionist Lionel Picton in the 1940s]. They have achieved engineering without mathematics, morality without moralising, agriculture without chemistry, health without medicine, sufficiency without trade. In the harsh and unpromising surroundings of the Hunza [valley], mastery of the art of life has been engendered by an unremitting agriculture.[18]

As farmers the Hunzakuts may indeed be in a league of their own. John Staley, who conducted field research in Gilgit and Hunza in the 1970s, certainly thought so.

Here are a people who live and thrive in one of the most
unpromising places on earth, who draw simultaneously upon
the resources of a vertical range of 4 miles [6.4 kilometres], and a
climate varying from permanent ice to permanent drought, who
can raise crops in a desert during summer and survive months
of freezing in winter, and who for centuries have provided for
themselves without dependence on the outside world.[19]

The outside world has since elbowed its way in. In the 1950s Hunza's
mule trails became jeep tracks, which became the Karakoram Highway
in the 1970s and which, after a long section was inundated in 2010,
became the Realigned Karakoram Highway in 2015.

Yet terraced fields stacked up hillsides like the treads of a staircase
and barely wide enough for a wheelbarrow still make for fastidious
farming. Weeding is done by hand. The harvest is cut with pocket-size
sickles and hefted home on bent backs. Apricot trees (there are said
to be thirty-nine varieties) double as trellises for grapevines. Livestock
do well on a mush made from mulberries. All waste is recycled, all
resources meticulously husbanded – and none more so than water.

Hunza's rainfall being minimal, arable production depends on run-
off. Meltwater from the winter's snows lasts well into summer, after
which irrigation comes courtesy of the mica-rich soup that is glacial
ice melt, aided by gravity and faultlessly angled aqueducts. The
glacier snout may be several kilometres from the parched terraces it
serves. In between may be landslides, sheer bluffs, deep canyons and
perhaps moraines or even another glacier. Ditched here, rock-cut
there, embanked and revetted, and gliding across cliff faces in long
timber troughs, the aqueducts cleave to the mountains' contours with
uncanny precision. This is engineering not just minus mathematics but
minus spirit levels and survey instruments, let alone machinery. And
if, for some insurmountable reason, a glacier's meltwater still can't be
conducted to the furthest fields, there remains the option of 'seeding' a
new glacier somewhere more convenient.

This miracle is achieved by artificial insemination (AI). Glaciers, say
the Hunzakuts, are gender-specific. Those with light or bluish ice that
produce a bounteous flow are female, and those coated with rocks and
soil that yield only a trickle are male. Naturally they have an affinity
for one another, and this mating instinct helps explain their erratic

Devotees undertaking Kashmir's most arduous pilgrimage snake towards the Mahagunas Pass before tackling the final ascent to Lord Shiva's icy lingam in the Amarnath Cave. The annual Amarnath *yatra* provides militant jihadists with an easy target and often necessitates an armed escort.

The great monastery of Lamayuru and its village announce the erstwhile kingdom of Ladakh. Travellers from Kashmir here enter the Tibetan world of sparse vegetation, minimal rainfall and the fervent Vajrayana Buddhism which once prevailed throughout Tibet.

In 1987, for the first time since Mao's Cultural Revolution, monks of the Tashilunpo monastery at Shigatse (Xigaze) were authorised to consecrate a giant mandala. Depicted in coloured grains of sand, a mandala becomes the focus of devotions for just a week. It is then swept away and must be recreated anew – a paradigm perhaps for the tenacity of Tibetan Buddhism.

An alluvial fan near Kargiakh in Zanskar. 'No one confronted by these contorted scarps could possibly question the instability of Himālaya or the one-time plasticity of its rocks … The thin air positively zings with lithospheric tension' (page 33).

Ladakhis await the arrival of the 14th Dalai Lama at Choglamsar in Zanskar in 1987. The visit of His Holiness was to celebrate the Kalachakra, a revered teaching tradition requiring the initiation of practitioners.

Known to Hindus as Mount Kailas and to Buddhists as Tisé or Kang Rinpoche, the most revered of Himālaya's sacred peaks rears above the shoreline of Manasarowar (Tso Mepham), a lake in western Tibet. The lake, at 4,590 metres asl, is traditionally supposed the source of four of Asia's great rivers, but its hydrography mystified all comers until the mid-twentieth century.

The source of the Indus River was traced to a patch of vegetation and a spring among rocks in the middle of nowhere. In 1908 when Sven Hedin reached the spot after a four-day march from Kailas-Manasarowar he claimed to have dined on fish scooped from the water by hand.

Gasa Dzong in north-west Bhutan is typical of Himālaya's many fortified monasteries. Besides their defensive capabilities, such perilously sited dzongs served as administrative centres and as religious repositories.

Tsaparang, a citadel and monastery fashioned from the canyons of the Satluj River in western Tibet, was the capital of the ancient kingdom of Gugé. Ravaged by drought and erosion, it was in terminal decline before being comprehensively vandalised during China's Cultural Revolution.

Belying Himālaya's reputation as a high-altitude desert, Bhutan and south-east Tibet attract the heavy precipitation that favours their unrivalled flora and fauna. Over twenty-five years the botanist Frank Kingdon-Ward collected thousands of rhododendrons ranging from dwarf ground-cover species to forest giants.

The temple complex of Pashupatinath in Nepal's capital of Katmandu attracts devotees of Lord Shiva from all over India as well as from the hills. Its terraced ghats along the Bagmati River echo those beside the Ganges at Varanasi (Benares).

The Durbar Square at Bhaktapur, 16 kilometres from Katmandu, forms part of a palace-and-temple complex dating from the fourteenth century and noted for its pagoda-style roofs. Sadly, Himālaya is no respecter of architecture. Major earthquakes in 1934 and 2015 left many structures beyond repair.

In Pakistan's 'Gilgit-Baltistan territory' an unmetalled stretch of road between Chilas and Gilgit belies its billing as part of 'Karakoram Highway'. Nanga Parbat, Germany's 'mountain of destiny', sublimely presides in the background, giving no hint of its 'killer reputation'.

Seventy-six kilometres long, the Siachen Glacier in the eastern Karakorams was claimed by both India and Pakistan until 1984. Indian troops then made a successful grab for the glacier itself. The heavy casualties sustained on the world's highest battleground were due more to the extreme conditions than the fighting.

'There with not a particle of cloud to hide it, stood the great peak K2, on the watershed of Asia!' H. H. Godwin-Austen. The skyline ridge on the right is the Abruzzi, by which it was eventually climbed.

West of Annapurna, Kagbeni on the Kali Kandaki River gives access to the carefully conserved Upper Mustang region of north-western Nepal. Once part of Western Tibet, Mustang was an independent kingdom until annexed by the Gorkha kingdom of Nepal in the late eighteenth century.

movements. John Staley, who took an interest in glacier lore during his 1970s sojourn in Gilgit and Hunza, reported that the Minapin glacier spilling from Rakaposhi, Hunza's 7,788-metre presiding peak, 'was known to have advanced ¾ mile in a single year – 1892'. It had continued to advance about 30 metres a year for the next twenty years, then went into reverse; but it advanced again in the 1940s and by 1961 was back where it had started in 1892. 'Subsequent observations suggest it may now [1970] be advancing once more,' reported Staley. Its rate of flow was put at 'almost a yard a day'. Another glacier advanced at a sprightly 15 feet (5 metres) per hour and kept this up for more than a day.

> The people of Kohistan [the hill country south of Gilgit] explain these surges as attempts by individual glaciers to reach a partner of the opposite sex ... Such dangerous behaviour can be countered by burying a piece of ice from the prospective partner in an advancing snout. 'The kiss' is enough and the advancing glacier will retreat again.[20]

This reminded Staley of other amorous glaciers. Like the nutritionist Picton, he was familiar with a book by Emily Overend Lorimer. Called *Language Hunting in the Karakoram* and published in 1939, the work told of the inter-war years spent in Hunza by Mrs Lorimer and her political agent husband, David. Both were notable scholars. They are said to have learned and recorded a different local language every year, much the most difficult being the scriptless Burushaski of Hunza itself. With its seven words for 'mother', its four genders, twenty-eight plural endings, numerous syntactical oddities and subtle meaning shifts, it was the sole surviving relic of India's pre-Aryan languages, the Lorimers believed, and 'entirely unrelated to its neighbours, or to any other known form of human speech, alive or dead'.[21] And so it remains, a 'linguistic isolate', in the words of a modern scholar, 'whose relationship to any other language has never been seen'.[22]

Uniquely the Lorimers gained some facility in this demanding tongue and put it to use quizzing the Hunzakuts on their practices and traditions. Emily Lorimer concentrated on domestic lore, David collected folk tales and beliefs. Though no proof is to hand, it was almost certainly the Lorimers' researches that first alerted Staley to the mysteries of 'glacier-seeding' by artificial insemination.

As Staley tells it, 'if ice from a male and female glacier is mixed together and is covered in snow in a suitable site, and if certain ceremonies are gone through, then a baby glacier will grow there'. As an example he cites the Minawar glacier visible downriver from Gilgit. It had been implanted twenty-five years previously and was now 'thriving'. Another AI glacier was seventy-five years old and still growing; and there were 'dozens more'.[23]

Similar accounts are still current today. They differ only in the details. Some hold that the ice brought from the male glacier must be sewn into a goatskin and suspended over the female ice so that its meltwater drips on to it. Others emphasise the importance of insulating the new ice with a good covering of sawdust or charcoal. Choosing a favourable site in terms of shade and gradient is obviously crucial; best is a cave from which a natural glacier may once have lolled, or a hollow choked with rock that's embedded in permafrost. Crucial too, of course, is the performance of the prescribed rites. And in all such cases it is as well to be on one's guard against testimony derived, not from first-hand observation, but from popular repetition of the reports one is seeking to verify.

Thanks to a marked revival in glacier-breeding, there is perhaps less danger of this happening in the twenty-first century than there was in the 1970s. According to a carefully researched submission by the Norwegian Ingvar Nørstergård Tveiten, Baltistan, as well as Gilgit and Hunza, is taking a new-found pride in its home-grown glaciers. Experience is being pooled, scientific corroboration adduced and grants made available. Much of the credit goes to the Aga Khan. Hunzakuts are usually Ismaili Shi'ites, as are the Wakhis from the Wakhan and the Tajiks of the Pamirs. As leader of the worldwide Ismaili community and a renowned philanthropist, His Highness is revered even by the region's non-Ismailis. Courtesy of the Aga Khan's development initiatives, schools, public health facilities and female entitlement are more evident in the Karakorams than in much of down-country Pakistan.

Since 1999 the Aga Khan Rural Support Programme has been championing artificial glaci-culture. As a result, Tveiten found that by 2007 nurturing glaciers was back in fashion: '18 out of 20 people [in Gilgit and Baltistan] regard glacier-growing as a feasible way of increasing water availability' and, with more abundant water, crop yields were improving and more land was being brought under cultivation.[24] Admittedly this is microdevelopment. The longest home-grown glacier mentioned by Tveiten had yet to reach 1 kilometre. None is ever likely to rival the tsunamis of ice spilling from K2 or inspire a programme for restoring the dwindling glaciers on the Tibetan plateau.

The potential of such local initiatives is still noteworthy. ICIMOD's 2019 report takes cognisance of some related endeavours in frozen-water conservation adopted in Ladakh. These are not strictly speaking glaciers, more 'ice towers' or, as the Buddhist Ladakhis prefer, 'ice stupas'. They are intended to replace the melt of vanishing natural glaciers but are not self-sustainable; they must be recreated every year. At night, in the depths of the Ladakhi winter, at an altitude of perhaps 5,000 metres, water is released from a suitable mountain lake to flow down a long pipeline, from the other end of which a sprinkler squirts it on to a previously levelled and waterproofed surface. As it spurts forth, the water freezes and the ice builds up into a brittle sculpture, then a stupa-like pyramid, then a fairy-tale castle. Come spring the castle starts to melt and its water is available just when the newly sown crops most need it.

In 2015 Sonam Wangchuk, a pioneer in this endeavour, was reported in the *Guardian* as having successfully stored several thousand litres in a 15-metre-tall ice stupa and to be planning another '80 to 90 stupas, each more than 30 metres tall'.[25] Four years later Elizabeth Kolbert in the *New York Times* noted plans for a Ladakhi ice tower. It is to be 'the height of a ten-story building', she says. 'In contrast to natural glaciers, which are shrinking rapidly all around the world, the artificial ones are proliferating. The Ice Stupa Project began with a single prototype [in 2013]; this past winter [of 2018], stupas were created in at least ten villages in Ladakh.'[26]

9

Sublime Deliverance

The perpetual snow-clad Peak of the Holy Kailas of hoary antiquity and celebrity, the spotless design of nature's art, of most bewitching and overpowering beauty, has a vibration of a supreme order from the spiritual point of view. It seems to stand as an immediate revelation of the Almighty in concrete form which makes man bend his knees and lower his head in reverence.

Swami Pranavananda (1949)[1]

The pitfalls in trying to learn an unwritten language are considerable. Fifty years before David and Emily Lorimer went 'language hunting' in the high valleys of the Karakoram and Hindu Kush, Dr Wilhelm Gottlieb Leitner, the German-born principal of Punjab's Government College, had attempted something similar. From the same mountain region he had enticed down to his bungalow in Kashmir a group of long-haired hillsmen, each supposedly representative of a different language group. Dr Leitner called them all Dards, a term derived by Herodotus and Ptolemy from Sanskrit tradition but not much used since and quite unknown to the Dards themselves. Undeterred, and with notebook at the ready, Leitner set about cracking the mysteries of what he called the 'Dardic languages'.

Basic vocabulary being a good place to start, he would point at a familiar object and ask each man in turn to identify it in his native tongue. The responses came thick and fast but were not easily reduced to the consonants and vowels that constitute words in an alphabetic Roman script. Nevertheless, and perhaps for the first time ever, columns of useful nouns in the Shina of Gilgit, the Khowar of Chitral, the Burushaski of Hunza and so on were being committed to paper.

It was exciting work and Leitner, easily excited when on the verge of a linguistic breakthrough, did not at first realise that something was wrong. Wrong, though, it was, for either the words he was transcribing just happened to be very similar or these languages used the same word for an extraordinary variety of objects. In fact, as he eventually realised, it was neither. The turbanned Dards were doing their best to oblige him; they just weren't used to the idea of pointing at things. All those nouns carefully listed in Leitner's notebook would prove to be the Shina/Khowar/Burushaski and so on words for finger, stick, pencil or pipe. They were not what the doctor was pointing at; they were what he was pointing with.

Forewarned, the Lorimers adopted a looser approach for their linguistic studies in the 1920s. Interviewees were encouraged to talk about their homeland, its traditions and taboos, while the Lorimers, after much repetition and clarification, recorded this testimony as best they could in phonetic prose. Meanings were often conveyed by the context; quirks of grammar, syntax and word formation might be revealed in the telling; and the mysteries of Burushaski's thirty-eight plural endings, twenty-eight 'pronominal prefixes' and four genders could be reserved for later more leisured study. As Emily Lorimer put it, 'When we eventually bade a sad farewell to Gilgit in 1924 to go "on leave pending retirement", we were comforted by knowing that we ... had enough material in manuscript and typescript to keep us happily employed for the rest of our joint lives.'[2]

Reviewing his transcripts back in Britain, David Lorimer was struck as much by their subject matter as by the complexities of their expression. Unwittingly his linguistic interrogations had yielded what amounted to a modest archive of local folklore. It warranted its own monograph, he thought. 'The Supernatural in the Popular Belief of the Gilgit Region' by Lieutenant-Colonel David Lorimer CIE duly appeared in July 1929. Here subscribers to the *Journal of the Royal Asiatic Society* could, for the first time, learn about the creatures both animate and supernatural that haunted the high valleys of the Hindu Kush and Karakoram.

Methodical as ever, Lorimer divided these denizens into 'Living Supernaturals' and 'Literary or Legendary Supernaturals'. In the former category were the *boyo* of Hunza, small furry goblins that often took the form of puppies and lived under tree stumps. Then there were the *phut* ('plural *phutu* or *phutants*') and the *de.u* (sic). In Lorimer's only examples the *phut* seems to have been fond of goats and the *de.u* to have been an ogre who frightened fruit-picking children into falling out of mulberry trees. Lorimer believed both creatures to be relics of pre-Islamic times, they being either demonised members of the indigenous society that Islam had displaced or spirits revered by that society. The same was probably true of the *boyo*, Lorimer suggested, but 'from what I have heard I very much fear that the *boyo* are as extinct as the dodo, though their extinction is of quite recent date'. So here was a whole new category – extinct Living Supernaturals. And as with the *phutu*, it hinted at a pre-Islamic past. 'The *boyo* now bear the stigma attaching to the gods of a superseded religion. "Boyo-worshipper" is now used in the sense of *heathen, pagan*.'³

Moving on to firmer ground, as he put it, it was a relief to come to the more everyday peris and djinns. 'We know a good deal about their characteristics and they do still on occasion enter into relations with human beings.' Djinns were acceptable to Islam; the word itself was Islamic and could be used in a generic sense for 'all beings of the apparitional order'. Peris were more specific, though less agreeable to Islamic orthodoxy, and they were more ubiquitous. Indeed they still are. As the mischief-making fairies of the heights, peris are said to frequent the snows of Hunza's fang-like Rakaposhi, of Chitral's throne-like Tirich Mir and especially of the massive buttress that is Nanga Parbat. When defeat or disaster overtakes climbers attempting any of these peaks, the peris are invariably to blame. Lorimer was even given a name for one of them. 'A peri called Madi is said to live on the mountain of Diamer (Nanga Parbat).'⁴

Peris may be male or female. They resemble human beings, they always wear green and they like to eat pilau. But they have only one eye (it's set vertically above the nose) and though good flyers and liable to carry away children, they can always be immobilised by a whiff of cowpat. Toddlers reared by peris make the best mountaineers. Female peris, however delectable, make problematic wives.

David Lorimer logged all this with obvious delight, but he avoided jumping to conclusions about its possible relevance to the rest of Himālaya. He knew nothing of Tibet and seems never to have travelled further east than Kashmir. His previous studies and postings had been in Persia (Iran). He was familiar with Islam, which he rightly supposed to have spread to the Gilgit region 'some 500 or 600 years ago'; and as a political officer he had doubtless passed exams in Hindustani and probably Sanskrit. He thought the word *de.u,* for instance, could be a corruption of the Hindu *devta* (goddess). But of Buddhist iconography he had little experience, and despite archaeological evidence of the Indus, Gilgit and Hunza valleys having anciently served as conduits for the spread of Buddhism, he never made the obvious connection between the mountain peris of western Himālaya and the personified mountains of eastern Himālaya – or in other words between the animate supernaturals of the Hindu Kush and Karakoram and the deified supernaturals of the Great Himālaya and Tibet.

In fact he records nothing at all about mountain worship or about pilgrimage. He may have thought all pilgrimages other than the *haj* were anathema to Islamic teaching. He must, though, have been aware of Amarnath, the mountain cave 3,888 metres asl in Muslim Kashmir to which Hindus from all over India still toil in their thousands. To devotees the great attraction here is another piece of ice sculpture, not in this case an artificial glacier or an ice stupa like those being pioneered for water storage but a plump upturned icicle, or ice stalagmite, the frozen product of a steady trickle of snow melt dripping from the roof of the cave. Devotees hail it as a miraculous manifestation of Lord Shiva's phallic lingam, and to coincide with its maximum annual tumescence the Amarnath *yatra* (pilgrimage) is officially scheduled for high summer when the snow-melt dribble is at its most dependable.

But if this 'ice lingam' is the great attraction for Hindu pilgrims, the great attraction for the militant separatists of Muslim Kashmir is the congregation on their mountain doorstep of so many Shiva-exulting Indians. Potshots aimed at the long straggling columns of inadequately clad pilgrims, and bomb assaults on their makeshift encampments, have become so commonplace that a massive military presence is necessary by way of deterrence. As a result the pilgrims' prayerful progress increasingly has the look of an armed invasion; and yet, despite all the troops, the whole *yatra* is still liable to be summarily suspended 'on security grounds'.

Pilgrimage and politics seem to exercise a mutual attraction. Where the faithful congregate, the state claims a presence; national flags flutter alongside prayer flags. David Lorimer would have known of the great spike of rock in western Tibet revered by Buddhists as Tisé and by Hindus as Kailas; all South Asia had known of Kailas since the time of the Vedas.

But he may never have heard of Tsari's Pure Crystal Mountain astride the distant Assam–Tibet border, let alone Kawa Karpo (Kawa Garbo) and Amné Machin (Anye Machen) along the Tibet–China borderland. Together, these four mountains rank as the holiest in the atlas of Tibetan Buddhism; all are home to tutelary deities, indeed they personify them; and yet all bar Tisé/Kailas stand on politically contested ground.

Mountain deities and the sanctity of their abodes are thought to predate Tibet's adoption of Buddhism. They are therefore usually credited to the sky-riding spirits of *bon*. According to Giuseppe Tucci, *bon*'s awesome pedigree of supernatural creations 'encompasses the *shi*, the *dmu*, the *gnod*, the *bgegs*, the *sri*, the *klu*, the *bdud*, the *the'u rang* and the *spun dgu*. Here we see a colourful world of gods and demons in which opposing beings co-exist ...'[5] Perhaps the *boyo* and *de.u* of Hunza, and of course 'Madi' of Nanga Parbat, hark back to the same colourful world long before the advent of Buddhism, let alone Islam. Installing one's gods at altitudes inimical to humans must always have been a convenient way of laying claim to otherwise neutral heights and attaining whatever strategic or psychological advantage accrued to them.

At 6,740 metres asl, Kawa Karpo is the highest of all these pilgrimage peaks. It commands the narrow but steep watershed between the upper reaches of the Mekong and Salween rivers in a remote corner of Chinese Yunnan. Both rivers rise in eastern Tibet and, by way of their parallel gorges, hint at a secluded but challenging route to the Tibetan uplands. Like that used by Bailey and Morshead from Assam, it eliminated the risks of being turned back by the British south of the Himalayan frontier or by the Tibetans north of it, although to reach Yunnan in the first place the traveller from India had to make a 3,000-kilometre detour by land, sea or both. Then he – or she – faced the challenge of circumventing Kawa Karpo, whose snow-streaked spurs seemed to wall off the western horizon and mark the end of the road.

Happily, for one as fearless as Mme Alexandra David-Néel, a one-time opera singer turned Buddhist scholar, Kawa Karpo was no dead-end: on the contrary it was the gateway to Lhasa. In 1923, as the Lorimers were preparing to leave Gilgit and the botanist Kingdon-Ward was turning his attention to 'the riddle of the Tsangpo Gorges', the fifty-five-year-old David-Néel was swapping Chinese travelling costume for the rags of a Tibetan beggar before joining the pilgrims performing Kawa Karpo's sacred *kora*, or mountain circuit.

Alexandra David-Néel and Yongden, her companion and
guru, reached Lhasa as pilgrims in early 1924. Despite
British objections David-Néel's account of their journey
created a sensation

Despite being officially designated as part of China, Kawa Karpo's *kora*,
like that round Tsari's Pure Crystal Mountain, bisects a frontier, here that
between Yunnan and Tibet. As such it suited David-Néel perfectly. With
her usual amanuensis, a young lama whom she called her son (and whom
she would later adopt), she simply slipped away under cover of darkness as
soon as the *kora* entered Tibet. The lama was Yongden, her spiritual guru,
and she was his *jetsunma* or 'reverend lady'. Each so delighted in the other's
company, it's hard to say who was making the decisions. Heading north
into Tibet, they paused only to pay a fond farewell to the mountain that had
served as springboard for their Tibetan journey. In so doing, David-Néel
acknowledged one of the rather few geographical features to get a mention
in her unusual travelogue.

The majestic Kha Karpo, towering in a clear sky lit by a full moon,
did not appear to me that evening as the menacing guardian of an

impassable frontier. It looked more like a worshipful but affable
deity, standing at the threshold of a mystic land, ready to welcome
and protect the adventurous lover of Tibet ... Before it our group
[of pilgrims] seemed a gathering of tiny animals crawling on the
ground. The sight was really crushing and did indeed remind us of
our nothingness ... I was lost in a trance of admiration.[6]

In truth, as recalled in her *My Journey to Lhasa*, the entire expedition
seems to have been construed as a pilgrimage and conducted 'in a trance
of admiration'. Their route led north-west up the Salween into Kham
and then west and south into Poyul and the Tsangpo–Brahmaputra
catchment. Showa, where ten years previously Bailey and Morshead had
been forcibly detained, gets a mention, as do Tsetsang and the Yarlung
valley. But the rest of their itinerary is open to conjecture. They travelled
mainly by night, their only compass had early been lost and neither of
them appears to have kept a diary. Hidden about her person, David-Néel
did have a watch, which explains occasional mention of the hour. Dates,
on the other hand, are even scarcer than place names. When, 'after four
months of tramping', she became the first European woman ever to enter
the city of Lhasa, Christmas had come and gone and the Tibetan New
Year festivities were under way. It may, then, have been February 1924,
giving some time in October 1923 as their likely date of departure from
Kawa Karpo.

Instead of places and dates, *My Journey to Lhasa* offers a wealth of
fantasy, anecdote and domestic detail. Hallucinations, divinations, mystic
encounters, feats of endurance, family disputes, miracles galore and
some leaden humour leave little room for geography. Money was never a
problem. David-Néel had means of her own plus the support, back home,
of Philippe Néel, the prosperous and extraordinarily understanding
husband to whom she stayed both married and hyphenated throughout
their long separations. Destitution was the pilgrims' choice. Begging
for their keep and charging for their blessings sustained their disguise
and suited their ascetic inclinations; it was neither an affectation nor a
necessity. The silver rupees sewn into *jetsunma*'s underwear stayed there
until required for the purchase of devotional texts.

The hardships were real enough, however; so was the risk of being detected
as foreign impostors. Luckily David-Néel seems to have been impervious to
fatigue and to have soon grown resigned to squalor. Unrecognisable as the

former diva of the Hanoi opera, in her Tibetan incarnation she dyed her straggling hair with black ink, smothered her complexion with soot, and slipped off into the night without complaint when banished from Tibetan hearths as an unprepossessing encumbrance. Yongden was everywhere respected; *jetsunma* was mostly ignored. As she herself explained, 'I am a tiny woman with nothing dramatic in my appearance.'[7]

On the other hand, few foreigners came better prepared for the rigours of a Tibetan winter. As a Buddhist recluse, she had lived many months in a cave in Sikkim, had made a previous foray across the Himalayas into southern Tibet and had lately undergone a crash course in *tum-mo*, that yogic discipline whereby practitioners 'harness the body's psycho-physical energy' to generate a frostbite-defying inner warmth. When the pair limped into Lhasa, it was not because of chilblains but because of incompetent shoe repairs.

To David-Néel, Lhasa was, as she put it, 'the threshold of sublime Deliverance'. It was as much spiritual goal as traveller's destination. She had met the (thirteenth) Dalai Lama during his exile in India and had been yearning to set foot in the holy city ever since adopting Buddhism in her twenties. For two months she and Yongden roamed the city's sites and pored over ancient texts. They lodged with well-wishers and dodged inquisitive officials by melting into the tide of pilgrims. Still incognito, they then headed south to Gyantse and Calcutta.

There the British authorities threatened to deport her for having contravened the ban on travel to Tibet. Yet by the time she returned to France in mid-1925 she was an international sensation. Published simultaneously in French and English, her travelogue enjoyed instant success. Fame as an intrepid adventurer was matched by professional recognition of her Buddhist scholarship and popular fascination with her spiritualist insights. In a house in Provence bought with the royalties from her books, she devoted most of the rest of her life to study, writing and lecturing.

Philippe Néel died in 1941, Yongden in 1955. Alexandra followed them in 1969, a month short of her 101st birthday. She had made only one more journey to Tibet. That was in the late 1930s and again with Yongden. Reaching Beijing by the trans-Siberian express, they had headed back into the hills but this time got no further than the prairie-land of Amdo (or Qinghai) in the north-east of the country. Warlords and civil upheaval during the Sino-Japanese War, and then the Second World War, made travel in China's borderlands almost impossible, added to which

David-Néel was now into her seventies and no longer fleet of foot. She spent the war years in retreat at Dartsedo (Tachienlu, Kangding) on the Sichuan–Tibet border. It was not far from Yan'an (Ya'an) in whose maze of loess canyons the young Mao and other stalwarts of the Long March were also sitting out the war as they set up China's communist government-in-exile.

Dartsedo was also as close to Kawa Karpo and the long-postponed completion of her 1923 *kora* as David-Néel ever got. Perhaps it was just as well. Tibetan concern for the inviolable sanctity of their mountain-gods – and their god-mountains – would be barely audible during the repressive Maoist era; pilgrimage itself was banned. But, come the more 'liberal' 1980s, the faithful were allowed to return to the mountain, albeit at times and in numbers approved by the authorities. They were joined by adventure tourists, honeymoon hikers and hard-wired climbers edging ever nearer to the sacred summits. Tibetan anxieties over their deities taking offence at such intrusions were about to be tragically vindicated.

In 1991, in a blatant exhibition of divine displeasure, an avalanche swept down the slopes of Kawa Karpo to claim the lives of seventeen Japanese and Chinese climbers. It was one of the worst mountaineering accidents on record. The fatality count was exceeded only by that of the previous year when some forty climbers had been buried by an avalanche on Lenin Peak (Avicenna/Ibn Sina Peak, 7,134 metres asl) in the eastern Pamirs.

Because of the Tibetan tragedy, and following more Buddhist remonstration, all further attempts to scale Kawa Karpo were forbidden. As of 2021 it remains one of Himālaya's few named peaks never to have been climbed. This respect for Buddhist sensibilities does not extend to the mountain's lower reaches. Here, with the snow cover and the glaciers in retreat, today's pilgrims are more likely to be mown down by quad vehicles and lycra-skinned mountain bikers than by avalanches.

So David-Néel never got to complete her circuit of Kawa Karpo. Nonetheless, and possibly as a result of some confusion between the two mountains, her biographers (and hence Wikipedia) credit her with performing the *kora* of that other sacred massif along the Sino-Tibetan borderland, Amné Machin. Of this shy and much misunderstood cluster of peaks she certainly knew and makes mention. Geographers somewhat optimistically consider Amné Machin a distant continuation of the Kun

Lun range. Before her 1923–4 journey to Lhasa she and Yongden had explored the possibility of tracking that range west into the Tibetan heartland, and in 1938 they had again been probing this Qinghai approach when war obliged them to take cover in Dartsedo. On neither occasion is there any evidence that they undertook a *kora* of Amné Machin; on either or both trips, though, they may well have caught sight of the mountain.

If so, they were not the first. Two years before David-Néel had first set off for Lhasa, a lame British general with the unlikely name of George Edward Pereira had confided to his diary of sporting rambles in China (mostly in search of what he called 'pandars') the news of 'something of extraordinary geographical interest – the great mountain Amné Machin'. Sadly Pereira had died on his next Chinese marathon, leaving publication of his travels to a friend and admirer. This happened to be Colonel Sir Francis Younghusband; and thus, in a nice twist of irony, it fell to the erstwhile leader of the British invasion of Lhasa and now chairman of the committee tasked with organising the conquest of Mount Everest to explain the 'extraordinary geographical interest' of Pereira's Amné Machin.

The mountain was interesting not just for geographical reasons. It must have been the most heavily guarded pilgrimage peak in all Tibet. Greasy-haired warriors of impressive physique patrolled its approaches, and once a year they and their far-flung brethren converged on the mountain in their tens of thousands. Mostly mounted and bristling with firearms, they looked more like Tibetan braves than Tibetan pilgrims; they were locals in so far as their nomadic habits allowed; and despite a preference for the Panchen Lama over the Dalai Lama, they were devout Buddhists. But as Goloks (Ngologs) they belonged to a fraternity of displaced clans united only in their bloodily attested reputation for inveterate brigandage and gratuitous butchery.

Foreigners invariably fled from Golok attentions. Even fellow Tibetans were in awe of their ferocious reputation. In the early twentieth century Chinese attempts to avenge Golok outrages occasioned a succession of 'Golok wars', and to this day Golok sensitivities have to be appeased by the well-signposted creation of a Golog Tibet Autonomous Prefecture. Amné Machin stands guard over it, 'rising like a threatening ogre turned to ice in the middle of the land of the marauding Ngolok'.[8] It is the Golok's mountain, and its god is the Golok's god. It seems highly unlikely that Alexandra David-Néel ever risked an encounter with such trigger-happy banditti, let alone a clamber round the mountain's sacred *kora* in their company.

Amné Machin's other claim to fame is that it stands alone. David-Néel, who had heard of it from Pereira himself, rightly referred to it as 'a veritable fairy-tale mountain rising in solitary splendour from the surrounding plateaux'.[9] Glaciers, she says, streamed from its summit like tresses, and black rocks stood like sentinels guarding its tutelary god (whom she wrongly understood to be a goddess). Pereira had been less poetic: the mountain simply 'towered above everything else in its snow-clad grandeur'. It was 30–70 miles away (50–113 kilometres) when he spied it and, given that he estimated his own altitude at the time at about 13,000 feet (4,000 metres), it must, he thought, 'be well over 25,000 feet high [7,620 metres asl]'. Younghusband agreed. At the risk of undermining his fund-raising for the assault on Everest, he conceded that 'rumour had it that [Amné Machin] was higher than Mount Everest and certainly it must be a giant'.[10]

The Golok people of Amdo (Qinghai) looked to the peak known as Amné Machin as their guardian deity and harried all visitors, including an Amercan colonel

This was confirmed by a succession of long-range observers, only one of whom braved the Goloks to get anywhere near the mountain itself. In 1930, in an article for the *National Geographic*, the Austrian-American botanist Joseph Rock again noted how Amné Machin towered above its surroundings. In fact it soared, he reported, to 'more than 28,000 feet [8,534 metres asl]'. During the war, US aircrews put it even higher. Overflying China at 29,000 feet (8,839 metres asl) they spied a mountain wrapped in cloud whose summit reared above them. If anything, Pereira and Rock had underestimated it.

But, of all these sightings, much the most convincing looks to have been that of 1949 by the affable Colonel Leonard Francis Clark of the US army. All the other estimates of the mountain's height had been guesswork; only the colonel, a big man with a taste for tall tales, had faced down the Goloks and actually measured Amné Machin.

As a member of the OSS (forerunner of the CIA) Clark hints at having spent an interesting war behind Japanese lines in China. After the war he returned to San Francisco, but he missed the adventure and soon cast about for more of the same. Amné Machin's 'mountain range of mystery' promised just that. He is adamant – and perhaps a little too adamant – that intelligence-gathering formed no part of his new brief; the expedition was undertaken entirely on his own initiative as a private US citizen, he says. But that didn't stop him calling in a few favours, availing himself of courtesy rides on China's civil aviation and securing for his expedition the military support and financial backing of General Ma Bufang, the Tungan (Chinese Muslim) warlord who held Qinghai on behalf of the Nationalist government-in-exile of Generalissimo Chiang Kai-shek (Jiang Jieshi). General Ma lent Clark the services of another Tungan Ma, one Solomon Ma, as interpreter and liaison officer, while 'Prince Dorje', a diminutive Mongol with a winning grin, impeccable Oxford English and a remote connection to Genghis (Chinggis) Khan, was recruited as the expedition's surveyor.

In retrospect Clark would describe his journey as 'a record of semi-starvation, incredible loyalties and disloyalties, some hardship, many disappointments and troubles, and perhaps a few successes'.[11] The food shortage is hard to explain. Clark's considerable appetite often left him hungry, though with 'eighty yaks, seventy saddle horses, a herd of pack mules and a few camels' the expedition was not exactly short of transport. Perhaps it was just too big. In addition to a considerable military escort, General Ma had insisted on burdening it with whatever scientific talent his Qinghai administration could muster. Hence the presence of sundry botanists, mineralogists and geologists plus a man with a geiger counter, though not perhaps the clique of disapproving fellow travellers whom Clark identifies as the expedition's 'court of law'.

Progress was slow and none too steady. Clark had to dash back to Xining, capital of Qinghai, on some unexplained errand – possibly for a square meal at the home of the town's German Catholic missionaries. Meanwhile the Tungan scientists cast about for any deposits that might reward General Ma with prospecting potential.

By late April the expedition had got as far as had Joseph Rock in the 1920s. On cue, the Goloks appeared just as Amné Machin loomed out of the cloud cover. A fierce engagement served to disperse the Goloks at the price of a few fatalities and the certainty of Golok retaliation. By the time the Goloks withdrew, the cloud had returned. It was also bitterly cold. Snow fell on and off, and talons of wind swooped from the invisible heights to play havoc with exposed complexions.

This was not the weather for measuring mountains. Regardless, and more with hope than confidence, the ever obliging Prince Dorje prepared to triangulate the great mountain. The first thing was to clear a kilometre of level ground to serve as the baseline. By measuring the angles between the baseline itself and sightlines to the mountain's still invisible summit, it should be possible to calculate the mountain's height and plot its location. That was the theory of triangulation anyway, and it was well understood by the bookish Dorje. But he was not a surveyor. The pint-size 'prince' had had no survey training and had never before measured a mountain.

Nor had he come across a theodolite, this being the instrument with which one measures the angles. Unfortunately the only theodolite available to the Clark expedition was a borrowed one which lacked the calibrations needed to measure seconds (fractions of angles, as of time, are expressed in minutes and seconds). Pinpoint accuracy being the essence of triangulation, this deficiency made a nonsense of the whole exercise. Even Clark admitted 'it meant heights could ... only be determined to within 2,500 feet [762 metres]'.

Dorje continued laying out his baseline anyway. On Saturday 7 May his perseverance was rewarded. The cloud lifted enough to afford a clear view of the whole massif. Which of the mountain's several summits he actually measured is less clear, but by the end of the day he had values for the all-important angles. It remained only to make the mathematical calculations that would turn the degrees and minutes into a measurement of elevation. He worked late into the night. At one end of their shared tent the little prince did his sums while at the other end the dinnerless colonel fretted. Eventually the tension became too much. Clark not only broke the silence but kept a record of what was said:

'How high is the old girl then?'
'Higher than Mount Everest' came the reply.
'Are you sure?'
'Quite. I've checked all the sets of figures twice.'
'All right. Now let's eat. I'm starved.'[12]

Dorje's computed height came to a very precise 29,661 feet asl (9,041 metres). At the time Everest was reckoned at 29,028 feet (8,808 metres). Amné Machin was not just the world's highest; it was comfortably the world's highest. But after factoring in the theodolite's generous margin of error, 'the mystery mountain' retained its mystery. It could be just 27,000 feet or it could be as much as 32,000 feet.

This was the highest ever claimed for Amné Machin. Next day the Goloks returned and the weather broke. Clark and his men beat a speedy retreat to the fleshpots of Xining without attempting any further measurements. 'It is difficult to know when fact ends and fiction begins in [Colonel Clark's] highly readable account,' says a sceptical John Town in the *Alpine Journal*. Plausibility had been sacrificed in the name of readability. For the sensationalists, the game was up. The 'few successes' of Clark's expedition boiled down to just one – that of having discredited the amateur explorer. At a time when Everest was about to yield to the professional assault spearheaded by Hillary and Tensing, the only way for Amné Machin now was down. It was a swift and ignominious descent. 'In 1960 the Chinese reported that they had climbed Amne Machin and that a survey had revealed it to be 7160m (23,491ft). The legend of Amne Machin had finally been laid to rest … or had it?'[13]

Twenty years later, among several Himalayan giants opened to Western climbers for the first time was Amné Machin. The Chinese authorities also took the occasion to announce a new survey of the mountain. This showed its highest point to be the most northerly peak. Whether this was the point triangulated by Prince Dorje remained unclear. The news may still have been enough to send the climbers rummaging among their maps, though even this seems unlikely. For the massif so often bedevilled by ignorance was now threatened by irrelevance. 'The bad news', says Town, 'was that the height of the mountain had dropped to 6282m (20,610ft).' Roughly speaking, it had been reduced by a third in the space of thirty years. All mountains push up and are pared down; few can have been cut down as abruptly as Amné Machin.

Some 2,000 kilometres to the east, on the other side of Tibet, the most famed of Himālaya's pilgrimage peaks tells a very different story. On first glimpsing Mount Kailas the Swiss geologist August Gansser declared it 'one of the most magnificent temples I have ever seen, a sunlit temple of rock and ice'. It was, of course, better known as 'the most sacred mountain in the world' and the sublime throne of the gods.[14] But what struck Gansser was Kailas's architectural quality, its near symmetry, its structured look of purpose-built intent.

A granite pediment clad in sedimentary conglomerate supported a gleaming dome of ice to produce a curvilinear profile like that of the *sikhara*, or tower, of a Hindu temple. Buddhists, on the other hand, who called the mountain Tisé ('the Peak') or Kang Rinpoche ('Snow Jewel'), imagined its rotundity atop a vaguely rectangular base as that of a giant stupa or chorten. More fancifully still, the always inventive *bonpo* read into the sequence of snowy striations across the mountain's flanks the outline of a 'nine-storey swastika'. Needless to say, the swastika in question was the reversed version; and *bonpo* making a circuit of the mountain did so counter-clockwise.

Either way, Kailas stood in marked contrast to the clustered peaks of other Himalayan giants like Gurla Mandhata, visible some 50 kilometres to the south of Kailas, or indeed faraway Amné Machin and Kawa Karpo. But to an Alpine geologist like Gansser, Kailas's detached location and architectural features bespoke a unique genesis. 'Strangely enough it consists of horizontally stratified conglomerate with erratic admixture,' says his translated travelogue; 'in the course of geological aeons these strata have been elevated many thousands of feet without any change in their horizontal lay-out.'[15] The Himalayan folds of the advancing Indian plate had intruded into the granite of the Tibetan (or Trans-Himalayan) crust and been heaved skywards to 6,714 metres asl;

yet, like a skilfully balanced stack of crockery, the strata miraculously retained their original alignment and sequence.

Clearly such a mountain must have been thrust aloft for some divine purpose. In time, Hindu and Buddhist cosmology had come to identify Kailas with Mount Meru, the world axis around which the planets circulate and the realms of gods and men are disposed. In devotional terms, Hindus know Kailas as the throne of Lord Shiva, its erect profile recalling that of other Shiva lingams like the ice statue in the Amarnath cave. Atop Kailas's largest of lingams the Great Lord sits in profound meditation attended by Parbati, his mountain consort, and both of them seemingly untroubled by the proximity of a mythical pleasure palace of later tradition called Alaka, 'a kind of celestial oriental Las Vegas' according to John Snelling.[16]

For Shiva and Parbati of the Hindus, Buddhists substitute Demchog ('Eternal Bliss') and Dorje Phagmo ('Thunderbolt Sow'), a pairing of guardian deities at one level or, more esoterically, of certain attributes of human nature as personified and deified in the practice of tantric meditation. Since the human attributes in question are things like rage, passion and violence, the divine pair vie with one another in ferocious embraces and gruesome accoutrements.

Artistic convention depicts Demchog as blue-skinned, his consort as red-skinned, and because he once had to subsist on a diet of nettles, the most revered of Tise's historically attested patrons is given skin tones of green. This was Milarepa, a beloved figure with a chequered career who played a key role in the reintroduction of Buddhism to Tibet in the eleventh and twelfth centuries CE. Milarepa's tantric austerities in a nearby cavern included mastery of the warming art of *tum-mo*, which must have been comforting, and inspired some of the sage's pithy, often cryptic, verses and aphorisms ('I need nothing, I seek nothing, I desire nothing').

Buddhist tradition also credits this Milarepa with the decisive rejection of the *bon* religion's claim to the sacred geography of Kailas/Tise. Before Buddhism's 'first coming' in the seventh century CE, *bon* had prevailed throughout western Tibet, and no one contested its sole proprietorship of the mountain. '*Ti se*', says Professor Tucci, '[was] *Bon ri*, the *Bon* Mountain, or *Zhang zhung bon ri*, the *Bon* Mountain of *Zhang zhung*.' It features in *bon* creation myths, it provided Zhang Zhung (Shang Shung and so on) with a royal heir when the succession failed, and it acted as a ladder down which Shenrap Miwo, the *bon*

Messiah, descended from heaven 'to accomplish his work of revelation and carry out his first mission'.[17]

Milarepa, who as an orthodox Buddhist no doubt thought all this nonsense, is said to have acquired Kailas for the resurgent Buddhism in the course of an epic trial of strength with a *bonpo* lama called Naro Bun Chon. Repeatedly worsted in a succession of magical challenges, the *bon* master proposed, by way of a final decider, an early-morning race to the top of the mountain. Like Kawa Karpo, Kailas/Tisé is too sacred ever to have been climbed. Neither of the contestants would contemplate such a sacrilege or stoop to anything quite so pedestrian. Nor did they need to, given their extraordinary powers of locomotion.

Naro Bun Chun was up early on the appointed day. Apparently unchallenged, he was soon to be seen sky-riding towards the summit astride his little *bonpo* drum. Meanwhile the saintly Milarepa had yet to emerge from his meditations. He was in no hurry. Perhaps he was confident of victory; after all, he whose followers would hand down the story could hardly be the loser. With a spell and a mystic gesture, the Buddhist halted his rival in mid-flight. Then, in a blink of the eye, he sped past him as he teleported on to the summit.

Naro Bun Chon accepted defeat. But by way of a consolation prize a compromise was reached: *bonpo* believers might continue to visit the mountain provided they made no trouble. Sven Hedin would meet a group of them performing the *kora* (known to Hindus as a *parikrama*) in 1908, and it's said they may still be met trudging doggedly against the press of Hindu and Buddhist pilgrims processing round the mountain in the orthodox direction.

10

Swede and Swami

Fifteen years ago, when writing my autobiography, the thought
of Manasarovar came to me. That thought had been hovering in
my mind for many a year. It was something which filled me with
delight and I had worked out many a plan for paying homage
to this wonder lake of Tibet and snow-covered Kailas nearby. But
these plans remained unfulfilled, for my journeys led elsewhere,
and I was filled with regret that I would never reach Kailas and
Manasarovar. Still the thought of them cheered my mind …
                    Pandit Jawaharlal Nehru (1949)[1]

The reputation of Mount Kailas as the most revered mountain in all
Himālaya owes much to its unexpected proximity to a large body of
water. Hindu temples customarily have access to a stone-lined pool, or
'tank', where worshippers perform their ritual ablutions. Kailas is no
exception. Here, at over 4,500 metres above sea level, in the rain shadow
of the Himalayas and amid the dust-blasted wastes of western Tibet, the
most improbable of 'temples' is complemented by a glassy expanse of the
clearest water. Eighty-eight kilometres in circumference, over 90 metres
deep and roughly circular, Manasarowar (to Hindus) or Tso Mapham
(to Buddhists) has been hailed as 'the holiest, the most fascinating,
the most inspiring, the most famous of all the lakes in the world and
the most ancient that civilisation knows'.[2] Pilgrims circumambulate it
just as fervently as they do Kailas itself; it takes a few days longer but
has some easy stretches along the shoreline. According to Burrard and
Hayden, the 1930s superintendents of the Indian Surveys, Manasarowar

is not just 'famous in Hindu mythology' as having been mentioned in both of the great epics but is 'the first lake known to geography'. It has been acknowledged as sacred since 'before the dawn of history; and as such it has remained for four millenniums'.[3]

Other authorities have been moved to write of this 'fairest of lakes' in more lyrical terms. 'She is majestically calm and dignified, like a huge bluish-green emerald or a pure turquoise set between the two mighty and equally majestic silvery mountains, the Kailas on the north and the Gurla Mandhata on the south.'[4] This comes from the flowery pen of Rev. Swami 1108 Pranavananda Maharaj, FRGS, the holder of a record even Milarepa might have envied in that Pranavananda (for short) had been semi-resident on Manasarowar for twenty-three years. During this time he completed, in all, twenty-five circuits of the lake and a like number of Mount Kailas. The swami was unquestionably the world's greatest authority on the region that he was the first to call Kailas–Manasarowar. A history of its exploration, which is really more a gazetteer, and a pilgrim's guidebook were his life's work. But he was not born to the mountains, nor did he die among them. He was called to them.

He was born Kanakadandi Venkata Somayajulu in 1896, a Telugu-speaking brahmin from what would become the south Indian state of Andhra Pradesh. He completed his education at a college in Lahore and worked for the railways before returning home as a Congress Party worker during the non-cooperation movement against British rule. From political activism he turned to religion and the natural sciences. He had received 'a call from the heights of the Himalayas in consequence of an internal urge for [the] search after Truth,' he says.

The call was mediated by the man who had become his guru. This was Shri 1108 Dr Swami Jnanananda DSc, PhD etc., a contemporary and also from Andhra, who had spent some years in the Indian Himalayas as a renunciate. According to Pranavananda, Dr Jnanananda had crossed the mountains on pilgrimage to Kailas 'wearing only a loincloth'. But Jnanananda was better known as an exponent of yogic philosophy and as one of India's foremost nuclear physicists (the DSc was from Dresden, the PhD from Liverpool). This example of ancient wisdom in the service of modern science appealed to Pranavananda; no doubt it explains his adoption of Jnanananda's honorifics (including that '1108', a sacred number in Hindu and Buddhist thinking and a nod, perhaps, to both men's mathematical competence).

No doubt, too, it was his guru's example that inspired Pranavananda to follow his example and head for the hills. In 1928, travelling as a pilgrim by way of Kashmir and Ladakh, Pranavananda made his first acquaintance with Kailas–Manasarowar. Thereafter he would return nearly every year until 1950, by which time India had gained its independence, Tibet had ceased to be a British responsibility and Hindu pilgrims were no longer welcome there. Pranavananda may have been relieved. He was by then in his mid-fifties, his hair had reached apostolic length and he must have been feeling the strain of the near-annual trans-Himalayan commute. Only twice did he pass what he calls 'a twelve-month' on the lake, overwintering in a nearby monastery. The lowest temperature recorded was minus 28 degrees Celsius on 18 February. 'It was so cold', he says, 'that the sputum of a person standing on the balcony [of his lakeside retreat] would reach the ground as solid ice.'

Swami Pranavānanda (of the Holy Kailas and Manasarovar)

After twenty-five-years' acquaintance, Shri Swami 1108 Pranavananda Maharaj, FRGS, was the indisputable authority on the Kailas-Manasarovar region

Throughout the years, Pranavananda's overriding objective remained 'the Truth' – or sometimes 'the Realisation of the ULTIMATE'. One can only hope he succeeded in finding it. Research has failed to reveal anything at all about his subsequent career. His pen seems to have been laid down for good in 1950; even the date of his death (in 1989) may be conjectural.

His work schedule in Kailas–Manasarowar must have been demanding enough in itself. It meant compiling a record of everything that could conceivably pertain to the region, from the mystic to the mythological, geology to geography, archaeology to ornithology, mineralogy to zoology, cultivation to climate. In his published works, each of the sacred region's dozen or so monasteries is described and its particular associations recorded; so are Kailas–Manasarowar's several varieties of waterfowl, its four kinds of bear (including the 'man-bear', a possible yeti contender), the few available options in terms of fuel, and one or two interesting plants (including a creeper called *thuma*, 'a marvellous specific for spermatorrhoea and an excellent aphrodisiac').[5] No tarn is left unvisited by the swami, no cave unexplored, no sizeable rock unexplained. The grasses prove few, the fossils plentiful. Pilgrims are advised to wrap up warm and bring with them all the drugs indispensable to life in such extreme conditions. Some thirty nostrums are listed, plus 'an enema can or syringe', 'a rubber catheter' and 'a hot water bottle'. No mention of a spittoon.

Of particular interest to Pranavananda was the hydrography of the region. If Kailas's fame owed much to Manasarowar, Manasarowar's fame owed much to its being the traditional source of no less than four great rivers: the Indus, Tsangpo–Brahmaputra, Satluj and Karnali. Sometimes all were said to flow from the snows of Kailas itself. Pranavananda was not so gullible as to take any of this literally; ancient geographies invariably favoured simplicity and symmetry over precision and scale. But he realised that all these rivers could well emanate from catchment areas in the vicinity of Kailas.

European geographers agreed. In the nineteenth century they had made river sources, most notably those of the Nile, subjects worthy of attention in the name of science. Would-be explorers rose to the challenge; august bodies like London's Royal Geographical Society (RGS) assessed each discovery; and the wider public feted the returning travellers as celebrities. The 'true' source of the little-known Oxus in the remotest eastern Pamirs kept the savants arguing for the best part of a century. Even Lord Curzon, the former viceroy of India and later British foreign secretary, had lent his considerable authority to the debate.

The idea of four of South Asia's mightiest rivers originating in the same unknown lake on the 'roof of the world' challenged geographical opinion even more than the source of the Oxus. But as early as 1812 the ubiquitous William Moorcroft had investigated the matter and apparently laid it to rest. Travelling in what proved a not very convincing disguise, he and Hyder Young Hearsey, an Indo-British adventurer, had set off from Saharanpur, crossed the Great Himālaya by the Niti Pass and reached Gartok (the windy settlement chosen for the remotest of Britain's Tibetan trade agencies a century later), from where they turned south-east for Manasarowar. There, despite carefully examining most of the lake's shoreline, Moorcroft found no rivers at all emanating from Manasarowar; 'not a break, nor any other appearance indicated the escape of any river or even of any small stream from it', noted Moorcroft, 'Manasarowar sends out no rivers to South, North or West.'[6] In fact the lake was, as Hearsey put it, 'perfectly insolated'.[7]

But this only raised the question of where, then, all those rivers did originate. One possibility was Rakas Tal (Rakshas Tal, Rawan Hrud), a lesser lake just to the west of Manasarowar. Local tradition insisted that it was from this little-visited body of water that the infant Satluj flowed. But again Moorcroft drew a blank. He could find no watercourse running out of Rakas Tal and none running into it from Manasarowar. The drainage of both lakes was becoming as much a mystery as that of the rivers.

Further investigation from British India then had to wait. Interest was stalled by the first Anglo-Sikh War in the mid-1840s and the Great Rebellion ('Indian Mutiny') in the late 1850s. In between, the need to establish the frontiers of the new state of Kashmir (which included Ladakh) brought two border commissions to western Tibet. Henry Strachey, a British officer whom Pranavananda would consider the region's greatest geographer, served on both of these and, like his brother Richard two years later, visited Rakas Tal to ascertain whether there was any sign of the Satluj issuing from it. There wasn't; but to the Stracheys' surprise there was a goodly stream gushing into Rakas Tal. It came from Manasarowar, and Moorcroft could only have missed it because it didn't exactly flow out of Manasarowar; it percolated down from it. Invisible to anyone following the lake's shoreline, it seeped through and beneath a wide embankment of shingle, not to emerge into its own twisting bed till some distance away. This being the case with Manasarowar, it was just possible the Satluj issued from Rakas Tal in similar fashion. Thus Manasarowar might indeed be that river's ultimate, or 'genetic', source. And if there were more such subterranean seepages, the great lake could yet be the source of the Indus and the Tsangpo–Brahmaputra as well. The possibilities were several, the problems far from solved.

No one paused to ask whether it was all worth it. Great minds were being distracted, young lives risked, local sensibilities offended, international relations endangered, and all because of an obscure geographical conundrum involving the vagaries of snow-fed rills and invisible seepings on the barely breathable heights of the Tibetan tundra. Yet somehow the absurdity of the quest only added to its appeal. Better still, its scientific plausibility afforded some welcome cover to all who chose to challenge Tibet's closed frontiers. Sportsmen intent on bagging Tibetan trophies, mineralogists excited by reports of Tibet's goldfields, spies on the lookout for gaps in the imperial defences, gun-runners arming distant pockets of discontent – all cheerfully professed an interest in Himālaya's hydrography.

In *A Mountain in Tibet*, Charles Allen's classic account of this whole saga, Ekai Kawaguchi, a Japanese monk with a poor sense of direction, vies with an unsavoury braggart called Henry Savage Landor in attracting the unwelcome attentions of the Tibetan authorities. From time to time, like wandering albatrosses, the Survey's long-distance Pundits pass quietly by, counting their paces as they come and go. Further north, where Tibet peters out in the empty quarter of the Changthang, a succession of international contenders files along the horizon. French expeditions like those of Gabriel Bonvalot and Prince Henri d'Orléans, or the unfortunate Dutreuil de Rhins (murdered in Jyekundo), are interspersed with swooping incursions by the unstoppable Nikolai Przhevalsky and his later Russian lieutenants.

From China in the 1880s William Woodville Rockhill, a Tibetophile member of the American legation in Beijing, had taken an interest in the Goloks when cutting across Qinghai and had named a peak that could well have been Amné Machin as Mount Caroline. The sportsmen (and sportswomen) are if anything over-represented. In 1895 Captain H. H. P. Deasy, a entertaining Dubliner after whom a gerbil, *Dipus deasyi*, was named, found whole hillsides in the Changthang alive with the dainty antelopes known as chiru; his companion 'who had experience in sheep-farming in America, was of the opinion that at least 15,000 were seen'.[8] A year later the Liverpool Littledales – St George Littledale, his sharp-shooting wife Theresa, a lanky nephew and a fox terrier called Tammy – got to 'within 48 or 49 miles [77–9 kilometres] of Lhasa'. Against all the odds, it was therefore this family group that drew nearer to the Tibetan capital than anyone before Younghusband, including the insufferable Landor and those unstoppable trailblazers, Nikolai Przhevalsky and Sven Hedin. Naturally none of them advertised their designs on the ultimate prize of Lhasa. Denying any such intention made it easier to concede defeat when they were forced to turn back – as they invariably were. Yet nearly all openly avowed an interest in Kailas–Manasarowar and its lakes.

For any but a devout Buddhist like Alexandra David-Néel, the Younghusband-led invasion of Tibet in 1904 dispelled the allure of Lhasa. Lord Curzon, the viceroy responsible for authorising the incursion, nearly apologised for it. 'I am almost ashamed of having destroyed the virginity of the bride to whom you aspired, viz, Lhasa,' he would tell Sven Hedin. But the main Younghusband expedition went nowhere near Kailas–Manasarowar and had added nothing to the mystery of the rivers. Nor, later that year, had Cecil Rawling and 'Hatter' Bailey in the course of their winter's journey through western Tibet to Toling–Tsaparang and Simla. Bailey was busy stalking a black wolf when they passed within sight of Kailas. He returned, bearing the carcase, just in time for a brief inspection of the two lakes. Neither man was impressed. Strachey's channel linking the lakes turned out to be bone dry, though 'they were assured by local Tibetans that it filled up when the snows melted in the summer'.[9] The same went for an old streambed leading out of Rakas Tal that might once have been the Satluj – no water.

It was beginning to look as if the source of the Satluj was lost to the Kailas–Manasarowar region. So too was that of the Karnali, which becomes the principal river of western Nepal before reaching the Ganges as the Ghaggara. Likely contenders as the remotest feeders of both these rivers descended from the Himalayan passes to the south, not from Kailas to the north, and veered away before either reached the lakes.

That left the Indus and the Tsangpo–Brahmaputra. Moorcroft in 1812 thought he had found the beginnings of the Indus north of

Kailas, but he may have mistaken a fast-flowing feeder of the river for the parent stream. Henry Strachey in 1846 had been nearer the truth when he suggested as a likely source of the Tsangpo–Brahmaputra a glacial torrent 80 kilometres to the east of Kailas. But neither of these discoveries agreed with local tradition, nor were they anything like conclusive.

The uncertainty cried out for an authoritative appraisal from Inner Asia's most experienced traveller. This in time would provoke a trial of strength as Herculean in its way as that between Naro Bun Chon and Milarepa. But, the wizardry being academic rather than supernatural, the tone would be more waspish and the disputants much less well matched. For the authority in question would be the great Dr Sven Hedin, heroic conqueror of all the deserts and mountains from Iran to Mongolia; and his disputant would be the bedraggled south Indian Truth-seeker, Swami 1108 Pranavananda Maharaj.

Hedin had first entertained the Tibetan challenge while at home in Stockholm. It was 1905. Younghusband had just stormed his way into Lhasa, but Curzon was still Britain's viceroy in India and it was he who urged Hedin to 'make one more big Central Asian journey before you desist from your wonderful travels'. Indeed Curzon, himself a notable traveller, had personally undertaken to facilitate Hedin's expedition. All within the viceroy's power was being arranged, including the 'requisite permits and protections'. 'A good native surveyor' had been detailed to accompany Hedin and another man was being trained to handle astronomical observations and meteorological recordings. An armed escort was also on the cards. Hedin's expedition was coming together of its own accord.

Hedin himself was more than ready. 'I must return to the freedom of the desert and hie away to the broad plains between the snow-clad mountains of Tibet,' he wrote.

> Three years had passed since my return to my own country; my study began to be too small for me; at eventide, when all around was quiet, I seemed to hear in the sough of the wind a voice admonishing me to 'come back again to the silence of the wilderness'; and when I awoke in the morning I involuntarily listened for caravan bells outside.[10]

Now forty-one and evidently under the spell of Rudyard Kipling, Hedin looked forward to an adventure that would begin and end in India. It would be the fourth of his great expeditions and the only one to be launched from British territory. He travelled east by sea and rail, taking to the desert only in Iran and Baluchistan. He had nothing left to prove in Central Asia. He had done enough already to eclipse all his contemporaries. He had also outwritten Herodotus and outdistanced his all-conquering hero, Alexander the Great. Thus Himālaya was not his first love. The interminable deserts of Inner Asia – the Karakum of Khorasan, the Gobi of Mongolia and expecially the Takla Makan of Xinjiang – preceded his affair with Himālaya. Even above the snow line he would ride whenever he could. As beasts of burden, his preference was for camels or mules rather than yaks or porters.

Hedin's second expedition in 1893–7 ('a jolly jaunt, a wild and whizzing expedition on horseback, by sleigh and carriage through all of western Asia') had brought him nearer the Himalayas.[11] Lingering in the Pamirs he had repeatedly failed to conquer Mustagh Ata, the 'Father of Ice Mountains', before embarking on the gruesome death march across the waterless Takla Makan that established his reputation as a stop-at-nothing adventurer. No doubt he exaggerated the horrors. As always he travelled alone except for locally recruited servants, guides and animal handlers. Whether Indian, Tibetan, Ladakhi or Turkestani, these men were well treated. They were unlikely to question his decisions or dispute his version of events; and unlike the beasts in their care, they knew the risks and were being well rewarded. The bones of two of his men lay among those of the camels that had expired of thirst in the shifting sands of the Takla Makan. Such losses might have plunged another traveller into paroxysms of remorse. Hedin offered not even an apology; rather were such deaths to be seen as testimony to the deprivations they'd had to endure and the magnitude of his achievement.

Only on his third expedition in 1899–1902 had he ventured into Tibet. By way of a warm-up, this jaunt began with another blistering traverse of the low-lying Takla Makan; then followed a sudden dash south up through the snow-filled skirts of the Kun Lun; and finally a lunge across Tibet's Changthang ('Northern Plains') until the intruder was halted well short of Lhasa. On the map his meanderings appear speculative, but they were never haphazard. Judging by his retrospective accounts, there was always a plan, always a carefully defined goal and usually a final triumph waiting to be embellished.

The publication of his travels was as carefully choreographed as the expeditions themselves. The scientific results of the second expedition (including 1,149 pages of maps) had appeared in eight volumes, and the findings of the third were already being prepared for the twelve shelf-bending tomes that would comprise *Southern Tibet*, his magnum opus. The superb maps and the author's own photographs, plus his sketches of people, places and hair-raising encounters which had somehow eluded the camera, added immeasurably to the value of all this paperwork. An aspiring surveyor and competent geologist cum naturalist, Dr Sven Anders Hedin belied his slight stature, suspect eyesight and rimless spectacles with a boxer's physique and the stamina of a fell-runner. Exploration in general was, he thought, being tarnished by too many amateurs. For Hedin it was a competitive profession, and he saw no shame in being both the ultimate competitor and the consummate professional.

Pranavananda, a wandering sadhu with no scientific training, scarcely qualified even as an amateur. How he funded his travels and who, if anyone, supported him is unclear. Hedin, on the other hand, was a meticulous organiser and an insistent advocate of his projects. Sweden's King Oscar II had been persuaded to lend his gracious support to the new venture. Emmanuel Nobel, the Russian oil baron and nephew of armaments magnate Alfred Nobel, headed the list of those subscribing a handsome £44,000 to equip the expedition. In India Younghusband stood ready with encouragement and introductions, while Curzon in London prepared to rally the RGS.

This last was news to Hedin, bad news. He was relying on Curzon's backing in Simla, but by the time he arrived there in June 1906 Curzon, the most outspoken of imperialists, was no longer in office. At odds with Lord Kitchener, his commander-in-chief ever since the Younghusband expedition, Curzon had resigned the viceroyalty and been succeeded by the steady-as-she-goes Lord Minto. The offer to Hedin of a native surveyor had accordingly been withdrawn, there was no more talk of an escort or an astronomical recorder, and on instructions from London the very idea of Hedin entering Tibet was firmly ruled out. Where British travellers were routinely being denied access, why should a Swede be admitted, never mind one who seemed to expect viceregal protection?

Sven Hedin in triumphalist pose with members of his
1906–9 Tibetan marathons

The expedition, which had come together thanks to Curzon, was now coming apart thanks to Curzon. In Hedin's eyes the British were proving more obstructive than the Tibetans. He appealed to Henry Campbell-Bannerman, the new British prime minister, but was again rebuffed.

In lieu of Tibet, there was, however, no objection to Hedin's returning to the Takla Makan and his other old stamping grounds in Eastern Turkestan (Xinjiang). So be it. Hedin let it be known that he would do just that; and since the most direct route from India to Turkestan was the ancient trade route across the main Himalayan ranges from Ladakh to Yarkand, he announced he would set off from Leh. There another official directive reminding him not to enter Tibet was either ignored or perhaps withheld by one of his supporters, probably Younghusband who was now serving out his time in Kashmir. On 13 August 1906 the expedition's thirty men, fifty-eight horses and thirty-eight mules filed out of Leh, following the Indus upstream. They could still have turned north for Turkestan but didn't. 'I watched my men pass along the road, and the whole world lay open before me,' wrote Hedin. 'Now I was free, out of reach of all that is called government; now I could rule myself.'[12]

A month later they were still heading east, now across a lunar wilderness of blood-red rock and azure salt lakes. Hedin knew it as 'the unannexed region of Aksai Chin'. 'Does the Maharajah of Kashmir lay claim to it, or the Dalai Lama, or is it part of Chinese Turkestan?' he wondered. Where the borders were so uncertain, 'I could move on eastwards without acting in direct opposition to the wishes of the English government.'

A flock of sheep – the expedition's larder-on-legs – was already much depleted, while the horses were succumbing to the poor grazing and biting winds; a few had expired; others carrying the collapsible boat in which Hedin proposed to explore the Tibetan lakes had to be replaced with yaks. The expedition blundered on.

October came and went; each day ended with a storm. The thermometer dropped to below zero Fahrenheit (minus 17.7 degrees Celsius). Captain Deasy's 'Antelope Plains' lived up to their name. Hedin himself seldom killed anything but his men easily shot the wherewithal for a kebab supper. Courtesy of several yak kills, beef was also plentiful. Otherwise they encountered only rodents and ravens. 'Of human beings, not a one.'[13] As if to relieve the 'monotonous solitude', there came instead a succession of dramas. 'On the Lake in a Storm', 'In the Land of the Wild Yak', 'Death in the Jaws of Wolves', 'Great Losses' – the chapter headings in Hedin's *Trans-Himalaya* read like those from a *Boy's Own* yarn. However repetitive the scenery, his readers were promised plenty of action. By now Turkestan was quite out of the question. It was Tibet or nothing.

New Year 1907 found them at 'Camp No. 99'. They had been on the move for four months, and of the ninety-six horses and mules that had set out from Leh only nine were still alive. But after an eternity of getting nowhere and meeting no one, help was at last on hand. The black tents of Tibetan graziers here dotted the hillsides. Yaks and guides could be hired, supplies replenished. By the same token, their presence was no longer a secret. Armed detachments were now shadowing their progress and the local governor had issued orders that they return to Ladakh.

Camp 100 was made on the shores of the Ngangtse-tso (Ngangce Co), about 400 kilometres west of Lhasa and one of several vast lakes here strung across the Tibetan plateau. Keen that his discoveries be ranked with those of the great African explorers, Hedin would call this 'the Lake Region' of Inner Asia in imitation of Burton and Speke's east African 'Lake Region'. But as Pranavananda would put it, 'Hedin's enthusiasm for fame seems to have got the better of him.'[14] Tibet was no Tanganyika. The lakes were mostly saline and, it being mid-winter, all were frozen solid.

Hedin still insisted on conducting a survey of Ngangtse-tso while he waited for the governor to arrive in person. The collapsible boat was hauled on to the ice; zinc runners acting as stabilisers turned it into a boat sledge; and the mast and sail made it what he called an 'ice-yacht'. Drawn upwind by horsepower and blown downwind by the raging gale, the contraption zigzagged across the ice before disappearing into the swirling mists. Measuring the lake was made more difficult by the white-out. The shoreline of encrusted salt proved indistinguishable from the lake's sheet of ice. And anyway, boreholes revealed the maximum depth as a mere 3 to 4 metres. The storms, which seem to have lain in wait for Hedin whenever he got in a boat, acquired added bite from splinters of salt mixed into the hail and snow flurries.

When the governor finally arrived he turned out to be the man who had halted Hedin's dash for Lhasa back in 1901. This was not necessarily bad news. Times had changed. The Dalai Lama was in exile in Mongolia, where he had fled during the Younghusband incursion. As was usual in such circumstances, his spiritual authority had devolved on to the Panchen Lama of Tashilunpo, the great monastic complex at Tibet's second city of Shigatse (Xigaze). Moreover this 'Tashi Lama' was on good terms with both the British and the Chinese (who had reasserted their interest in Tibet the moment Younghusband withdrew). Hedin carried a letter of introduction to the Tashi Lama from a fellow lama in Leh, and the obstructive governor was naturally in awe of someone so well connected. Better still, Shigatse lay only a ten days' march from where Hedin was encamped beside the Ngangtse-tso.

Two further visitors decided the matter. One was an emissary from the Tashi Lama's brother whom Hedin took to be inviting him to attend the New Year ceremonies in Shigatse. The other was a postal runner bringing him his first newspapers and letters in nearly six months. The mail had been forwarded from Gyantse, along with champagne and other delicacies, by its acting British trade agent, the menagerie-mad 'Hatter'

Bailey. Hedin had never met Bailey but would have known of his larger-than-life reputation from mutual acquaintances like Younghusband and Rawling. Things were looking up. Notwithstanding doubts about the status of an unauthorised Swedish national on the loose in Tibet, it was now as clear to the Tibetans as it was to Hedin himself that the British in India, unlike the government in London, were not implacably opposed to his expedition. He broke camp and headed for Shigatse, confident of winning the Tashi Lama's support for his onward plans.

Until now Hedin had failed to reveal his ultimate destination even to his followers – let alone to his readers. There were in fact several options, one of which he had already discounted: to the attractions of Lhasa he claims to have become indifferent. Younghusband had stormed the gates. Journalists and others embedded with Younghusband's 'mission' had laid bare the city's 'medieval squalor' and its 'monk-ridden' intrigues. Shigatse and Tashilunpo, on the other hand, were less known and well worth the several chapters he would award them in his *Trans-Himalaya*.

But to a geographer–explorer, far more inviting were the great white spaces marked on the map as 'Unexplored'. Hedin had always been obsessed by these cartographical blanks and was delighted to be riding into just such a vacuum en route to Shigatse. In his mind's eye he would soon score out the map's 'Unexplored' and scrawl his own coinage. The name he eventually chose would be 'Trans-Himalaya':

> this region [Tibet north of the Tsangpo–Brahmaputra] presented
> the grandest problems which remained still unsolved in the
> geography of Asia. There must exist one or more mountain systems
> running parallel with the Himalayas and Karakorams; there must be
> found peaks and ridges on which the eye of the explorer had never
> lighted … and from their southern margins voluminous rivers must
> flow down, sometimes turbulent, sometimes smooth.[15]

It was not quite true that no explorer's eye had ever lit upon this mountain system. As Pranavananda would rightly observe, Pundit Nain Singh had passed this way in 1866; so had Rawling, Bailey and colleagues in 1904. As Hedin must have appreciated, anyone approaching Tibet's second city from the north could hardly miss such a lofty mountain barrier. The 'problem', as Hedin saw it, lay in recognising the range's significance. For if most of South Asia's great rivers originated not in the Himalayas but north of them, might not this, 'my very own Trans-Himālaya', be the greatest of Asia's watersheds?

From the frozen lakes and the plateau country of the Changthang, the expedition climbed steeply to three passes of between 5,350 metres and 5,500 metres asl, then dropped down towards the Tsangpo–Brahmaputra valley. Winter reigned north of the passes; to the south it was almost spring. Trees, the first since Ladakh, were in bud. Fields were being sown, and the mercury edged above freezing. The range they had just crossed should, thought Hedin, be a continuation of that reported north of Lhasa. There it was named after a peak, the Nyenchentanglha. But given that its length and height served as a major climatic and hydrographic divide, Hedin preferred his own, grander coinage: 'I called it the Trans-Himālaya because it was on the other side of and beyond the Himālaya.'[16] His achievement lay in having identified what he thought a fitting rival to the Himalayas themselves.

> I venture to describe this geographical problem that I have
> succeeded in solving as one of the finest, perhaps the most striking,
> of all [the] problems connected with the surface of our earth that
> awaited solution … It is one of the greatest and grandest watersheds
> of the world, for from its northern flank the water flows down to
> the undrained lakes of the plateau, and from its southern flank to
> the Indian Ocean.

As his greatest discovery to date it called for a burst of purple prose and some not untypical triumphalism.

> In silent meditation my eyes swept from the rocky crests, brightly
> lighted by the moon, down to the dark shadowy depths of the
> valley, where there were only wolves crouching in their holes. It
> seemed as though all belonged to me, as though I had marched into
> this land a conqueror at the head of victorious legions, and had
> crushed all opposition.[17]

The opposition was not entirely crushed. Though deeply impressed by the young Tashi Lama – and in no doubt that the regard was mutual – Hedin failed to win the Lama's support for what was his preferred option: a return to India by way of the Tsangpo–Brahmaputra. At the time that river's course across southern Tibet and through the Great Himālaya was still a mystery. Kintup's report had been forgotten; Rawling and colleagues, who were to have followed the river downstream when returning from Lhasa in 1904, had been redirected upriver to Gartok and Gugé; and Bailey had yet to try his luck by working up the Brahmaputra's tributaries from Assam with Morshead.

Hedin had projected the solution of the river's route down to India as the crowning achievement of his whole expedition. But it was not to be. Neither the Tashi Lama nor the Lhasa authorities would hear of him continuing east and forcing a passage through the Himalayan 'Great Wall'. It would complete a traverse of the entire country that was almost as provocative as Younghusband's invasion. He must go back to Ladakh by the way he had come.

Hedin would not have been Hedin had he taken this lying down. He agreed to return to Ladakh if he had to, but not by retracing his route across the Changthang. Instead he would follow the Tsangpo–Brahmaputra west to its supposed source in Kailas–Manasarowar and, from thereabouts, track the infant Indus to Ladakh. He claimed this was a more direct route (which was debatable), and that he wouldn't deviate from it (though he would).

The sources of the great rivers were at last emerging as prime objectives. He would resolve that ancient conundrum about the drainage of Manasarowar; and if he could locate the genetic sources of the rivers in the catchment of his beloved Trans-Himālaya, it would add immeasurably to his claims for that range. Suddenly the expedition had a renewed sense of purpose. 'I have only one goal, the north of the Tsangpo, where most important discoveries await me.'[18] With the Tibetans providing transport, guides and two officious 'mandarins' to monitor his movements, Hedin left Shigatse on 27 March 1907.

Four months later he reached the headwaters of the Tsangpo–Brahmaputra at a place called Shamsang (Samsang). Here, in open country about 120 kilometres short of Kailas, a decision had to be made as to which of three feeder streams constituted the main river. Pranavananda would insist that he should have explored each in turn, but to this Hedin's Tibetan escort had objected as being a deviation from the agreed routing.

Relations were already strained to breaking point. No sooner had the expedition left Shigatse than Hedin had begun pushing his luck. First he had headed north again into the Trans-Himalayan highlands. April and most of May had been passed in exploring the western extent of the Trans-Himālaya and of the 'Lake Region' beyond. His minders, desperate to redirect him to the prescribed route, threatened to withhold supplies. An appeal to Shigatse won him time but no concessions. The expedition was ordered back to the well-travelled east–west road up the Tsangpo–Brahmaputra valley.

Hedin's mood was not improved by the sudden death of Muhammad Isa, his head man and probably the most experienced expedition organiser in the whole of Himālaya (he had been *caravan-bashi* for Younghusband, Rawling, de Rhins and Deasy among others). Hedin was genuinely upset by this loss and arranged for a headstone to be erected over his grave. Then, again contravening orders, he crossed to the southern bank of the river. In June he set off on another unauthorised excursion, this time into the Great Himālaya. Camp 184 had actually been pitched beyond the crest of the mountains and well within Nepal.

Now at Shamsang he was back on the river again, within striking distance of its sources, and more closely chaperoned than ever. 'The force of circumstances had forced us to leave behind us step by step ever larger areas of unknown country to the north. I was vexed but I would at any rate endeavour to do all that was possible in my hampered condition.'[19]

Hedin's decision as to which of the Tsangpo–Brahmaputra's feeders should be regarded as the parent stream would prove controversial. Among the many who would contest it none would do so more bitterly than Pranavananda. But Hedin did the best he could under the circumstances. Pressed by his minders to move on, he launched his collapsible boat at the confluence of the tributaries and, from moorings midstream, took measurements of the volume of water in each. To his satisfaction, the Kubi-tsangpo from the south-west was found to discharge three times the combined volume of the other two. This seemed conclusive. Though it hardly placed the source in the Trans-Himālaya, it nicely complemented the historical record. Strachey had reported on one of the other feeders, and both Pundit Nain Singh and Captain Ryder, one of Rawling's surveyors, had investigated the other one. The Kubi-tsangpo was not just the largest of the three; it was Hedin's because no one else had laid claim to it.

On 13 July, from Camp 200 at 15,991 feet (4,874 metres) asl, Hedin and three guides rode up to the glaciers which they took to be the ultimate source of the Kubi-tsangpo. The setting was the usual chaos of mud, moraine and icy tarns, but the backdrop 'could not have been grander or more magnificent'. Hedin could hardly contain himself: 'Holy and thrice holy are these mountains, which from their cold lap give birth and sustenance to the river celebrated from time immemorial in legend and song, the river of Tibet and Assam, the river *par excellence*, the son of Brahma.'[20] Generations of 'black Tibetans', he supposed, had heard its waters 'roar[ing] between the two loftiest mountain systems of the world, the Himālaya and the Trans-Himālaya'. Assamese tribesmen had watered their fields 'with its life-giving floods and drunk of its blessed water'. Great cities had grown up in its Bengal delta.

> But where the source lay no one knew. Three expeditions had determined its position approximately but none had been there. No geography had been able to tell us anything of the country round the source of the Brahmaputra ….
>
> But here it is, here in front of three glacier tongues, that the river so revered by the Hindu tribes begins its course of some 1800 miles [2,900 kilometres] through the grandest elevations in the world and from which its turbulent volumes of water roll first east, then southwards, cutting a wild valley through the Himalayas and finally flowing south-westwards through the plains of Assam. The upper Brahmaputra, the Tsangpo, is truly the chief artery of Tibet … while its lower course is surrounded by the most fruitful and populous provinces of Assam. The Brahmaputra is therefore one of the noblest rivers of the world …[21]

One down, then; and one – possibly two – to go. It was the perfect start to his hydrological discoveries. But, before tackling the Indus and the Satluj, there were Kailas and Manasarowar. 'By forbidden paths, making as many discoveries as possible in spite of the Mandarins and the Lhasa government', Hedin continued west. The excursion to the source of the Tsangpo had cost 110 rupees in bribes and fees. Since his 'Mandarin' minders were supposed to be escorting the expedition as a whole, another way of evading their attentions was to divide up the party. Half the men were therefore sent on ahead to Ladakh, leaving just thirteen men, four horses and a dozen yaks. By detaching himself even from these for short explorational forays, Hedin counted on evading surveillance.

On 20 July they crossed the watershed between the Tsangpo catchment and the Manasarowar basin 'over country never visited by a European before'. Pilgrims began appearing from nowhere. A glacier-fed spring whose stream flowed towards the lake was said to be the traditional source of the Satluj. It was also said to have miraculous healing properties; 'it is a Lourdes in miniature'. But, doubtful as yet about any link between Manasarowar and Rakas Tal, Hedin pressed on. He was straining for a first glimpse of the great lakes when his men suddenly dismounted and threw themselves on the ground. They had spied the unmistakable profile of Kailas.

Hedin's homage to 'the holy mountain lifting its summit under a cupola of eternal snow' was no less heartfelt; like Manasarowar it 'had long been the subject of my dreams', he announced. But the moment was sullied by his identifying the peak as 'where Siva, one of the Hindu trinity, dwells in her paradise among a host of other deities'.[22] This could be a misprint, though the feminine 'her' is applied to the Great Lord elsewhere in Hedin's account. It may serve as a reminder that Hedin, though game to grapple with the rites of Tibetan Buddhism, knew little of India and next to nothing of the devotional fervour that brought Hindu pilgrims toiling through the Himalayas to such a bleak but god-favoured arena.

Manasarowar, 'my beloved lake', called for launching the boat. Overnight the wind dropped and they were able to row the 13 kilometres to the opposite shore, taking soundings as they went. Hedin sat on a cushion by the tiller wrapped in three shirts, plus 'a leathern vest, Kashmir boots and an Indian helmet'. They had been told to expect disaster since trespassing on to the sacred waters was accounted a sacrilege. They returned the following day without mishap, having found a maximum depth of 82 metres.

Rowing and riding round the southern shore, they continued past the termination of the rivulet they had followed from the 'Lourdes in miniature' and which Hedin was now inclined to accept as the 'genetic' source of the Satluj. The stream seeped into the lake through a raised bed of aggregate like that reported by Strachey as obscuring the drainage of Manasarowar into Rakas Tal. This was promising. They repaired to nearby Tugolho Gonpa, the same monastery in which Pranavananda would pass two winters and from whose balcony his sputum would hit the ground with an icy tinkle. But this was August. The weather, though changeable, was almost warm.

The next monastery was Gossul Gonpa on the western shore. To a mariner less storm-prone than Hedin it would have looked an easy sail. So it was at first. Then out of nowhere came a fusillade of hailstones, followed by such torrential rain that the boat became a bath in which the crew risked being 'drowned like cats'. Hedin used his Indian helmet to bail. Finally a hurricane whipped up waves 'as high as the billows of the Baltic in stormy weather'. Through spectacles awash with spume, Hedin saw little. They eventually squelched ashore, soaked to the skin and badly shaken.

The storm was the prelude to a long circuit by land and lake of both Manasarowar and Rakas Tal. It detained them for a month, during which they repeatedly ignored orders to rejoin their by now frantic escort. Hedin, of course, was busy with his detailed survey of possible channels exiting the two lakes. But painstaking measurements of ground and water levels, and much stomping about in swamps and fetid puddles, revealed nothing new. 'In the year 1907 no water flowed from the eastern into the western lake,' he reported. In fact, to do so it would have had to flow uphill since the bed of the connecting channel was 2 metres higher than the water level of Manasarowar. Only after exceptional rainfall and heavy snow melt might the lake overflow sufficiently to wet the channel. There was, though, ample evidence of underground seepage. It was hard to quantify, but it was enough to convince Hedin that Manasarowar did, however discreetly, drain into Rakas Tal.

Much the same conclusions were prompted by the exploration of Rakas Tal itself. Again there was no detectable flow out of the lake. But, on following the streambed reported by Strachey to the north-west, Hedin found stagnant pools, then freshwater springs and finally another Lourdes-like grotto. This, too, his guides declared to be the traditional source of the Satluj. So the river now had four interconnected sources: Lourdes 1, Manasarowar, Rakas Tal and Lourdes 2. Assuming that, like Hedin, one accepted subterranean filtration as a valid form of flow, it could just about be argued that in Lourdes 1 he had indeed identified the remotest 'genetic source'.

Next came the inevitable *kora*, or *parikrama*, of Mount Kailas. Their minders objected vehemently. The mountain was out of bounds to foreigners and anyway time had long since run out; they must head off to Gartok and Ladakh immediately. Hedin listened, then slipped away regardless. With just five followers and minus the boat, he threaded his way round the mountain past rows of chortens and massive granite boulders amid a throng of pilgrims hailing from as far away as Kham and Amdo. If any were Indian Hindus, they were few. Nearly all those identified were Tibetan Buddhists whose mantra *om mani padme hum* hung in the sharp air like the drone of swarming bees. Most of the pilgrims expected tangible rewards for their exertions – protection for their families and flocks from disease and predators, good pasturage, more rain, forgiveness for sins. 'My pilgrimage was of no value at all because I was riding,' says Hedin. To gain any benefit one had to walk or, better still, progress by prostrating oneself, stretching and retracting the body along the rocky trail like a looper caterpillar.

Hedin's only reward was some good news of the Indus. From Kailas the river's source could be reached in just three days. Hastily concluding the last stage of the 50-kilometre circuit, he found a guide and immediately struck off to the north-east. The 'Kailas Range' was crossed by a pass of 5,630 metres asl and the Trans-Himālaya by one of 5,466 metres. Dinner at Camp 236 was fish, hauled from the river by hand.

As at the headwaters of the Tsangpo–Brahmaputra, three tributaries contended for the paternity of the Indus, but in this case there was no need for the boat; they were all wadeable. He didn't even measure their discharges. A passing trader with a caravan of pack-bearing sheep simply directed him to the one which was the generally accepted source. Here, in open ground at the base of a slope of detritus, a slab of white rock protruded. Around it were a scatter of engraved *mani* stones, three cairns and a small votive shrine. And at its base, noted Hedin, 'several springs well up out of the ground forming weedy ponds and the source stream, which we had traced upwards, and which is the first and uppermost of the headwaters of the mighty Indus'.

Because the area was infested with robbers, no foreign traveller was known to have been there. It was as if 'the discovery of the Indus source

had been reserved for me and my five Ladakhis'. The moment cried out
to be milked; and who better to do the honours?

> Here I stood and saw the Indus emerge from the lap of the earth.
> Here I stood and saw this unpretentious brook wind down the
> valley and I thought of all the changes it must undergo before it
> passes between rocky cliffs, singing its roaring song in ever more
> powerful crescendo down to the sea at Karachi ... Here I stood
> and wondered whether the Macedonian Alexander, 2,200 years
> ago, had any notion of where its source lay and I revelled in the
> consciousness that, except the Tibetans themselves, no other human
> being but myself had penetrated to this spot ... Not without pride
> but still with a feeling of humble thankfulness, I stood there,
> conscious that I was the first white man who had ever penetrated to
> the sources of both the Indus and the Brahmaputra.[23]

A pioneer so comfortable with his own conceit asks to be ridiculed.
Howsoever humble, the 'poor Swede', as he liked to characterise
himself, would face a barrage of critics the moment, eighteen months
later, he reappeared in Europe. Indeed so hostile was the response to
Hedin the braggart that Hedin the explorer would be largely written
out of the record. In his 1955 *Abode of Snow*, a classic 'History of
Himalayan Exploration and Mountaineering', Kenneth Mason, retired
superintendent of the Indian Survey and Oxford professor of geography,
would manage to avoid any mention whatsoever of Himālaya's most
determined explorer or of Trans-Himālaya, his dearest discovery.

The eighteen-month time lapse between news of Hedin's achievements
reaching London in late 1907 and his presentation of them in person in
1909 is explained by his having again disappeared into the unknown. In
fact, no sooner had he left Tibet than he had slipped back into it. From the
source of the Indus, the expedition had repaired to Gartok and continued
on to Ladakh as planned. But there, instead of riding into Leh in triumph,
Hedin had announced his intention of reviving his original ploy of visiting
Chinese Xinjiang. On this understanding, employees and well-wishers
brought out from Leh the transport, provisions and sacks of rupees needed
for a new expedition. He himself didn't actually go to Leh. Nor did he
reach Xinjiang. Instead, after marking time in the eastern Karakorams, he
wheeled about and headed east again for Tibet's empty quarter.

This sequel to his greatest Tibetan journey took six months. Except that it meant crossing the Changthang in the depths of winter and so merited his likening it to 'a polar journey', the itinerary approximated that of the previous marathon. Again the horses and mules died, again the storms came thick and fast (one lasted a week), again the Tibetan authorities blocked his progress, again he led them a merry dance through Trans-Himālaya, and again he checked the outflow of Manasarowar before being shooed back to India, this time to Simla by the route taken by Rawling and Bailey in 1904.

'I could not return without having done all that was humanly possible,' was his only explanation for what he rated 'the hardest journey I ever accomplished'. It did flesh out his thinking about Trans-Himālaya. Only now did 'the great idea' of calling it Trans-Himālaya come to him, and only now did he concede that it was in fact 'a family of ranges'; 'there was no question of a single continuous range, but ... a whole collection of ranges quite independent of one another existed'.[24] More of those 'Unexplored' voids on the map were also traversed, but without any great discoveries. Charles Allen thought it likely his real goal still lay either in 'the golden portals of Lhasa itself' or in the Pemako gorges where the Tsangpo becomes the Brahmaputra. Both chimed with the objectives of his previous travels; so did his reluctance to divulge them. But the Tibetan authorities proved as keen to get rid of him as ever. For a warmer welcome he looked to India and London.

From Gartok, to where he had repaired after locating the source of the Indus, Hedin had written to the secretary of the RGS. 'I have made the most important and splendid discoveries that were left to be conquered on earth,' he declared; a whole issue of the Society's prestigious journal must be devoted solely to them; his sister in Stockholm, acting as his agent, would invoice the Society for the publication rights. She did; but when told payment was out of the question, she became abusive. Hedin's braggadocio was one thing, his family's chutzpah another. The RGS was as horrified by the suggestion of contributors to its journal being paid as it was by Hedin's ranting.

An abstract of his proposed presentation then alerted the London geographers to the possibility that his discoveries might not be quite as earth-shattering as he contended. Tom Longstaff, an army doctor and climber who had just set a new altitude record by scaling the 7,210-metre Trisul in Kumaon, rubbished his claim to have discovered the Trans-Himālaya range. Others merely objected to his calling it Trans-Himālaya. A former surveyor-general cast doubt on the accuracy of his survey work, old-timers like Godwin-Austen and Henry Strachey were unhappy about his ignoring or disparaging the labours of previous travellers, and nearly everyone wanted greater clarity about his river sources.

These objections were not allowed to mar his formal address to the RGS in early 1909. With Hedin sensibly subdued, the occasion proved a personal triumph. A packed audience was spared the spectacle of Himalayan giants trading blows in public. Two weeks later, at Hedin's insistence, a more intimate inquest was convened for what was billed as a 'Discussion of Geographical Problems Concerning Tibet'. Again Hedin delivered a masterly and even modest résumé of his work, after which each of his critics in turn aired their concerns. Most were conciliatory, acknowledging his unequalled record and applauding his undoubted courage while regretting his hasty assessments. His Trans-Himālaya, it was pointed out, had been known since the 1870s and surveyed in 1904. A party of British sportsmen had in all probability preceded him to the Kubi-tsangpo source of the Tsangpo–Brahmaputra. And the underground seepages of the Satluj were neither convincing nor relevant. He might claim just one 'first', the bleak spot three days' ride from Kailas where he had 'stood and saw the Indus emerge from the lap of the earth'.

Hedin felt understandably aggrieved and said so; his attempted tact and restraint had been met with ill-founded accusations from armchair amateurs and outright betrayal by old friends. Though himself an RGS medallist and honorary member, he wanted no more to do with London or the Society. He left soon after for Germany and Sweden. Five years later he burnt his boats with the British establishment by lending his support to Kaiser Wilhelm in the First World War. The RGS removed his name from their membership. Hedin would lead more expeditions but never return to Tibet or India. His last misjudgement came in the Second World War when he endorsed Hitler's Third Reich. He died in 1952, having never met, and probably never heard of, Swami 1108 Pranavananda Maharaj.

Hedin's 'Trans-Himālaya' range had been effectively demolished by the London geographers. The term is now to be found only in a historical context – as here. Pranavananda ignored it; unlike Hedin, the swami was not particularly interested in Tibet's vast landscape. He could, though, claim an intimate acquaintance with western Tibet's hydrography. With Hedin's mountain discovery discredited, it remained for Pranavananda to 'explode', as he put it, Hedin's findings regarding the four great rivers of Kailas–Manasarowar.

The swami did so at length in 'New Light on the Sources of the Four Great Rivers', a vigorously argued chapter in his *Exploration in Tibet*. The footwork seems to have been done in 1936–7 when he visited all the headwaters specifically to compare them. What immediately struck him was the difficulty of determining which of a river's tributaries would lead one to its true source. Did one look for the longest feeder, or that with the largest flow of water, or that which local tradition held to be the parent stream? If length was the decider, was it the distance in a straight line from the feeder's remotest source to its confluence with other feeders that counted, or did one have to include in one's calculations all the wiggles and bends of what were often wayward rills? And if water volume was the decider, how to compare flows which, depending on snow melt and precipitation, varied from year to year, season to season and even hour to hour. In August 1907 Hedin, like his predecessors, had found the old streambed leading out of Rakas Tal that might have been the Satluj to be 'stone dry'. In October 1942 Pranavananda found it a raging torrent and too deep to ford. Did a river with only an intermittent discharge count as a river? And how, short of mounting a year-round vigil, did you measure that discharge?

Pranavananda might have been more sympathetic to Hedin's predicament had he not been convinced that Hedin had taken advantage of these uncertainties. The Swede 'would have served the cause of truth better', wrote the swami, 'if he had frankly admitted the difficulties of deciding upon suitable and consistent criteria for fixing the sources of these rivers'. Instead he distorted the facts with defective logic 'in a desperate struggle to achieve the coveted honour of being the first and original discoverer of the sources'.

It was the swami's contention that Hedin's discoveries were often retrospective, as with Trans-Himālaya, or prescriptive as in this case. Having first selected a tributary source to which no previous traveller had penetrated, he then chose whichever of the various criteria best suited its claim to be considered the only true source. 'Enthusiasm for fame had got the better of him.' His conclusions contradicted one another, huffed the swami. They were palpably contrived.

By giving preference to the quantity of water in the case of the Brahmaputra, tradition in the case of the Indus, and far-fetched tradition and length in the case of the Sutlej, he has not hesitated to sacrifice mercilessly all consistent, reasonable, and uniform procedure which has to be adopted in dealing with such important problems.[25]

Hedin was wrong about other things too. Manasarowar didn't freeze over in an hour; it took days. And in Rakas Tal there were actually three islands, not Hedin's two. Pranavananda was confident of such matters because he too had sailed across the sacred waters. His first boat – and the first on the lakes since Hedin's – had been an inflatable presented by a well-wisher in Bombay. This was replaced by 'a galvanized steel sailing dinghy-cum-motor boat' paid for by the Maharajah of Bhavnagar. After long delays because of the difficulty of transporting it over the Himalayas, in August 1947 this *Jnana Nauka* was eventually launched 'into the turquoise-blues of the celebrated celestial lake amidst deafening cheers of pilgrims, Tibetans, and Bhotia traders'. But the launch coincided with India's independence and was soon to be followed by Maoist China's first assertion of its authority in Tibet. It was the end of Pranavananda's affair with Kailas–Manasarowar.

Writing in 1950 the swami hoped the boat was still there. He looked forward to the day when Indian pilgrims would again be welcomed to the region and he foresaw a great future for trans-Himalayan aviation. Seaplanes could land on the lake; and between it and Kailas there was an excellent site for an airstrip. 'Pilgrims and tourists who love adventure and mountain-trotting [trekking?]' might still come on foot or pony-back; but, for the less fit, the Kailas–Manasarowar Air Service would operate day trips from, say, Delhi offering a dip in the holy lake and home again for dinner.

But this was all wishful thinking. Not until the late 1970s were pilgrims from India readmitted to the region. Access was improved with the opening of routes via Nepal and Sikkim, and occasional helicopter tours were authorised. But under the terms of the Beijing–New Delhi agreement, pilgrim numbers remained regulated. Successful applicants were often selected by lottery, tour operators could be unreliable and, since the whole concession was at the mercy of Sino-Indian relations, it was, and still is, liable to be withdrawn at short notice.

11

Sages and Heroes

This centre of heaven,
This core of the earth,
This heart of the world,
Fenced round by snow,
The headland of all rivers,
Where the mountains are high and the land is pure.
O country so good,
Where men are born as sages and heroes,
To this land of horses ever more speedy,
He chose his mode and came.
      Ninth-century Tibet as described in a Dunhuang document[1]

Although Kailas–Manasarowar is sacred to both Buddhists and Hindus, how frequently Hindus found their way from India through the mountains and across the high plateau has been questioned. Sven Hedin writes of thousands of pilgrims having performed the *koral parikrama* of Kailas every year since time immemorial, but it seems those he encountered in 1907–8 were mostly Tibetan Buddhists. This will not have surprised Alex McKay. In a 1998 study of pilgrimage in Tibet, Professor McKay notes the absence in the Kailas–Manasarowar region of the structural features typical of other places of Hindu pilgrimage – ashrams and dharmsalas, temples, ghats, shrines, statuary. Until the 1930s neither Kailas nor Manasarowar seems to have been 'institutionalised as a site of pilgrimage', and 'this raises the question of how many Hindu pilgrims actually visited the region' – and indeed, who they were. McKay believes that few if any caste Hindus (for example, family groups) undertook the journey and that, until the 1930s, Hindu pilgrims were mostly renunciates, either sadhus or yogins, plus the occasional man of science like the truth-seeking Swami Pranavananda.

The swami himself, while usually so minute in such matters, is disappointingly vague about pilgrim numbers. But his publications helped promote the attractions of Kailas–Manasarowar, and in this they may unwittingly have served the interests of British India. For according to McKay, popularising and facilitating Indian pilgrimage to Tibet became imperial policy during the last decades of the raj.

After the 1904 Younghusband expedition the British could largely discount Russia's interest in Tibet. Not so China's. In 1909–12 Chinese troops repeatedly entered eastern Tibet to nibble at the frontier in Kham and support assertions of China's suzerainty over the whole country. These were the incursions which obliged 'Hatter' Bailey to abort his first attempt to trace the Tsangpo–Brahmaputra through the mountains and which persuaded the Dalai Lama, who had barely returned from his Younghusband-induced exile, to flee Lhasa once again, this time to India. That was in 1910. In 1911 the last emperor of China's Qing (Manchu) dynasty was swept aside in favour of the first in a dynasty of revolutions (nationalist, communist, cultural and so on). Initially an unsteady Chinese republic was good news for Tibet. The Dalai Lama returned from India, and while China's nationalists battled with warlords and communists for control of the heartland of the erstwhile Celestial Empire, Tibet's nationalists enjoyed a golden age of qualified autonomy, cautious modernisation and avowed neutrality.

This interlude ended in 1933 with the thirteenth Dalai Lama's death (or 'attainment of nirvana'). To 'pay posthumous tribute to the deceased', China's nationalist government in Nanjing unexpectedly sent a representative to Lhasa; and after a long search for the next incarnation, another Chinese official attended the enthronement of the four-year-old Llamo Dhondrub as the fourteenth Dalai Lama. Each of these Chinese missions left an accredited representative in Lhasa. There was much to sort out. Fighting had again broken out on Tibet's eastern frontier, the Panchen Lama was planning to return from his self-imposed exile in China with a large escort of Chinese troops, and the uncertain nature of the whole Sino-Tibetan relationship cried out for clarification. Having appointed themselves guardians of Tibetan autonomy, the British decided to follow suit by installing their own representative in Lhasa – just as Younghusband had proposed thirty years earlier. The man chosen was Hugh Richardson, a Tibetan scholar as well as a political officer, and he would stay on till after the British had left India. 'Although the arrangement [for a resident British official] was undefined and considered temporary, yet from 1937 it was to remain virtually unchanged until the Chinese invasion of 1950.'[2]

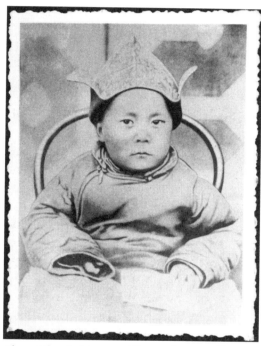

Llamo Dhondrub was aged four at the time of his recognition as the 14th Dalai Lama

In the context of this Sino-British rivalry in Lhasa the British took up the cause of Indian pilgrimage to Kailas–Manasarowar. Trails were improved, protection afforded where necessary and some basic facilities provided. 'The British patronage led to a dramatic increase in the numbers of Indian pilgrims to the Kailas region after the 1920s,' says McKay.[3] By emphasising the customary and devotional nature of the journey and its importance to the wider Hindu public, not just renunciates, the object was to impress on the Chinese that India too had legitimate interests in Tibet. It was another case of pilgrimage serving politics – or nowadays of politics regulating pilgrimage.

There was nothing unusual in Hindus and Buddhists reverencing the same sites; it had long been so. Sectarian awareness, though not unknown in pre-Islamic India, had rarely been exclusive or confrontational. In so far as labels like 'Hindu' and 'Buddhist' can usefully be applied to anyone in the fifth century BCE, the young Siddhartha Gautama had himself been a Hindu. The great temple at Bodh Gaya (in Bihar) marking the spot where he achieved enlightenment as the Buddha has spent much of its 2,000-year history in Hindu hands or under Hindu management. Likewise, the rock-cut cave temples that honeycomb the cliffs of western India were hewn by adherents of both faiths, seemingly within the same generous time span of c.200 BCE–200 CE. Since Hindus recognise the Buddha as the last of the nine avatars of Lord Vishnu, Bodh Gaya's much reconstructed temple still attracts Hindu devotees as well as Buddhist pilgrims.

So do less august locations. In the foothills of the western Himalayas near Mandi in what is now the Indian state of Himachal Pradesh, the boating-size lake of Rewalsar draws Hindus, Buddhists and even Sikh devotees. Tradition holds that Lord Shiva declared Rewalsar's once wooded shoreline ideal for meditation; history confidently has it that Gobind Singh, the last of the ten Sikh gurus, lodged on the lakeside while soliciting support against the Mughal emperor Aurangzeb; and between these visitations, Buddhist texts record the appearance here in the eighth century CE of a pivotal personality in Tibetan history.

One school of Tibetan Buddhism regards this apparitional figure as 'the second Buddha'. All Himalayan Buddhists still revere him as Guru Rinpoche or 'the Precious Master'. And among several contenders, Rewalsar is widely supposed the magic lake of his first manifestation. Besides being one of the few, albeit fancifully, documented figures in early Tibetan history, the Precious Master commands devotion as the yogin and teacher who introduced the peoples of Himālaya to the doctrines, practices and institutions of Vajrayana ('the thunderbolt vehicle'), otherwise Lamaism or Himalayan or Tibetan Buddhism.[4]

The details of his manifestation are confused. Many pilgrims regard the little lake of Rewalsar as the Precious Master's actual place of birth, but according to David Snellgrove, Giuseppe Tucci's one-time pupil, 'this does not accord with the literary accounts'. These prefer some version of the story that the Guru Rinpoche flew into the then lakeless Rewalsar from his homeland in Swat, a valley renowned for Buddhist endowments and scholarship and now in northern Pakistan. He had come to Rewalsar to offer instruction to a devout local princess who was about to take the Buddhist vows. But when the visitor's intentions towards the princess were misrepresented, the king, who was the girl's father, took swift action. Princess Mandarava was thrown into a pit full of thorns, and her Precious Master was 'taken to a wild place' and there burnt at the stake.

> The smoke did not clear for ten days so the king went to see the cause. The place had changed into a lake, from the centre of which emerged a lotus, and upon it was a boy of eight years. The king was abashed. Padmasambhava [literally 'the one born of a lotus'] manifested himself and accepted the offer of the kingdom and Mandarava as his bride.[5]

The Precious Master known ever since as Padmasambhava then flew on to Nepal. There he was intercepted by Santarakshita, another apostle of Indian Buddhism, bearing an invitation from the Tibetan king Trisong Detsen (Trhi Songdetsen, Khri sron lde brtsan). Trisong Detsen, a long-reigning sovereign (c.756–c.800 CE) whose extensive conquests resulted in what scholars confidently call a Tibetan 'empire', saw Buddhism as affording an endorsement of his rule plus a useful arsenal of supernatural powers with which to confound the machinations of his presumably *bonpo* opponents. For a consummate magician like Padmasambhava, here then was the perfect patron. Forgoing his own kingdom in the foothills, but not forgetting his consort and Santarakshita, the Precious Master accepted the Tibetan ruler's offer. Like Columba a century earlier when the Irish saint had set sail across the Minch to convert the Gaelic Scots, Padmasambhava winged his way over the mountains to begin his mission to Tibet.

His sojourn in 'the land of snows' may have been brief, but among the many myths that grew up around his ministry certain features deserve attention in that they anticipate the pattern of later Buddhist propagation. As a miracle-worker, Padmasambhava is described as defeating those who, by way of opposing the king's extravagant Buddhist tendencies, were sabotaging the 765 CE building of Tibet's first monastery. The monastery was at Samye, between Lhasa and the Yarlung valley, and though Santarakshita seems to have been responsible for its layout being a replica of one of Bihar's great monastic establishments, it was Padmasambhava who decontaminated the site of baleful influences and destructive demons so that construction could go ahead.

With the assistance of Santarakshita, the Precious Master is also said to have amassed and translated from Sanskrit into Tibetan numerous Buddhist scriptures and commentaries. Among them were examples of those Indian texts classed as tantra, 'practice' (as opposed to sutra, 'teachings') and from whose esoteric formulae, rituals and liturgies the Master derived his awesome powers. In a sense he was not only promulgating Tibetan Buddhism but defining it, for the greater emphasis given to tantra is one of the features that distinguishes Himālaya's Vajrayana Buddhism from the mainstream Mahayana ('Great Vehicle') once practised in India and the Theravada or Hinayana ('Lesser Vehicle') which still prevails in Sri Lanka and South-east Asia.

Finally, and as if to make up for what may have been his hurried expulsion from Tibet, Padmasambhava used his powers of foresight to lay out across the length and breadth of the country what sounds like an elaborate treasure hunt. Known as *terma* (*gter-ma*), meaning 'hidden texts' or 'spiritual treasures', the prizes could indeed be scriptures but could also be devotional aids such as mantras and talismans. And they were always well hidden, sometimes encased in the ice of a glacier, sometimes deep in the ground or in the bed of river. From a listing compiled by Matthew Kapstein, they might be sought 'in the mountains, ravines or woods' of almost anywhere in either India or Tibet. They could be in a vase, on a pillar or somewhere about the house. One was found within the nipple of a statue of a female deity, another was uncovered 'beneath a yak-like boulder'.[6]

The *terma* were Padmasambhava's legacy, a legacy that preserved and substantially augmented his patchy life story and incorporated enough additional miracles and textual feats to ensure his enduring centrality in the development of Tibetan Buddhism. Future generations had only to find a yogin empowered to discover such *terma* to enjoy the insights of the Precious Master's hitherto unrecorded teachings and wrestle with his Delphic responses to issues of the day.

It goes without saying that *terma* were impossible to authenticate. Some were clearly contrived to serve a latter-day cause or laud a new teacher; nearly all have since been classed as apocryphal. They were also slow to appear and were often credited to Padmasambhava's disciples and later followers rather than the Guru Rinpoche himself. The most important were not discovered until the 1300s, by when Tibet had long been subject to its second infusion of Indian Buddhism. Yet because they 'also contain original texts and liturgical and magical elements' which Tucci thinks analogous to those found in other Vajrayana schools, some may be neither anachronistic nor contrived.[7] And doubtless the great majority were accepted as no more far-fetched than, say, the airborne mahatmas and frost-proof ascetics vouched for by Alexandra David-Néel. As Snellgrove observes, 'to non-believers belief all too often appears a mere superstition. It is rational to believe nothing, but always difficult, if one believes, to know where to draw the line, and so the Tibetans do not draw the line.'[8]

Just as no Tibetan draws the line between belief and superstition, so no clear distinction is made between history and myth. King Trisong Detsen is real enough. Chinese sources confirm his conquests, including the occupation in 763 CE of the city of Chang'an, the imperial Tang capital, and later of Silk Road oases like Dunhuang (in Gansu province), where the copious documentation discovered in the early twentieth century amply attests Tibet's first encounter with Buddhism.

In combining Buddhist patronage with Tibetan empire-building, Trisong Detsen was conforming to ancestral tradition. His lineage, Tibet's first and only dynasty, is traced back to the fertile Yarlung valley (between the Tsangpo and Pure Crystal Mountain) and to a day early in the seventh century CE when one of his antecedents had died. Such a thing had never happened before. The Yarlung kings were immortal. For untold generations they had begun their reigns by abseiling down from heaven and had ended them by being hauled back up again. A celestial rope, or in some accounts a lofty mountain, served as the means of ascent, and the kings were of course deities – all, that is, except for the last divine incumbent. His all-too-conventional demise – he was murdered – had opened the way for a more prosaic father-to-son succession, from which, following another regicide, there had emerged the thirteen-year-old Songsten Gampo (Sron btsan sgam po).

As the age at which able-bodied boys were supposed capable of controlling a horse, thirteen had come to be reckoned old enough for the succession. And thanks to his father's tireless campaigning there had been much to which to succeed. 'Prince Songsten was ... heir not only to a divine heritage but to the largest kingdom Tibet had ever seen.'⁹ The young sovereign had dealt swiftly with insurgents and asserted his authority over subordinate clan leaders. A fortress near the Tsangpo's Kyichu tributary known as Rasa, and later as Lhasa, served the Yarlung kings as a strategic headquarters between the productive valleys of the Tsangpo–Brahmaputra catchment and the barren uplands of Changthang. Beyond, by a combination of marital alliances and timely interventions, the young Songsten had reduced to vassalage one by one the indeterminate kingdom of Zhang Zhung in the west, that of Katmandu in Nepal to the south and various peoples in eastern Tibet.

The king of Nepal is said to have provided Songsten with a bride. So too, after an exchange of hostilities and missions, did Taizong, the first emperor of China's illustrious Tang dynasty and an almost exact contemporary of Songsten Gampo. 'Henceforth, neighbouring states – even haughty China – were to deal with Tibet on an equal level,' writes Christopher Beckwith in *The Tibetan Empire in Central Asia*, 'and [they] actually refer to the *btsanpo* either by his Tibetan title or, in unofficial writings, with Chinese terms meaning "emperor". The spread of Tibetan power left them little choice.'¹⁰

Since Tibetan sources invariably acknowledge Songsten as the nation's founding father, it is to be expected that tradition would award him other nation-building achievements. He is said to have adopted and promoted Buddhism, constructed the first Buddhist temple and introduced the first Tibetan script, and perhaps he actually did all of that. The temple was the Great Jo, or Jokhang, still Lhasa's holiest shrine and the repository of a Buddha statue brought from China by Princess Wencheng (Kongjo in Tibetan). Wencheng was Songsten's Chinese bride, and in her wake had come Chinese agronomists, scholars and calligraphers. The latter introduced the still largely illiterate Tibetan court to brush and ink, though the new script itself was modelled not on China's written characters but on the Indian scripts used in Kashmir and among the Newari people of Nepal. Perhaps it was Songsten's Nepali queen who urged the sending of a young Tibetan over the Himalayas to learn as many languages as he could and compose an alphabet from them. Songsten himself is said to have been the first to master the new medium and appreciate its value in administering an empire.

Though Songsten may not have been the law-making Solomon of later tradition, his example as 'the great empire-building king who was also a compassionate Buddhist ... became a model to which all subsequent Tibetan rulers aspired. Down the centuries, through to the twentieth century, the Tibetan ideal of government was a union, not a separation, of Church and State.'[11]

The ideal proved, up to a point, ideal. Taking Songsten Gampo as its founder, Himālaya's first and only home-grown empire lasted nearly 200 years and spanned, east to west from Xi'an to Afghanistan and Samarkand, over 40 degrees of longitude. But it was a loose and untidy affair, liable to contraction and neither exclusivelyTibetan nor entirely the creation of the Yarlung dynasty. Its seventh-century expansion had been the work of a subordinate clan, the Gar, whose members served successive emperors as ministers and military commanders. The troops involved were as often local allies and recruits from the mostly Turkic or Mongol peoples along the Silk Roads as Tibetan-born Tibetans. And the distances were so great that, however mobile the court and effective the emperor, their directives left much to the initiative of individual governors and commanders. Campaigning was almost continuous. There is little evidence of a *pax tibetana*. Tibetan penetration was as likely to shatter peaceful commerce along the Silk Roads as protect it.

A century and some seven kings later, the lustre of Trisong Detsen's reign (756–97) was the exception rather than the rule. Blessed in matters spiritual with the guidance of Padmasambhava and Santarakshita, Trisong found his temporal prospects greatly boosted by a favourable turn of events in China. The Tang dynasty was facing its greatest ever crisis. In 755 China's imperial troops had been routed, and the elderly emperor forced to flee his capital, by an outlawed courtier of non-Han descent called An Lushan who commanded one of the empire's frontier armies. To deal with this challenge, loyal troops were hastily recalled from far-flung garrisons in Sichuan, Gansu and Xinjiang, whereupon 'Turks, Tibetans, Arabs and others took advantage of this retraction to dismantle the empire's entire western extension'.[12]

A year later envoys from Wakhan and Shignan in the Pamirs were paying homage to the Tibetan court in 'a reassertion of Tibetan influence in the heart of Central Asia'. Content with this assurance of safe passage to what Beckwith calls Tibet's 'colonial empire' in what is now Uzbekistan, Tibetan arms were first directed towards China itself and the north-western approaches to Chang'an (Xian). But by the time the Tibetans entered Chang'an, albeit briefly, in 763, An Lushan was long dead and his rebellion had collapsed. A Tang emperor had been restored, and the task of reclaiming the oasis cities along the Silk Roads was being entrusted to the Celestial Empire's latest allies, the formidable Uighur Turks.

It was therefore in competition with the Uighurs and, from the west, another new contender, the Arab-led forces of an advancing Islam, that the great conquests of Trisong's reign were made. The oasis cities fell to the Tibetans like dominoes – Hami (Yizhou) in 781, Dunhuang (Zhazhou) in 783 and Kucha (Qucha) and Khotan (Hotan) in the early 790s. Beckwith likens the Tibetan expansion into Central Asia to the Arab eruption into West Asia; both were unexpected and they were almost simultaneous. In a couple of decades the Silk Roads north and south of the Takla Makan were relieved of their Chinese commanderies and garrisons, and 'the whole region effectively removed from direct Chinese control'. Neither the Song nor the Ming would recreate the Tang's mastery over the western desert. 'It was to remain free of China until after the Manchu conquest nearly a thousand years later.'[13]

Trisong Detsen seems not to have taken to the battlefield in person. His achievement lay in keeping Tibet and its clans sufficiently united to undertake these conquests and in ordering and exploiting the resultant empire. There was nothing contradictory in the idea of a Buddhist emperor. A 'wheel-turning' *Chakravartin*, or universal ruler, was expected to support the *sangha*, the monastic community, which in turn upheld the sovereign. Trisong would have been aware of India's imperial precedents, from Emperor Ashoka of the Maurya dynasty who distributed the Buddha's relics in the third century BCE to Emperor Kanishka of the Kushana dynasty who convened the fourth Buddhist council in the second century CE.

Trisong may even have seen military expansion, however sanguinary, as actually serving the cause of Tibetan Buddhism; for the Central Asia he brought under Tibetan control was already part of the wider Buddhist world. Its thriving monastic communities and its seemingly inexhaustible supply of sacred texts, plus the learned monks to expound them, were just what neophyte Tibet needed.

Mahayana teachings had reached the oasis cities along the Silk Roads, and then China itself, all of 500 years before the 'conversion' of Songsten Gampo, let alone that of Trisong Detsen. They had come from India by way of Afghanistan and the Pamirs of Swat and the Hindu Kush passes. The first archaeological evidence of Buddhism in Himālaya is to be found not in Tibet but in rock-cut inscriptions along this trail of dissemination through Baltistan, Gilgit and Hunza. When Trisong Detsen organised a new programme of translations in a massively ambitious project rated as 'among the greatest achievements of the medieval world', the texts from which the translators worked were as often of Chinese or Central Asian provenance as of Indian.[14]

So too were the monks engaged in the work; and therein lay trouble. In the course of time the Mahayanist teachings favoured in China had diverged from those current in India. In Tang China, and probably under the influence of contemplative Daoism, a new school of thought within orthodox Mahayana offered an alternative to the hard slog towards nirvana by way of endless rebirths and the accumulation of merit. Known as Chan (and in Japan as Zen), the new school emphasised a 'sudden path', short-circuiting the 'gradual path' of good works and ethical conduct with the promise of an instantaneous realisation, attained through meditation, that Buddhahood is within us. Neither school precluded the other; the Indian also enjoined meditation, and the Chinese variant also applauded good works. But 'Chan/Zen and the art of minimal engagement' was obviously attractive and was passionately promoted.

The whole controversy is said to have become so heated that blood was shed. More in keeping with Buddhist tradition was the idea of holding a formal debate, as duly convened at Trisong's new Samye monastery in c.792 CE. The debate is said to have been presided over by the sovereign himself. A Chinese monk called Moheyan (that is, Mahayana, the name being an assertion of his school's orthodoxy) led off for the 'sudden path'; Kamalashila, an Indian monk and disciple of Santarakshita, put the case for the 'long slog'.

As is the way with Tibetan history, the outcome of the famous Samye debate would be disputed. 'According to the Chinese tradition, the king [Trisong Detsen] would seem to have decided in favour of the Chinese,' says Tucci.[15] Later Tibetan tradition, on the other hand, is unanimous in awarding victory to Kamalashila. 'The Chinese lost and were forced to leave the country, in no gentle fashion,' says Rolf Stein, Tucci's counterpart at the Paris Collège de France. 'The king proclaimed that only the doctrine that had been maintained by the Indians was to be recognised in Tibet.'[16]

Ever after, the texts and teachers sought by the Tibetans would indeed be overwhelmingly from India. Buddhism would remain true to its Indian roots throughout Himālaya, and it was to India that Tibet would look for the inspiration and direction behind the second wave of Buddhism's diffusion in Tibet.

More immediately the country's history was rendered as obscure as ever by the use of two competing chronologies and a shortage of references in China's dynastic histories. The Samye debate looks to have been one of Trisong Detsen's last public engagements. His immediate successors benefited from an 822 treaty with China, while they let slip his conquests and took up his pious initiatives rather too enthusiastically. 'More of Tibet's wealth was spent on Buddhist projects than ever before.' The revision and translation of texts went on apace. Lhasa's streets filled with monks, 'new monasteries were under construction throughout the empire, and the cost of maintaining them fell upon the local population'.[17]

A backlash was probably inevitable. The culprit is traditionally held to have been Langdarma (Darma Wudunten), a boorish libertine who, on succeeding to the throne after the 841 murder of his saintly brother, outlawed Buddhism, expropriated the monasteries and dispersed the monks. A year later Langdarma himself was murdered, supposedly by the abbot of Samye. His sons then disputed the succession. Into all this strife between Buddhist and anti-Buddhist factions abetted by a variety of royal contenders there rode Tibet's ever restless clan leaders and landowning chiefs. Tucci dismisses the ensuing chaos with a mighty huff.

> The political unity of Tibet began to crumble; an era of decay
> and dissolution commenced. The state of anarchy lasted for two
> centuries, and led to Tibet being split up into a number of little
> states. The intellectual contacts with India were broken. Even the
> royal tombs were desecrated.[18]

It was the end of the Tibetan empire, the end of the Yarlung kings and the end of Tibet's first engagement with Buddhism.

How to account for the peculiarities of Himalayan Buddhism has taxed the Western mind no end. It has been supposed, for instance, that the physical environment may in some way be responsible. Himālaya's terrain is so awesome that it positively enjoins an imaginative, even credulous, mindset. That was surely the case with *bon*, many of whose mountain cults and airborne sprites were accommodated by Lamaism. As already noted, Tucci thought it was true of Himalayan Buddhism as a whole.

> One can understand why this people lives so intensely in its religion, almost entranced and lost in visions of its own. The landscape is the natural background for Lamaism, for its rites, for its demons; all is gigantic and mysterious, infinite and sad.[19]

Where a sudden shift of cloud can turn the ethereal into the savage, anything can happen. Mountains invite communion with the supernatural; wildernesses encourage a contemplative state of mind; isolation makes for introspection. The play of light stimulates metaphysical speculation, and the bare rock seems ready-made for the anchorite. Lack of oxygen could induce a light-headed euphoria even in impetuous imperialists like Younghusband.

It may not always have been so. The climate of Gugé in western Tibet is believed once to have been as moist as that of eastern Tibet. Nor is it true of Himālaya as a whole. Favoured peripheral valleys like those of Kashmir and Katmandu certainly invited early settlement. In Chitral, Gilgit, Himachal, Uttarkhund, Sikkim and Bhutan, corridors of trade and cultivation snaked invitingly into the mountains. But elsewhere, across the length and breadth of Ladakh and Tibet, an average precipitation of just 46 centimetres a year and an altitude of over 4,000 metres asl combined with the harsh terrain to make life as testing as possible.

Arable farmers struggled with a growing season of barely three months, while the scant grazing obliged stockbreeders to adopt nomadic pastoralism. A monotonous diet and minimal domestic comforts offered no compensation. Home for the herdsman was a yak-hide tent thick with yak-dung smoke. Diseases of the eye were among the commonest, as were hearth-side burns. Scalding was rarer only because at 4,500 metres asl water boils at 16 degrees Celsius lower than at sea level. Without a pressure cooker, rice won't cook. Barley, the staple, has to be roasted and ground into flour before becoming digestible as *tsampa*. Clothing was traditionally of woollen homespun and was shared with lice and fleas which could on no account be killed; hygiene was unheard of, soap was for foreigners, and ablutions of even the most perfunctory nature were reckoned a risk to health.

> One must remember [wrote Snellgrove in the 1950s] that the Tibetans are still living (or were until a few years ago) in a world vastly different from our twentieth century. Their world is still flat. Virtue practised for its own sake will still gain its certain reward. Learning and sanctity are still worthy of universal regard ... In short the Tibetans still believe in their religion.[20]

Seeing Tibetan Buddhism at first hand, members of the Younghusband expedition had been more critical and less inclined to make allowance for the challenging terrain and the medieval outlook. Dr Waddell, the expedition's most knowledgeable Tibetanist as well as its medical superintendent, marvelled how Padmasambhava was revered 'notwithstanding his grotesque charlatanism and uncelibate life'. The Lamaist liturgy, according to Waddell, was 'a silly mummery of unmeaning jargon and magic circles'.[21] Prayer was so repetitious it could be left to wind-fluttered flags, revolving prayer wheels and a single muffled mantra of mind-numbing inconsequence. In Tibet the Buddha himself had somehow become one of many Buddhas, most of whom could be depicted as hideous monsters. Superstition and witchcraft had eclipsed meditation and compassion. Tibetan Buddhism was just 'a priestly mixture of ... mysticism, magic and Indo-Tibetan demonolatry overlaid by a thin varnish of Mahayana Buddhism'.[22]

Perceval Landon, special correspondent of *The Times* with the Younghusband expedition, had been equally dismissive, though from a different perspective. 'There is ... a belief that the Buddhism of Tibet is a lawful descendant of the Buddhism which the Master preached beneath the peepul [trees] of Bengal,' wrote Landon. Obviously, over time, this pristine Buddhism had become 'encrusted in ritual and adorned by traditions'. The same could be said of early Christianity, whose 'flame

Some members of the 1904 Younghusband expedition had been in South Africa for the Boer War. Younghusband himself had monitored the Jameson Raid, and Perceval Landon, the Times *correspondent in Tibet (here standing), had edited a local newspaper with Rudyard Kipling (seated on table)*

yet burns as steadily today as ever it did'. But the flame of Buddhism in twentieth-century Tibet was barely guttering. It no longer had the faintest resemblance to 'the plain austere creed which Gautama preached'.

> It is doubtful if the great Founder of Buddhism would recognise in its forms or formulae any trace of the purity and sobriety of his own high creed ... Under a thin mask of names and personifications suggested by the records of the Master, a system of devil-worship pure and simple reigns in Tibet; the monkish communities spare no effort to establish their predominance by fostering the slavish terror which is the whole attitude towards religion of the ignorant classes.[23]

It was a 'religious tyranny'. Tibet was in thrall to its monks. Meeting their expectations in terms of produce and labour, and paying for their expensive intercessions, were all that stood between the toiling masses and a host of demon-induced misfortunes. Spread over endlessly reprised existences in ever more wretched incarnations, these might be relieved only by spells of detention in one of the mandala's hellish realms. 'Not even the darkest days of the Papal States, nor the most bigoted years of Puritan rule in New England, not the intolerance of German Calvinism, not Islam itself can afford an example of such utter domination.'

So, according to Landon and most other foreign observers, in Himalayan Buddhism the fundamentally humane and compassionate teachings of the Buddha had been 'corrupted', 'distorted' and 'debased'. And for the resulting 'travesty' it was not the terrain or the lifestyle that was responsible; it was the monastic orders.

There were, though, some obvious objections to this sweeping diagnosis. For one thing, it overlooked the fact that the monks and the masses were one and the same. About a seventh of all Tibetan males were in holy orders. They were not unwillingly inducted, they were spared the annual obligation of public labour, their families were proud to see them go and they could usually return home if they wished. Moreover monasticism was not peculiar to Himalayan Buddhism. Along with the Buddha and his teaching (*dharma*), the monastic community (*sangha*) was one of the three fundamental 'refuges', or 'gems', of all shades of Buddhism. Its institution dated back almost to the time of the Buddha, and Buddhism was inconceivable without it.

But that was long ago. The Buddha had lived in the fifth century BCE. Not till eleven centuries later, a time lapse roughly equivalent to that between Christendom's first Crusades and the latest Gulf War, did the Master's teachings reach Tibet. In so long an interlude, just as Christendom's teachings were moulded by schisms, reformations and counter-reformations, so were Buddhadom's teachings. And if the result in Tibet was not 'what the Buddha had preached under the peepul trees of Bengal', the supposed 'corruption' was down to Buddhism's long maturation in India, not to its belated exposure to Himālaya's harsh terrain and the mysterious fantasies of *bon*.

In India 'one must conceive of Buddhism as an active living tradition … continuing to enrich and adapt itself for 1500 years or more', says Snellgrove. It did so in tandem with India's many non-Buddhist cults and doctrines which are now recognised as Hindu. Hindus and Buddhists were 'one and the same people'. They inhabited the same north Indian cultural milieu and their beliefs were 'analogous and sometimes identical'. There was much cross-fertilisation. Buddha figures became as numerous and influential as Hindu deities. For devotional focus, individual identities were accorded to the myriad pre-incarnations (bodhisattva) of the historical Buddha and were even bestowed on the attributes and epithets with which he was credited. Both belief systems awarded their deities fearsome aspects, both paired them with female consorts, and both equipped them with significant accoutrements, plus the limbs to wield them.

Both were also alert to the popular appeal of a class of renunciates well represented in post-500 CE Bengal whose ritualistic practices and meditational aids would be embodied in tantric yoga. Some of these practices, especially those of a sexual or carnal nature, were considered distasteful by many Buddhist scholars and would appal later observers like those of the Younghusband expedition. But, as affording another shortcut towards enlightenment, this tantric 'mantra-vehicle' (*mantrayana*) was adopted by Buddhists as well as non-Buddhists. It entered mainstream Mahayana at a time when Buddhism in India was already in decline but before the last great monastic establishments there were destroyed, and their communities dispersed, by the Muslim invasions of Bihar and Bengal in the late twelfth century CE.

Even then Nepal and Kashmir were spared. Kashmir's Buddhist heritage was merely reprieved; it would be all but obliterated in the fourteenth century. Katmandu fared slightly better. The pagoda style and iconography of the Katmandu valley's surviving Buddhist shrines and temples still 'represents the India of a thousand years ago'. It was, however, the last flowering of Buddhism south of the Himalayas.

Hence it is the Buddhism of Tibet, 'at once the purest and most debased of Buddhist traditions', which Snellgrove recognises as the 'direct descendant of Indian Buddhism'.[24] Vajrayana, far from being an aberrant development, should be seen as proof of Mahayana being a still living and evolving tradition. Thanks to the Herculean feats of acquisition and translation performed under Tibetan patronage, Tibet would remain the main repository for knowledge of early Buddhism and its texts – until, that is, the Tibetan diaspora of the late twentieth century.

Nearing Lhasa on 2 August 1904, leading members of the Younghusband expedition had pushed ahead to claim the first sighting of the 'Tibetan

Vatican'. There was no danger. The city was reported to be undefended. Scarcely a shot had been fired since the expedition crossed the Karo La east of Gyantse. But Edmund Candler, the maverick *Daily Mail* correspondent who'd been wounded at the Hot Springs encounter, chose to take the day off. He wanted to write up his notes in peace. The byways around the township of Nethang looked inviting. He didn't let slip that the excursion was also by way of a pilgrimage.

It was a fine morning, 'almost tropical' he thought, as he rode amid a swarm of 'hoopoes, magpies and huge black ravens'. The fields were bisected by clear-running streams bordered with flowers, and 'in a grove of poplars where doves were singing all day, I found Atisa's tomb'. The tomb — or more accurately the stupa/chorten — was within a barn-like building that smelled of hay and was mercifully devoid of the monks' 'spurious apparatus of terrorism' — garish frescoes of ghouls and she-devils and 'hideous grinning devil-masks'. It was the cleanest and simplest place of religion Candler had seen in Tibet. But then, he mused, 'this was the resting-place of the Reformer, the true son of Buddha, who came over the Himalayas to preach a religion of love and mercy.'

> I entered the building out of the glare of the sun, expecting nothing but the usual monsters and abortions ... But as the tomb gradually assumed shape in the dim light, I knew that there was someone here, a priest or a community, who understood Atisa, who knew what he would have wished his last resting place to be; or perhaps the good old monk had left a will or spoken a plain word that had been handed down and remembered these thousand years — that there must be no gods or demons by his tomb, nothing abnormal, no pretentiousness of any kind. If his teaching had lived, how simple and honest and different Tibet would be today.[25]

Presumably Candler had been picking the brains of the knowledgeable Dr Waddell on the subject of Atisa. Waddell too calls him 'the reformer', indeed the instigator of 'the Lamaist Reformation'. And it is that reformation which history recognises as the great achievement of the conclusive 'second wave' of Buddhist diffusion in Tibet.

In the tenth century, during the chaos that had engulfed the crumbling realm of the later Yarlung kings, monasteries in the central provinces of U (dbUs) and Tsang (gTsang) had been sacked or abandoned, their monks dispersed for want of patronage. One group had repaired to Amdo in the far north-east, from where their disciples returned to central Tibet in the early eleventh century. Soon after, in the extreme south-west, a distant descendant of the Yarlung kings, having conquered and reunited the provinces of Ladakh, Zanskar and Purang, formed the kingdom of Gugé. It corresponded to the supposed heartland of the erstwhile Zhang Zhung and prefigured the later Gugé kingdom where the Jesuit father António d'Andrada would be so well received.

Under the first of these Gugé dynasties Buddhism had been reinstated and the task of attracting Indian masters and authoritative texts began anew. A promising scholar called Rinchen Zangpo was one of several despatched to the monastic communities in Kashmir. On his return with a baggage train laden with board-bound books of Sanskrit text, Rinchen Zangpo founded new temples, including that at the today spectacularly crumbling Toling, and another, still standing, at Tabo (Dabo) in Spiti (in Himachal). He organised their consecration, oversaw their embellishment with frescoes and bronzes and above all immersed himself in the work of translation.

Universally hailed as 'The Translator' (*lotsava*), Rinchen Zangpo laboured to a ripe old age and at some point seems to have visited the great monastic complex of Vikramashila in eastern Bihar. If so, it may have been he who suggested luring away Vikramashila's leading light, an Indian monk who had previously instructed the Buddhist communities in Sumatra and was renowned for his 'philosophical acuity, mastery of tantric practices, and personal qualities of humility and compassion' – in other words the saintly fifty-eight-year-old known as Atisa. Travelling by way of Nepal, Atisa arrived in Toling in 1042 to a royal welcome, stayed there for three years and then removed to Lhasa, where he joined with the ordained returnees from Amdo and made Nethang his eventual home. He died there in 1056.

And 'never before in the thousand years since the good monk was laid to rest at Nethang', wrote Candler, 'had a white man entered his shrine'. Even the monks who had charge of the shrine seemed ignorant of his achievements. 'Merely a name to a few dry-as-dust pedants', supposed Candler as he bent over his notebook in a courtyard bordered with hollyhocks and snapdragons. 'Everything human he did is forgotten ... Here the unsavoury guardians of Atisa's tomb watch me as I write, and wonder what on earth I am doing among them, and what spell or mantra I am inscribing in the little black book that shuts so tightly with a clasp.'[26]

In central Tibet as in Gugé, Atisha used his intimate knowledge of Buddhism's Sanskrit texts and commentaries to contextualise and sometimes marginalise the more outrageous tendencies of tantric practice. Tantra's practitioners might not be monks at all; they were often individual miracle-workers who promised sensational results in the here and now as well as a 'shortcut' to enlightenment, like that promised by Chan/Zen. In his best-known composition, *The Lamp for the Path to Enlightenment*, Atisha countered these tendencies by providing the uneducated with simple injunctions on how to follow 'the gradual path'. On more tutored novices he urged the importance of following the advice of a spiritual guide or guru. 'Just be kind,' he told the simple layman; 'the teacher is more important than the text', he told the aspiring novice. Monks must be celibate. He had a special reverence for Tara, the female aspect of bodhisattva Avalokiteswara and handmaiden of Boundless Light in the creational story of the monkey-king.

Like other leading lights in the intellectual ferment of the second diffusion, Atisa insisted on 'a solid tradition of teaching'. The authority of a doctrine was to be 'guaranteed through direct transmission from master to student'.[27] Hence the new schools resulting from these ministries depended on an apostolic succession from their original founders. Such doctrinal linkages were then, in the course of time, 'institutionalised into monastic organisations'. A monastery's prestigious teachers attracted ever more followers, and its lay supporters, who were often scions of a powerful local clan, provided the endowments which would augment the monastery's spiritual authority with economic might and political clout. The scene was set for Tibetan history to 'alter its course and become the history of religious groups and sects'.[28]

Tucci provides a listing of these eleventh- and twelfth-century monastic foundations – 'Puton was founded in 1040, Sakya in 1073, Thil in 1158, Drigung in 1179' and so on. At about the same time Marpa, a cantankerous contemporary of Atisa, shuttled between Tibet and Nepal to bring back the Bengali tradition of mystical verse. This

was perfected and popularised by Marpa's devoted disciple, Milarepa –
he who worsted his *bonpo* rival in the ascent of Mount Kailas. The
followers of Marpa and Milarepa formed the Kagyu (Kagyu-pa) or
'whispered transmission' school, one of the four main orders of Tibetan
Buddhism, which itself generated a host of subsects, among them that
of the originally Kham-based Karma-pa.

When Chinggis Khan and his Mongol hordes burst upon the scene
in the 1220s, it was the Karma-pa who vied with the Sakya-pa (those
of the Sakya monastery) for the support and influence on offer under
the new dispensation. Summoned to the Mongol court, a Karma-pa
endeavoured to convert the future Kublai Khan in 1255 and actually
succeeded with the emperor Möngke. By the time Möngke died five
years later, Tibetan Buddhism was in high favour at the Mongol court.
Kublai Khan, Möngke's brother and now the emperor (of China,
Manchuria, Mongolia and much more beside), inherited this preference
for Buddhism but followed the example of their uncle in accepting the
authority of the Sakya-pa. Mongol favour thus proved a mixed blessing.
Tibetan forces, though defeated more than once, were spared the mass
slaughter meted out by the Mongols elsewhere. On the other hand,
competition among Chinggis Khan's numerous descendants led each
to favour a different monastic school, so embroiling the monasteries
in the Mongols' interminable succession disputes and militarising the
monks.

The Sakya-pa emerged from these struggles pre-eminent. As
patriarchs and wonder-workers at the Mongol (or Yuan) court, and
as governors exercising what Stein calls the Mongols' 'fairly loose
suzerainty over Tibet', they held their own against rival monastic
houses from 1270 until 1330. But as Mongol rule in China began
limping towards extinction at the hands of the first Ming emperor, so
their Sakya-pa surrogates in Tibet fell prey to twenty years of strife in
which monastic allegiance determined loyalties and the combatants
were often monks.

The eclipse of the Sakya-pa meant the Karma-pa again basked
in Mongol favour. The third Karma-pa Lama officiated at the 1333
enthronement of Toghon Temur, the last Mongol emperor of China,
and founded a Tibetan temple in Dadu, the Mongol capital that
would become Beijing. In 1358 his role as tantric instructor to the
emperor and palace tutor to the imperial progeny passed to the fourth

Karma-pa – or rather to his reincarnate self; for 'the Karma-pa Lamas have the distinction of being the longest line of reincarnate lamas in Tibet' and may have been the first to 'consciously claim' this *tulku* (incarnate) status.[29] Derived from the belief that the early kings were Buddhist masters, indeed bodhisattva, who had opted for a 'corporeal existence', the idea of reincarnating lamas is peculiar to Vajrayana and was soon adopted by schools other than the Karma-pa.

The ninth in the Karma-pa succession founded the Rumtek monastery in Sikkim, and here the sixteenth took up residence after fleeing Tibet in 1959. But while continuity is the great advantage of reincarnation, controversy is its great corollary. On the death of the sixteenth in a New York hospital in 1981, two candidates with such confusingly similar names that they are best known by their initials were identified as successors. Each attracted powerful support, each was duly consecrated and each has since been more successful in discrediting himself than in discrediting his rival. Press coverage of their monkish followers warring over the possession of Rumtek has won neither of them any sympathy. The claimant known as OTD has been recognised by the Dalai Lama and also Beijing (which some regard as suspicious). He made his 'escape' from Tibet in 1999 but has since acquired Dominican Republic citizenship and spends most of his time in the US. TTD, on the other hand, left Tibet for India in 1994 and is still there; but in 2017 he got married, which, given the insistence of Atisa and others on monkish celibacy, would appear to rule him out. He now has a son, while OTD is said just to have 'many girlfriends'. Scandals, mutual recriminations and an occasional attempt at rapprochement dog their days. Meanwhile Rumtek remains under lock and key with an armed guard of Indian troops to pre-empt any further monastic fisticuffs.

A 'good old monk' remembered only by 'dry-as-dust pedants' would not have approved of these goings-on. Nor, surely, would Edmund Candler. Like the ghouls and she-devils of the 'tawdry' frescoes, they were the sort of thing that gave Tibetan Buddhism a bad name. But Candler, despite his high regard for Atisa, had done him a disservice: in suggesting Atisa's name was forgotten he had hopelessly underestimated the old monk's legacy.

The Kadam-pa school, founded by Dromton, Atisa's leading disciple, to preserve his teachings, survived a century of 'loose' Mongol suzerainty over Tibet (c.1240–c.1350) and a takeover bid by the Sakya-pa, only to rise again to an unheard-of prominence. In the early fifteenth century it emerged as the 'new Kadam-pa', otherwise the Ganden-pa or more commonly the Gelug-pa (Geluk-pa), the 'official' yellow-hatted order that would be headed by the Dalai and Panchen Lamas and is nowadays much the largest of all the monastic orders.

This transformation was the achievement of the most intellectually brilliant of the monastic founding fathers, a man from Tsongkha in Amdo known inevitably as Tsongkha-pa (1357–1419). As a child Tsongkha-pa's promise had supposedly been acknowledged by the fourth Karma-pa; and as a teenager he found his way to central Tibet to continue his studies, initially under Sakya-pa tutelage. His formidable debating skills and his appetite for the niceties of Buddhist exegesis attracted followers and led him to other monasteries. One was the Kadam-pa monastery of Reting (Radreng) north of Lhasa founded by Dromton. He went into retreat there and in 1403, with a portrait of Atisa looking over his shoulder, he composed for his 'new Kadam-pa' order a great two-part work of synthesis. It was to become what Tucci calls 'a *Summa* of Lamaist doctrine as [Tsongkha-pa] saw it' and 'a kind of bible ... to which [his followers] constantly refer in the form of excerpts, explanations and commentaries'.[30] It included little in the way of ritual or doctrinal innovation. Like Atisa, Tsongkha-pa did not exclude tantric practices and rituals, and like Atisa 'he insisted once more on the need for monastic discipline and the gradual path (of morality, etc.) for the generality of men and even as a preliminary to liberation'.[31]

Well organised and enjoying the patronage of one of the great landowning clans, the new school was an instant success. Gelug-pa monasteries, all of which would grow to colossal dimensions, were founded in quick succession – Ganden (1409), Drepung (1416), Sera (1419), Chamdo (1436), Tashilunpo (1447). But, throughout the fifteenth century, rivalry with the Karma-pa and the other schools continued, as did strife between the central provinces of U (including Lhasa) and Tsang (including Xigaze/Shigatse with Tashilunpo). 'There were many wars' (Stein), indeed 'more than a century of uninterrupted guerrilla warfare' (Tucci).[32]

The great Gelugpa monastery of Sera in the outskirts of Lhasa was founded by Tsongkhapa in 1419. It was destroyed by the Chinese in 1959 but has been partially rebuilt, while the monastic community was relocated to near Mysore in Karnataka

As with all Tibetan Buddhism's founding fathers, Tsongkha-pa's reputation owed much to the calibre and devotion of his disciples. Prominent among these were the abbots of the new monasteries, and it was thus that, in 1578, Abbot Sonam Gyatso of Drepung monastery travelled to Mongolia to meet with Altan Khan, the leader of one of the confederations into which the now China-less Mongols had dissolved. As a descendant of Kublai Khan, Altan Khan had already raided China with a view to re-establishing Mongol rule there. He now hoped to enhance his prospects by reprising the patron–priest relationship forged by Kublai 200 years earlier.

The plan succeeded. Altan Khan had the satisfaction of retaining Abbot Sonam Gyatso's highly auspicious presence and of establishing, through the khan's patronage of the Gelug-pa, what could be construed as a nominal suzerainty like that which Kublai had exercised through the Sakya-pa. The Tibetans too were gratified. The Gelug-pa now had the assurance of a powerful supporter in their domestic vendettas. Tibetan Buddhism welcomed the khan's prohibition of animal sacrifice and embraced the opportunity for Buddhist proselytisation throughout

Mongolia. Abbot Sonam Gyatso was personally rewarded with gifts and favours, among them a Mongolian title – that of Dalai Lama. The original title was actually more of a mouthful, but 'it was quickly shortened by the Tibetans' till it was no more than a rendering of the abbot's Tibetan honorific of 'gyatso', 'dalai' being its Mongol translation and 'ocean' being the English for both of them.[33]

Although he was the first to hold the title, it soon transpired that Sonam Gyatso was not in fact the first Dalai Lama; he was the third. To give the title a prestigious pedigree, and because the abbot was already credited with two pre-incarnations, his recognition as Dalai Lama was, as it were, backdated. Thus two previous abbots were posthumously elevated to the same dignity, so taking the first manifestation of the Dalai Lama back to the 1390s and opening the possibility of some association with the great Tsongkha-pa and the founding of the Gelug-pa.

And there was another twist, possibly foreseen, possibly not. When in 1588 the third Dalai Lama (Abbot Sonam Gyatso) 'entered nirvana', two candidates for the succession were identified. One was a Tibetan, the other a Mongol. The Tibetan looked well qualified, but politics argued for the Mongolian candidate. He would cement the Mongol–Gelug-pa alliance, he could be given a Tibetan name – and was – and he happened to be the grandson of Altan Khan. That settled it. After a prolonged stalemate, the thirteen-year-old Mongol now known as Yonten Gyatso was escorted to Lhasa and officially recognised as the fourth Dalai Lama.

His reign was brief and inglorious. Still in his twenties, the only ever non-Tibetan Dalai Lama died in 1617 amid mounting opposition to the Mongol–Gelug-pa alignment. This came from another Tibeto-Mongol alignment, that between the Karma-pa order with their Xigaze-based patron, the ruler of Tsang, and some Mongol allies of their own choosing. 'So Tibet implicated foreign powers in her own struggles and was accordingly involved in the rivalries that were breaking down Mongol unity.'[34]

In and around Lhasa, Mongols now threatened to outnumber monks. Recognition of the next Dalai Lama was stalled and fighting looked inevitable. To add to the confusion, Ming rule in China was also breaking down. From the Celestial Empire's far north-east, the Jurchen, a Tungusic people with shaven forecrowns, long pigtails and a history of China incursions, were bearing down on Beijing. They called themselves Manchu (hence their homeland became known as Manchuria) and, once installed in the Forbidden City, they would adopt the dynastic title of Qing (Ch'ing).

If the Gelug-pa were to maintain their primacy in such turbulent times, they needed an exceptionally gifted leader. The 'Great Fifth', as Tibetans call him, was just that. As a child, this Lobsang Gyatso displayed such precocity that his failure to identify any of his predecessor's personal effects was not enough to disqualify him. And in a notable act of reconciliation, the honour of ordaining him as a monk went to the abbot of Tashilunpo who, with some dextrous backdating like that by which Sonam Gyatso had become both the first and the third Dalai Lama, was now acknowledged as both the first and the fourth Panchen Lama. 'From that time onwards [whoever was] the elder of the two hierarchs, the Dalai or the Panchen Lama, always initiated the other.'[35]

The Great Fifth's greatest achievement is reckoned a reunification of Tibet that would last until the mid-twentieth century. With the assertion of Lhasa's authority in Amdo, Kham, Bhutan and what had become Ladakhi-ruled Gugé, more of Himālaya was brought under unitary rule than at any time since Trisong Detsen 900 years earlier. The feat was suitably symbolised by the construction, begun in 1645 on a site hallowed by association with Songsten Gampo, of the cascade of quintessentially Tibetan architecture that is the Potala palace (or fortress cum temple cum seat of government). Flanked at no great distance by the Sera, Drepung and Ganden monasteries, the Potala completed a monumental ensemble that was as much about Gelug-pa ascendancy as about Tibetan integrity. And just as the defeat of the Great Fifth's domestic and doctrinal foes owed much to his regent and then minister, so the political triumphs of his reign were once again down to the Mongols.

They were different Mongols, though. Gushi (Gusri) Khan and his Khoshut (Qoshot) troopers were not only recent Buddhist converts but avid Gelug-pa loyalists. From Kokonor in Amdo they swept into Lhasa in 1637 to revive the patron–priest relationship and thrust their services on the young fifth Dalai Lama. Four years later, by when they had completed the bloody conquest of dissident pockets in Kham, they returned to central Tibet to rout the Karma-pa and his Tsang patron. Despite the Dalai Lama's supposed doctrinal neutrality, Kagyu and Sakya monasteries were converted into Gelug-pa houses; writings critical of Tsongkha-pa's philosophy were banned and destroyed; monks not slaughtered during these operations were imprisoned or dispersed. 'And so it was', writes Tibet's latest historian Sam van Schaik with just a hint of irony, 'that, in 1642, the fifth Dalai Lama, Lobsang Gyatso, became the ruler of Tibet.'[36]

It was rulership of Tibet with Mongol strings attached. The Dalai Lama might rule a reunified realm but Gushi Khan, backed by his fearsome 'bannermen', reigned over it as *dharma* king'. His days, however, were numbered. By mid-century there were emerging new contenders for the spiritual sanction and territorial suzerainty afforded by the reincarnate leadership of the now pre-eminent Gelug-pa. One such interested party was China's new Manchu/Qing dynasty. In 1654 the Dalai Lama accepted an invitation to visit Beijing. The meeting of the ten-year-old emperor and the now middle-aged Great Fifth was conducted as between two sovereign rulers. *Pace* the revisionist school of Chinese history, it scarcely signified the Dalai Lama's submission either to China, most of which the Qing had yet to conquer, or to the Han Chinese, who were still in awe of their Manchu overlords.

These things were imminent but had to wait on the outcome of another Mongol intervention. The Great Fifth attained nirvana in 1682, yet supposedly lived on in a meditational trance for another fifteen years. The sixth Dalai Lama was therefore a dissolute teenager by the time he was called on to succeed. Refusing ordination, he dabbled in verse and took up archery. Questions were asked about his suitability as a spiritual leader and about the legitimacy of his selection. Lajang Khan, Gusri Khan's grandson and now the Mongol leader, even consulted the Qing in the person of the great Kangxi emperor.

The emperor offered no objection to the youthful misfit in the Potala being deposed, asking only that he be sent to Beijing. Acting on this suggestion, Lajang Khan and his Mongols kidnapped the fun-loving sixth. He escaped, but then surrendered when Mongol forces stormed Drepung monastery. Under escort and far from well, he was hustled north and east. It was November 1706 when, somewhere in the vicinity of Xining, the unlikely sixth Dalai Lama died, possibly from natural causes.

Back in Lhasa, the Mongol leader Lajang Khan now both reigned and ruled. He arranged the assassination of the former regent – he who had engineered the Great Fifth's delayed demise – and he touted as the real sixth Dalai Lama a twenty-five-year-old monk who may have been his own son. This was a reincarnation too many and too cynically political for the Tibetans. They rose against the Mongols and called in the Dzungars, the Mongols' inveterate enemies. Lajang Khan died in battle with these intruders, and his candidate as Dalai Lama was cast aside. A boy of the right age and the necessary accomplishments to be the seventh Dalai Lama was found in Kham, well away from Lhasa where the Dzungars had inaugurated a reign of terror.

The Kangxi emperor in Beijing needed no further invitation. Lajang Khan, his surrogate in Tibet, was dead, the Mongols had been dispersed and the Dzungars were the Qing's *bêtes noires*. Scooping up the young seventh Dalai Lama, in 1720 a Qing army marched on the Tibetan capital. It encountered little resistance and was generally welcomed. The seventh Dalai Lama was enthroned in the Potala and a Qing-approved minister appointed to supervise him. 'Tibet was formally deprived of its autonomy,' says Tucci, 'a situation that had never arisen before even in the days of the Mongols.'[37]

> The walls of Lhasa were pulled down, a Chinese garrison was installed at Lhasa, and [eastern] Kham annexed to the Chinese province of Sichuan. The Chinese protectorate that was to last till the end of the Qing dynasty (1912) was established.[38]

The authority of the Dalai Lamas remained intact. So did the integrity of most of what was once Tibet. The price paid was that of Tibet's sovereignty.

12

Gold Dust and Yak Tails

And he gave him handfuls ... of precious dust, gold, silver, copper, iron etc.; [and bade him] scatter them in the Abode of Snow and they shall become treasures in the earth and shall be found in mines, and after a season these thy progeny, become men, shall subsist by this precious gold, silver, etc., and after a time they shall open these precious mines.

From the *Mani-ka-bum*[1]

Everything has its price. Just as preserving Tibet's integrity and upholding the Dalai Lama's authority meant mortgaging the country's sovereignty, so in earlier times the adoption of Buddhism had been funded by depleting the country's natural resources. These were limited but by no means negligible; some, like wool and salt, were plentiful; others were rarer but of exceptionally high value. Musk pods, for instance, obtained from the slaughter of Himalaya's scant population of musk deer, were possibly the world's most precious animal product. Until an artificial substitute was developed in the twentieth century, perfumers depended entirely on the shy musk deer's odoriferous excretions to fix and retain the beguiling properties of their preparations.

Almost as valuable – and, until a machine gun was first mounted on a Toyota pick-up, more abundant – was *tus* (pronounced 'toosh'), the underwool of the Tibetan antelope or chiru. Adulterated with *pashm* (pashmina, from the undercoat of Himalaya's domesticated goats), this shahtoosh ('royal *tus*') was reckoned the finest and softest of all known fabrics and still commands exorbitant prices. But, like the musk deer, the dainty chiru must die for its yield to be realised. Both species were nearly exterminated in the twentieth century, and despite CITES protection, they remain endangered and susceptible to poaching.

Pilgrimage apart, the great incentive to travel in Tibet had always been trade. Indeed to a Tibetan the two are inseparable. Pilgrimage without trade, they say, is like tea without butter; it's a wasted opportunity. The butter was home-made. It came from the large herds of yak grazed by *Drokpa* pastoralists in the Changthang and was traded so freely within Tibet that it constituted a form of currency. Rents paid in butter kept the lamps burning in countless monasteries and lesser domiciles, so ensuring a greasy black patina on every surface and imparting a rancid aroma to the entire population.

Butter was readily obtainable; the tea was more problematic. None was grown in Tibet, and British attempts to interest the Tibetans in the leaf teas introduced in Assam and Darjeeling got nowhere. From at least the eighth century CE Tibetans had preferred the shavings pared from blocks of compressed black tea. This so-called brick tea was produced in China specially for the Tibetan market because it travelled better, and it too sometimes served as currency. It came from Ya'an in Sichuan, and judging by contemporary accounts it sustained the biggest trade in a single commodity anywhere in Himālaya. In 1888 William Woodville Rockhill from the American legation in Beijing, having failed in an attempt to reach Lhasa, crossed from Tibet back into China at Ta-chien-lu (Dartsedo, Kangding), 'the main emporium between China and Tibet'. From here the tea caravans funnelled up through the mountains into Kham. The trail was so jammed with porters staggering under towering stacks of brick tea that Rockhill's progress against the flow was halted. He reckoned the volume of the trade at '10–13 million pounds a year' but had no idea of its value, though he presumed it greatly exceeded that of Tibet's exports by the same route of mostly musk and rhubarb.[2]

Rockhill makes no mention of gold. By the late nineteenth century Tibet's gold was evidently less abundant than it once had been. Herodotus had known of it and had famously told of how it was mined by 'gold-digging ants'. These creatures were unusually furry for insects; they lived in burrows and grew to about the size of foxes, all of which has been taken as a reference to the still plentiful marmot colonies of Himālaya's high pastures. How much gold the marmots' ankle-twisting excavations actually yielded is unknown; but, by more conventional panning or mining, gold dust and nuggets were certainly obtained in quantities sufficient to gild religious structures and add sparkle to wall paintings and texts, as Tucci discovered at Toling.

Much of the gold was exported. 'The Tibetan chronicles abound in stories showing devotees busy amassing large quantities of gold in Tibet before setting out for India in quest of books and oral teachings,' says Rolf Stein.[3] Rinchen Tsangpo, the great 'Translator', took gold dust to Kashmir to pay for the texts he hauled back to Gugé, and the Bihar monastery of Vikramashila was reconciled to Atisa's departure by a similar inducement.

Something like a price list – the 'gold standard', perhaps, for religious inductions – emerges from the life story of Marpa, guru of the teleporting Milarepa and the acknowledged inspiration, along with Milarepa, of the Kagyu-pa school. A born tearaway, the young Marpa had converted part of his birth inheritance into yaks with a view to trading cattle for some basic tantric training. When quoted '15 yak-cows for each of the four initiations of *Nairatmya*', he changed his mind. Reasoning that 'without gold you don't get religion', he returned home, exchanged the rest of his inheritance for half a kilo of gold dust and set off for India. The gold was essential, he says, since 'if for the sake of religious teachings, one goes to India without much gold, it is like drinking water from an empty bowl'.

Marpa's investment paid off. A short stay in Nepal was followed by twelve years' instruction from Naropa, the tantric master at the great university cum monastery of Nalanda in Bihar. It cost Marpa his half-kilo but empowered him with a tantric pedigree that ensured his own teachings and intercessions would be handsomely rewarded. He returned to Tibet, amassed a considerable fortune by exercising his miraculous powers and attracted a wealthy following, among whom was the owner of a gold mine. Then he headed back to India for a repeat performance, this time with 1.4 kilos of gold. Religion was proving good business. After the second trip Marpa acquired land and properties in Tibet. When a supplicant of limited means applied to him for instruction, Milarepa – for it was he – was so discouraged at the sight of the by now corpulent tantric master with his rolling acres and myriad servants that he contemplated suicide.

'Frequent references in the biographies of this period to Tibetan gold mines suggest that Tibet was awash with gold in the eleventh century,' confirms Sam van Schaik.[4] Indeed, to that resurgence of tantric Buddhism in Tibet known as the second diffusion, 'gold was vital'. And so it continued. It contributed to the growth of the monastic orders and their great monasteries in the fifteenth century, featured in the embellishment of the Potala in the seventeenth century and was no doubt responsible for attracting the last Mongol warlords in the early eighteenth century.

When in 1774 Governor Warren Hastings of the Honourable East India Company despatched a British mission to Tibet, gold was high on the list of Tibetan commodities to be investigated. Seventy years later in 1847, the Scottish traveller Dr Thomas Thomson resumed the quest for Himalayan gold when serving on a commission sent to establish an Anglo-Tibetan border. A close associate of Joseph Dalton Hooker, Thomson was also a botanist but doubled as the border commission's mineralogist when he investigated gold workings on the upper Indus. The site was just downstream from Khapalu in Baltistan, the gold was being panned rather than dug and 'the produce seemed to be very trifling': 'I purchased for a rupee (paying, I believe, a good deal more than the value) the produce in gold dust of one man's labour for three weeks. I suppose, however, he only worked occasionally.'[5] More promising, and more indisputably Tibetan, were the gold workings of Thok Jalung, an encampment about 150 kilometres north of Kailas–Manasarowar in the Changthang, Tibet's empty quarter (actually more like its empty three-quarters). In 1867 Nain Singh, the first of the Survey of India's intrepid pundits, was directed to locate this impossibly remote spot and report back. He would not disappoint. Nor would Thok Jalung, although its mines still took him by surprise.

Crossing a wind-blasted desert in the middle of nowhere, Nain Singh found his solitary pacing being interrupted by an eerie singing. It was more a chant than a song, with 'a great number of voices singing together' and yet more providing a chorus; and it was coming from under the ground. Evidently the miners, like the marmots, sheltered from the icy wind by living in their diggings. Their tents were pitched in pits 2 metres below the surface and, this being August, there were about 300 of them. In winter, Nain Singh was told, there would be twice as many because the work was safer when the ground was frozen. The soil was being dug with long-handled spades and panned in a wide trench 8 metres deep through which a stream was directed. In total, the workings extended to 'over a mile [1.6 kilometres]'. Their yield was described as 'large' and the finds as 'occasionally very heavy'. One nugget seen by Nain Singh weighed nearly a kilo. For gold dust the traders from Shigatse who were camped nearby were paying the equivalent of 30 rupees an ounce (28 grams).[6]

Nain Singh heard of other workings in the vicinity. Most were abandoned, but north of Hedin's Trans-Himālaya the auriferous sands were said to extend east as far as Lhasa. Further afield, in 1903 Cecil Rawling met a nomadic couple in the northern Changthang near what Deasy had called the Antelope Plains. The nomads seemed to be acting as purveyors of meat and dairy produce to the local gold miners, whose excavations were extensive and still being worked. They appeared to be profitable, 'but all attempts to ascertain the amount of gold extracted were fruitless'.[7] Not far from Manasarowar, more gold mines were reported by Swami Pranavananda – who else? – but here operations had been 'discontinued when small-pox broke out among the miners: a blight attributed to the wrath of the presiding deity of the mines'.[8] This seems to have been in the 1930s. Then, as now, the gods did not take kindly to excavations unsanctioned by the religious authorities.

'Xizang', the preferred name for today's Tibet Autonomous Region, translates as the 'treasure-house of the West'. The name is apt. The mineral treasure of the Tibetan plateau is now known to include not only gold but silver, lead, zinc and, more importantly, large deposits of copper, chromium, lithium and uranium. Under Chinese direction, the first industrial mining got under way in the 1990s. Three thousand potential sites were identified with a possible yield of 80 million tonnes of copper, 2,000 tonnes of gold and 30 million tonnes of lead and zinc.

By 2013 a map based on satellite imagery and obtained by the Canadian environmentalist Michael Buckley marked about a hundred sites where large-scale mining was already disfiguring the landscape.[9] In the same year one such site, a vast gold and copper mine at Gyama (Jiama) not far from Lhasa which even an official website described as a 'post-apocalyptic ... portrait of devastation', was suddenly bulldozed by a landslide. According to *The Economist*, 'at least 83 of the mine's workers' were killed. Officially the tragedy was attributed to 'a natural geological disaster' involving the movement of glacial boulders; it had nothing to do with unsafe mining procedures or environmental damage, said the press release. But how come the miners' camp was sited in the path of these unstable boulders? 'Critics of Tibet's mining frenzy felt vindicated.' The wrath of the mine's presiding deity had made its point.[10]

Today the ores and aggregates of the plateau all flow east into China. So do the pelts and hides of the grasslands and the timbers of Kham and Pemako. Those benefiting from all this bounty are overwhelmingly Han Chinese plus a few foreign investors. The machinery, the expertise and most of the labour come from China; the processing and the profits accrue to China. Local involvement seldom extends beyond marginal and often illegal activities like traditional gold-working, casual hunting and some much publicised foraging.

For to twenty-first-century Tibetans the most rewarding extractive activity is said to be scavenging for a different kind of 'Tibetan gold'. Known as 'caterpillar fungus' or in Tibetan as *yartsa gambu* ('summer grass, winter worm'), this is not a mineral but an epiphytic fungus whose spores infect the larvae of a species of upland ghost moth. In spring, after consuming all but the yellowish outer shell of the moth larvae, the spores produce a dark fungoid spike which, growing from the head of the larva, eventually protrudes from the ground cover just enough for keen eyes to detect it. Prised from the soil intact (that is, with the spike still attached to the remains of the larva) this least prepossessing of the vegetable kingdom's productions has the distinction of being 'the world's most valuable parasite'. Its medicinal and restorative properties may not be universally recognised but that only inflates its mystique in the Chinese market. 'The biggest and most attractive pieces sell for $140,000 per kilogram,' says one authority. 'The global market is now worth $5 billion to $11 billion.' Eastern Tibet (Qinghai/Amdo and adjacent areas of Gansu and Sichuan) attracts the most foragers, though substantial actvity is reported elsewhere in upland Tibet and in Nepal. 'In Bhutan, [*yartsa gambu*] accounts for a significant amount of the gross domestic product.'[11]

Oddly, and despite the caterpillar fungus having been prized for centuries, the spectacle of foraging families advancing on all fours across the immensity of the Changthang seems to have escaped the attention of earlier observers. Perhaps they mistook the fungi foragers for *kuth* diggers. The ginger-like root of this plant of the high pastures, otherwise *kut*, *putchuk* or *Saussurea costus*, appears to have been in demand even in the Prophet's Arabia and is still credited with medicinal, culinary and aromatic properties in India and China. Or it may be that caterpillar fungus went unnoticed in the past because it was less plentiful then. Mycology is mysterious enough, added to which recent studies suggest that the life cycle of the ghost moth is particularly sensitive to climate change. By 2020 the annual yield of the most favoured sites was said to be declining fast, in part due to excessive foraging and grassland mismanagement but also, and less reversibly, to rising temperatures. In other words the bounty of the 2010s may have been just that – a bonanza resulting from a favourable but historically infrequent combination of human and climatic factors. In a lyrical requiem for Tibet's wildlife, Michael Buckley says of the Changthang: 'This is a home where the wild yak no longer roam, where the musk deer and the antelope no longer play'[12] – and where the spores of the world's rarest parasite wait in vain for the cold-loving caterpillars of the Tibetan ghost moth.

On a map the arc of small states slung along the outer Himalayas brings to mind a sagging string of prayer flags. Stretching from Bhutan and Sikkim in the east through Nepal, Uttarkhund and Himachal, and on to Kashmir and Gilgit–Baltistan in the west, it makes one wonder whether the political configuration of southern Himālaya was designed to complement its geology. Along an invisible fault line, irreconcilable ideologies collide like continents. Opposing concepts of responsible rule meld into one another. Maoists jostle for power with monarchists, and aspiring legislators pay handsomely for endorsement by inert deities. Civil upheavals are as frequent as seismic ones. Here a government can be toppled by a real landslide. Over the centuries earthquakes have flattened more temples than iconoclasts.

In the politically unstable landscape, flags fray and history itself gets distorted. Exhibiting a disconcerting plasticity, those state formations strung along the mountains prove far from permanent. Much of what was once Tibet, for instance, is now no longer Tibet, with fewer Tibetan-speaking Tibetans living within the Tibet Autonomous Region than outside it. As Bailey and Morshead discovered in the dripping forests of the lower Tsangpo–Brahmaputra, Himālaya's borders had never been set in stone. They expanded and contracted with the seasons, with the flow of trade and the requirements of pilgrimage, and with the fiscal reach of the great monasteries and the local governors.

Trade and taxation being indicative of the whereabouts of frontiers, they had much to do with Governor Warren Hastings's decision to despatch the first ever British mission to Himālaya. In 1773 an emissary from the Panchen Lama at Shigatse had unexpectedly appeared in Calcutta. He had come to appeal for clemency on behalf of Bhutan, a Buddhist kingdom once part of Tibet and still vaguely subordinate to it, whose troops had invaded neighbouring Cooch Behar. Sandwiched between Bhutan and British Bengal, Cooch Behar had asked the Company for help in dispersing the Bhutanese, and by the time the Tibetan intermediary (he was actually a Hindu) reached Calcutta, the invaders had been expelled. Cooch Behar had placed itself under Company protection, the Company pocketed half its revenues, and peace returned to the Duars, a term for the 'gated' floodplains of north Bengal as intersected by the outermost spurs of the Himalayas.

But the unexpected intermediary from Shigatse had not come alone, nor was he without an interesting proposition. With him was a red-robed lama bearing gifts which included a small bag of gold dust and an invitation from the Panchen Lama himself: would Hastings care to send a reciprocal mission to Shigatse? The invitation may have been oral and it was unclear whether it originated with the Panchen Lama or one of his intermediaries. Also, if it was meant to humour Hastings into withdrawing his troops from Cooch Behar, it had been overtaken by events; the troops had already pulled out. But the offer still stood and it was presented in the most glowing terms. Tibet, said the lama, 'had mines of gold, silver and all sorts of metal'. There were skilled craftsmen and there was a busy entrepôt trade in items like 'shawls and spices from Kashmir, corals and pearls transported by Nepali traders, [and] porcelain, tea and silk from China'.[13]

This was music to Warren Hastings's ears. The Company in the 1770s was still primarily a trading concern, its most valuable monopoly being that of supplying the British and American markets with tea from China. Western demand for tea seemed insatiable but its eastern supply was by no means straightforward. Only at the port of Canton (Guangzhou), and even then on highly disadvantageous terms, did the Celestial Empire authorise foreign purchases, added to which, after a decade of private profiteering in Bengal, the Company's coffers were short of the silver needed to pay for this investment.

Several solutions had already been aired, of which the supply of Indian-grown opium to the coastal creeks of the South China Sea would prove to be the most promising. But opening a trans-Himalayan trade route also had its champions, especially in British Bengal and most notably in the person of Governor Warren Hastings. A favourable response had therefore been sent to Shigatse, and in May 1774 George Bogle, a twenty-seven-year-old Scot who was one of Hastings's protégés, accompanied by a friend, Dr Alexander Hamilton, and an inordinate quantity of baggage, boarded one of Calcutta's rivercraft to tack north up the Hughli towards its junction with the Ganges and the road to Bhutan.

On a good day, from the banks of the Ganges the travellers might actually have made out the snowy profile of the Great Himālaya peeking above the northern horizon. Tibet, and so Qing China itself, were not as remote from the Company's India as the long sea voyage to Guangzhou suggested. The land of gold and musk, and a back door into the tea gardens of inland China, lay within sight of the seat of British power in Asia. However hard of access, Himālaya could no longer be ignored.

'What a road for troops,' sighed Bogle as he and Dr Hamilton climbed into the hills up rain-lashed trails better suited to fell-running than trade. They were accompanied as guides by the two men from Shigatse, though even their enthusiasm was flagging. A letter from the Panchen Lama expressed surprise that the mission was coming at all. The Lama was now denying having ever issued an invitation and was asking that the mission return the way it had come. Bogle insisted he had his orders and must press on, at least as far as the Bhutanese capital. A similiar letter from Shigatse addressed to Hastings in Calcutta was said to have been neither opened nor forwarded. Bogle claimed that, if apprised of its contents, he would have been honour-bound to turn back and so become complicit in the mission's failure. This doubtful logic was the same as that later invoked by Moorcroft, Bailey, Younghusband, Hedin and countless others when defying instructions to return. Undelivered missives, unopened orders, unaccountably delayed directives would become clichés of Himalayan travel.

Bhutan brought the travellers some relief. The division of powers between Bhutan's hereditary ruler, its spiritual authority, its temporal leader and its actual leader was thoroughly Tibetan and so, to newcomers, impenetrable. But, during two months spent in the Bhutanese monastery of Tashi-cho-dzong, Dr Hamilton's medical skills won friends who were happy to convey to the Panchen Lama some assurance about the mission's pacific intentions. Trade in the case of Bhutan would never amount to much; the country was too poor, the trails too perilous. But good government and sensible laws made for contented subjects. The people of Bhutan were a delight, thought Bogle, 'industrious, faithful, hospitable, honest, grateful and brave'.[14] Unfettered by caste, and uncorrupted by the greed and self-promotion found in more sophisticated societies, they led the young Bogle to question his own Enlightenment values. With his generally favourable observations of Himalayan society both here and in Tibet proper, the notion of Himālaya as a halfway haven to the stars blessed with idyllic valleys and otherworldly values took root in Western minds. A road as bad for trade as it was for troops was good news for Bhutan's autonomy. Of all those state formations strung along the Himalayan suture zone, only Bhutan remains a sovereign kingdom in the twenty-first century.

How nearly it succumbed may be judged from the map. From Paro in Bhutan, it took Bogle and his companions less than a week to climb on to the Tibetan plateau at Phari. This was the grim fortress where, 130 years later, officers of Colonel Younghusband's expedition would live like 'wild men of the hills' as they choked on yak-dung smoke and stuffed chapatis in their pockets to keep them from the dirt. Younghusband would be well aware of the Bogle expedition. From Phari to Gyantse he would follow in Bogle's footsteps, as would all subsequent Lhasa-bound traffic from India. Had it not been for Bogle's poor opinion of the road, the Younghusband expedition might even have approached Phari through Bhutan instead of through Sikkim. But it didn't, and so Bhutan was spared any taint of colonial occupation. When a trade agreement had finally materialised, it specified that no Europeans be permitted to cross the frontier.

Sikkim, on the other hand, would meet a fate like that of the ghost moth caterpillar. Inclined to crumble, assailed from without and infected from within by Nepali migration, Sikkim now exists in little more than name and nostalgic remembrance. It too was once part of Tibet and, since the seventeenth century, had been an independent Buddhist kingdom, albeit at the mercy of its neighbours. In return for surrendering the Darjeeling district, it was acknowledged by the British as a sovereign protectorate and later by the Republic of India. 'It was not, and never had been, part of political India' – until, that

The fort at Phari as photographed by the Younghusband expedition in 1904

is, 1975 when an embattled Mrs Gandhi sent in the Indian army, stage-managed a doubtful referendum and steamrollered a bill through the Indian parliament. Wrenched from that string of prayer flags dangling along the Himalayas, the no longer autonomous principality was appended to India as the least of the two dozen administrative units that comprise that republic's constituent states and territories.[15]

Near Phari, the Bogle mission's lama was seen to be improvising his own flag and preparing a libation. Since they were now in Tibet, indeed were the first Britons ever to enter it, Bogle presumed the offering was some kind of welcome. He was a trifle disconcerted to learn the little ceremony was not in fact in his honour but in that of the mountain deity who presided over the Phari plain. This was Tsheringma, the divine persona of the colossus of Chomolhari (Jomolhari) from whose icy flanks the travellers were keeping their distance.

Ten years later, in 1784, Chomolhari attracted interest from another quarter. Sir William Jones, judge, polymath and close associate of the by then Governor-General Warren Hastings, while scanning the horizon from the Ganges at Bhagalpur in Bihar, noted a barely visible but shapely summit probing the ether in the nor'-nor'-east. Chief Justice Jones took its bearings as best he could and asked around when he got back to Calcutta. It was not Everest or Kangchenjunga, neither of which was known at the time nor are they visible over the intervening ranges. But, thanks to Bogle's information, Jones was pretty sure it was 'Chumalary' on the Bhutan–Tibet border. 'From the most accurate calculations that I could make,' wrote Jones, 'the horizontal distance at which it was distinctly visible must be at least 244 British miles [393 kilometres].' A sighting so remote could be explained only by the mountain's extraordinary elevation. And this being the case, Jones was prepared to chance his arm. There was now, he declared, 'abundant reason to think that we saw from Bhagilpoor the highest mountains in the world, without excepting the Andes'.[16] Sixty years before anyone would be able to prove it, and a century before anyone would attempt to climb them, the primacy of the Himalayan peaks was first asserted thanks to George Bogle and Sir William Jones.

Encouraged by the news that the Panchen Lama's ambivalence over their mission was the result not of a change of heart but of pressure from Lhasa plus a smallpox outbreak in Shigatse, Bogle and Hamilton pressed on to Gyantse. There they turned west to follow the Tsangpo–Brahmaputra upstream to where the Panchen Lama had quarantined

Chomolhari, as seen from Phari. Sighted from Bhagalpur on the Ganges in 1784, Chomolhari provided the first evidence that the Himalayas were 'the highest mountains in the world, without excepting the Andes'

himself in a side valley; and thence, when the smallpox panic was over, they entered Shigatse and its great monastery of Tashilunpo.

The entire winter of 1774–5 was spent as guests of the Lama. Bogle would look back on the months whiled away in their airy quarters at Tashilunpo as 'an enchanted interlude'. It was 'the most peaceful period of my life', he wrote in a letter to one of his sisters in Scotland. With Hamilton and the Lama's nephews he played chess and a local version of quoits. Hunting had to be discreet, but they were welcome to join in the New Year festivities. They were welcome too to enjoy the company of a trio of 'well-looking females' who were also part of the Lama's extended family and with one of whom Bogle may have fathered a child.

But the Lama himself was the main attraction. He spoke some Hindustani, was refreshingly candid and had a great sense of fun. His curiosity about the outside world was matched only by Bogle's about Tibet. For hours on end the two men sat closeted together, Bogle venturing an account of the British constitution and the Church of England and the Lama an explanation of his relationship with Lhasa. As so often, this relationship was complicated by the Dalai Lama being a minor and his regent being answerable to the imperial Qing government in the persons of its two Lhasa ambans, or residents. Any commercial arrangements would have to be approved by the Lhasa authorities, who would then have to refer them to Beijing; and in neither place was there any hope of Bogle being admitted to plead his case in person.

That didn't mean the mission had failed, although, in the stimulating company of the Panchen Lama, Bogle may have felt mention of musk pods and gold dust almost indecent. The Lama was planning to write a pilgrim's guide to the mythical utopia of Shambhala (or Shangri-la). Based on a Buddhist text, this work would benefit greatly from Bogle's impressions of entering Tibet, and they in turn relied heavily on the Lama's insights. After such a meeting of minds, the two men could hardly bear to part. For the edification of his sister, Bogle composed a heartfelt valediction to Tibet which must rank as one of the more unusual utterances by an advocate of closer trade links:

Farewell ye honest and simple People. May ye long enjoy that Happiness which is denied to more polished Nations; and while they are engaged in the endless pursuits of avarice and ambition,

defended by your barren mountains, may ye continue to live in
Peace and Contentment, and know no wants but those of nature.[17]

The return to Bengal went without incident. It was hoped that Bogle
would make a second visit but in the end it was Dr Hamilton who
battled his way back up to Bhutan in 1776 and again in 1777. Usually an
upstanding Captain Hastings to Bogle's inquisitive Poirot, Hamilton
minus Bogle lost patience with the Bhutanese, made no further progress
with a trade treaty, never re-entered Tibet and died in the Duars.

The Panchen Lama was still hoping that something might come
of Bogle's visit. To that end he got as far as Beijing, but there, while
trying to interest the emperor in British plans for Tibetan trade, he was
finally overtaken by smallpox in 1780. Bogle himself died a few weeks
later in a freak drowning accident, aged thirty-four. Only Governor-
General Hastings soldiered on. Hamstrung by his council in Calcutta,
undermined by government regulation in London and sidetracked by
the possibility of substituting Vietnam for Tibet as a back passage into
inland China, he never quite gave up on the Bogle initiative.

Hence in 1783 news that Tashilunpo had rediscovered its Panchen
Lama, albeit now dressed in swaddling clothes and accompanied by his
parents, prompted the despatch to Tibet of Samuel Turner, a Hastings
nephew. Turner too was delayed in Bhutan but was rewarded in Tibet
when he found the Lama at eighteen months as endearing as Bogle had
found him at forty-something. Turner thanked him for all the kindness
he had shown Bogle; the infant, with a gravitas beyond his single year,
solemnly acknowledged the compliment. If he couldn't actually put it
into words, 'it was not to be inferred he cannot understand', said one
of his monastic minders.[18]

The Turner mission – Turner too was accompanied by a doctor –
appeared to have been a success. Reincarnation had its uses. In postulating
a continuity of personnel, it offered some consistency of policy. Trade
with Tibet was still on the cards, and even after his departure from India
in 1785 Hastings remained wedded to the idea. Some 'Kashmir goats'
sent from Tibet by Bogle had mostly perished in the heat of Bengal,
but of two yaks secured by Turner, one, a bull, even survived the voyage
to England. 'Robust but bad-tempered, the yak lived comfortably on
Hastings's estate,' writes Kate Teltscher; 'he fathered many calves, gored
a coach horse and had his portrait painted by George Stubbs.'[19]

Hastings's parting initiative, a duty-free trade mission to Tibet for Bengali merchants only, 'met with some success', says Teltscher. The Bengalis' wish list again featured 'gold dust, silver, wool, musk and yak tails (for fly whisks)', and their reception in Tibet was encouraging. But Hastings's successor as governor-general had more pressing concerns than Himalayan trade, added to which Nepal and Tibet were about to fall prey to convulsions that would put paid to formal intercourse across the central Himalayas for over a century.

In despatching the missions of Bogle, Hamilton and Turner via Bhutan, one of Warren Hastings's objectives had been to ensure they circumvented the 800-kilometre-long tangle of mountain-studded contours which the maps show as Nepal. Traditionally trade between Tibet and India in this sector of the Himalayas had indeed been handled by Nepalis, notably the Newars of the Katmandu valley. Merchants as well as master craftsmen, the Newars drew on their ancient links with both Hindu India and Buddhist Tibet to act as middlemen in exchanging the produce of Hindustan's sweltering plains for the yield of Tibet's frigid plateaux. Cereals, spices, sugar, textiles, coral and cowries travelled north; musk, gold dust, wool, salt and borax (used as a flux) went south. Newar specialities like carvings and metalwork went both ways, as did gold and silver; ever since Tibet started using its own coinage it had looked to Katmandu to actually mint the coins.

But the scale of all this commerce may have been exaggerated. In a standard account of the 1814 British invasion of Nepal, historian John Pemble dismisses trans-Himalayan trade as merely 'a peddling business' and sees Nepal's pre-eminent role in it as 'an attractively simple theory – but dangerous and a sure enticement to error'. The 'theory' to which Pemble refers holds that chronic instability in Nepal in the late eighteenth century interrupted the pre-existing trade and thereby obliged participants – British, Bengali and Tibetan – to explore other routes, like that via Bhutan. If, though, the trade was of little consequence, why bother relocating it? The need to bypass Nepal falls by the wayside.[20] There must be another explanation.

In revisionist mode, Pemble offers the idea that the antiquity and value of the trans-Nepal trade was in fact a fabrication. It had been dreamed up by Hastings and Bogle to justify their endeavours to find a Himalayan market for British woollens and a possible back door into China. It had little to do with musk pods, yak tails or even gold dust. Tibet's only dependable export of value was its annual yield of shawl wool, and this being destined for Kashmir (indeed for cashmere), the trade led west, not south, and was handled by Ladakhis, not Nepalis.

Who exactly were the Nepalis, anyway? In the late eighteenth century their country had yet to take shape. (Their nation, arguably, has still to do so.) Spreadeagled across the Himalayan glacis, knee-deep in the lowlands of the malarial Terai and with out-thrust arms clutching at the mountains, the peoples of southern Himālaya's tight valleys were barely aware of one another, let alone Nepal. Even today, with its five-sided flag (or 'double pennant'), its inconvenient time change (fifteen minutes ahead of Indian Standard Time), its 129 languages and its appetite for constitutional change, Nepal struggles to assert a distinct identity.

For most of history, the country's actual configuration bore no resemblance to the neat parallelogram shown on a modern map. Usually it was much smaller, occasionally much bigger. Until the eighteenth century 'Nepal' as a toponym was reserved for a cluster of minuscule city states in the fertile Katmandu valley. The wealth and prominence of these statelets was down to the industrious Newar (or Nepar and hence 'Nepal') who had migrated here probably from Assam or Tibet. They were not the valley's first settlers, but, receptive and accomplished, they made the valley their own. The Buddha himself had been born in the nearby Terai and had lived and taught in the neighbouring Indian kingdoms. Naturally the first Newars conformed to the undifferentiated Hindu–Buddhist practices prevalent at the time. They retained an attachment to Buddhism long after it fell out of favour in India, although ultimately most adopted the cults and caste distinctions of Bengal's tantric Hinduism. Naturally, too, they attracted the attentions of India's Muslim conquistadores – and paid the price for their idolatry. The characteristic pagoda profiles and the elaborately carved woodwork of the brick and timber temples at Patan (Lalitpur), Bhadgaon (Bhaktapur) and Katmandu itself may be typical of a style once common in northern India, but nearly all the surviving examples in Nepal date from the reconstruction necessitated by Muslim desecration and destruction in the fourteenth century.

Only to this nucleus of small but highly developed Newar city states, each with its ruling house, was the name 'Nepal' originally applied.

In the hills beyond Katmandu's valley, dozens of other small state formations, many no more than fortified villages, proliferated under the rule of often émigré rajahs of the Rajput (*kshatriya*) caste displaced by the Muslim conquests in India. One such statelet was Gorkha, west of Katmandu and midway between there and Pokhara. Established in the late sixteenth century, the Shah rajahs of Gorkha had gone on the offensive in 1736. Overawing or conquering immediate neighbours, by 1767–8 the troops of Gorkha's Prithvi Narayan Shah (reigned 1743– 75) were spilling into the Katmandu valley. The Newar states either surrendered or were brutally besieged. With Katmandu as its royal capital, a still modest but enlarged Nepal (because it included Gorkha and Gorkha's earlier conquests) came into being. Gorkha's Shah dynasty now ruled from Katmandu. Their Gorkhali dialect became Nepali, the official language, and their *parbatiya* ('hillbilly') brand of Hinduism became the state religion.

Prithvi Narayan Shah is usually credited with 'the unification of Nepal' since, says Pemble, 'these conquests marked the birth of the Nepalese state'.[21] In reality the submission of the Newar cities just prompted more conquests. From cradle to coming of age, the infant Gorkha state continued to grow prodigiously. East of Katmandu, in the 1770s Gorkha's forces pushed on through the hills to Sikkim, much of which was occupied despite putting up a long resistance. In the west it was a similar story, but with more accommodations and much more substantial results. The Karnali River was crossed and then the Mahakali (today Nepal's western frontier). This brought Gorkha's sway to the independent hill kingdoms of Kumaon (capital Almora) and eventually Garhwal (capital Srinagar), which together now comprise the Indian state of Uttarkhund (capital Dehra Dun). The first was reduced to vassalage, the second ravaged and subjected. Continuing west in leaps and bounds, Gorkha troops then crossed the headwaters of the Yamuna to reach the banks of the Satluj and overran most of what is now Himachal Pradesh. Among those here forced to flee was the incumbent rajah of Sirmur, whose fossil-rich section of the Siwaliks would so handsomely reward Hugh Falconer and his fellow palaeontologists thirty years later.

There was talk of pushing on to Kulu and even to Kashmir, but by the end of the eighteenth century the Gorkha advance was faltering. Thanks to the lengthening catalogue of disputed successions, palace massacres and incessant intrigues for which the Nepalese court would become notorious, blood was being shed and heads sent rolling as freely in 1790s Katmandu as in 1790s Paris.

Even the army was losing its aura of invincibility. When rumours of Prithvi Narayan Shah's declining health had reached Shigatse back in 1775, Bogle and the Panchen Lama had paused their discussion of world affairs to savour the prospect of a peace-loving Katmandu resuming its role in the Tibet–India trade. But this never happened. No sooner had Turner departed the scene than Katmandu, taking advantage of the Panchen Lama's minority and citing a string of minor grievances, sent troops to occupy the Tibetan frontier and command the road to Shigatse. The young Panchen Lama was rushed off to safety in Lhasa, from where the Dalai Lama appealed for help to the Qianlong emperor in Beijing. Meanwhile both Shigatse and Katmandu secretly sought the support of Calcutta. This was not forthcoming, but by 1789 the threat of Chinese intervention had done the trick. The Tibetans agreed to pay off the Nepalis, the Nepalis agreed to send a tribute mission to Beijing, and the frontier districts and trading rights of old were restored.

Two years later the agreement broke down when the Tibetans reneged on their payments. Nepali troops crossed back into Tibet and this time kept going all the way to Shigatse. Tashilunpo, second in sanctity only to Lhasa, was comprehensively sacked by the Hindu Nepalis. 'Every temple and mausoleum, every shrine and hall, offered up its contents to the Gurkhas. The great monastic city was soon stripped and gutted, plundered and desecrated.'[22]

Retaliation was again swift and, for Tibet, almost as painful. After crossing the plateau in the depths of winter, in 1792 a two-pronged Chinese invasion, 17,000 strong, overhauled and defeated the loot-laden Nepalis, then pursued the survivors across the mountains to within a day's march of Katmandu. The Chinese called a halt only after extracting Nepali promises of non-interference in Tibet and of the reinstatement of the five-yearly tribute missions once sent to the Ming emperors in Beijing.

So Gorkha's proud record of conquest had been blotted; the Qing dynasty had notched up its sole trans-Himalayan success, and Tibet was left in no doubt about who was its saviour. A hundred porters were needed to carry the recovered plunder back to Tashilunpo, where the young Panchen Lama was reinstalled. In return he bound himself never again to engage in cross-border initiatives. Such trade as there had been between Tibet and Nepal now ceased. Beijing had made good its claim to Tibetan sovereignty and in Lhasa the Dalai Lama was more closely chaperoned than ever. 'Tibet hadn't just been protected; it was swallowed up ... Only as the Qing fell in the twentieth century [would] the Dalai Lamas emerge again as great leaders.'[23]

Trade between Tibet and Nepal was stifled, but the East India Company still nursed hopes of trade between India and Nepal, added to which there was a clear need of some understanding about the India–Nepal frontier. A Nepal not much bigger than the Katmandu valley could be ignored, but a Nepal that stretched the entire length of the central Himalayas from the Tista River (roughly Darjeeling) to the Satluj (roughly Simla) was a very different matter. East to west, Gorkha's conquests now marched with the Company's territories and dependencies all the way from Bengal to the Punjab; and in the Punjab the no less militant Sikh kingdom based on Lahore was already causing the Company grave concern. Being contained by two such formidable neighbours was anathema to the strategists in Calcutta; just being encroached upon by them had to be accounted an occupational hazard.

This was especially the case with Nepal. The difficulty of communication across the grain of the hills left Gorkha's far-flung commanders in Kumaon and Garhwal free to make their own mischief while, to the south in the unhealthy Terai, tentacles of Nepali authority tended to advance and retreat with the passage of the seasons and the hatchings of the malarial mosquito. As the nineteenth century dawned, Calcutta's relationship with Katmandu had less to do with trade and more with territory and its taxation.

Nepal's ill-judged adventures in Tibet had appeared to offer the Company an early chance of restraining the warmongers in Katmandu. To that end in 1792–3, under the pretext of mediating a truce with the advancing might of the Qing army, Captain William Kirkpatrick had led the first ever Company deputation to reach the Nepali capital. He arrived too late. Terms for a Sino-Nepali ceasefire had just been agreed. The one thing now to be avoided was upsetting the Chinese by hinting at Company involvement in Nepal. 'Kirkpatrick and his party were then sent packing, having gained little more than some second-hand information about the country and its people, but in the knowledge that they had at least "removed the veil that had so long interposed between the two countries".'[24]

The veil stayed removed largely due to a new and sanguinary power struggle in Katmandu. In 1800, the just installed rajah, evincing early signs of madness, abdicated in favour of his two-year-old son and fled the capital to seek sanctuary or self-redemption – and probably both – in British territory at Benares (Varanasi). A rival faction then snatched power in Katmandu and, in return for the Company's cooperation in detaining the ex-rajah in Benares, a new chief minister signed a treaty promising peace, trade and an exchange of representatives. Accordingly in 1802 Captain William Knox, who had commanded Kirkpatrick's escort, returned to Katmandu as its first British resident. He came with an impressive staff that included surveyors, a military observer and a botanist. Spacious quarters in a garden north of city were put at their disposal. They looked set to stay. Then another palace convulsion removed the new chief minister, negated the treaty and left Knox and his staff marooned under what amounted to house arrest.

They held out till the end of the year before being withdrawn. Francis Buchanan-Hamilton, the botanist, used the time to amass over 800 plants new to science, and being also an antiquarian with some experience of Burmese Buddhism, he made another altogether extraordinary discovery: among the Newar were to be found still knowledgeable descendants of ancient India's long forgotten Buddhist past. It was like coming across a huddle of apostles tending their nets on the shores of Galilee. From texts purchased at considerable expense from 'an old Buddha residing in the city of Patan', and from this same Amritananda's exposition of them, Brian Houghton Hodgson, the most prolific of Himalayan scholars, would twenty years later embark on what he called 'a full and accurate investigation of this almost unknown subject'.[25] Translator Rinchen Zangpo would have approved; Marpa might have queried the cost. But it was worth it. Buchanan-Hamilton's chance discovery and Hodgson's dogged endeavour ignited an explosion of scholarship which would result in the re-evaluation of ancient India's archaeology and the rehabilitation of Buddhism as India's greatest contribution to civilisation.

But if the Knox mission was really about peace and trade, it failed dismally. The long-sought trade with Nepal never materialised. The peace, such as it was, was shattered when the Nepali conquests resumed in Garhwal. By 1806, in the midst of another palace bloodbath, Katmandu acquired a new chief minister, indeed a war leader, in the person of the redoubtable Bhimsen Thapa. Bhimsen held the British in contempt. He didn't believe they would fight to uphold their claims to disputed sections of somewhere as unhealthy as the Terai. The British didn't realise that the low-lying Terai was in fact as dear to the Nepalis as their mountain valleys. 'The Tarai estates were the main source of income for the Gurkha government,' says Pemble; indeed they were commonly 'assigned to the army in lieu of payment'. Gorkha's war machine had been fuelled by land-grabs in the Terai and by raids beyond it. Relinquishing an interest in such tracts was therefore unthinkable.[26] With both parties confident that the other would not fight, they bluffed and blundered their way to war.

As wars go, the 1814–16 Anglo-Nepalese confrontation was short (about fifteen months), discontinuous (because of a lull during the malaria season) and extensive (because of what was theoretically an 800-kilometre front). It was also notably ferocious with heavy casualties, a few spectacular engagements and numerous instances of conspicuous bravery. The British failed to take advantage of their numerical superiority and greater firepower, while the Nepalis were hampered by poor communications and the early loss of Kumaon. Since both sides emerged with a better appreciation of the other's fighting qualities, this mutual regard would prove as valuable as anything in the 1816 peace treaty of Sagauli.

The treaty pledged 'perpetual peace and friendship' between the late combatants provided Nepal renounced all claims to Sikkim, to the disputed Terai territories and to Kumaon and Garhwal. Other lowland areas administered by Nepal were ceded to the Company, which was to compensate those Nepalis who lost the revenues of these areas. Ministers to oversee these arrangements were to be exchanged, which meant the British installing a resident in Katmandu; and Nepal was not to employ or admit any other foreigners. There were nine articles in all. Not one mentioned trade, although reinstating the former rajahs of Kumaon and Garhwal as feudatories of the Company left open the possibility of a direct trade in shawl wool between Tibet and these now British dependencies.

Pace Pemble, that there was indeed such a trade is indisputable. William Moorcoft had reported on it at length during his expedition from Garhwal to Kailas–Manasarowar in 1812. But the traders themselves would prove a greater asset than their trade, just as their fellow Nepalis would prove more useful as recruits than as allies. Sagauli guaranteed Nepal's status as a sovereign state and a neutral buffer between British India and Qing-controlled Tibet. But the arrangement depended on Katmandu not embarking on another round of Quixotic conquests, and this could be forestalled only by finding new employment outlets and alternative sources of income for the country's ever expanding and partially militarised population.

Before the war was won, some 5,000 Nepali troops had switched sides to fight with the British. Their courage and stamina, their cragsmanship and their willingness to dispense with the caste and dietary taboos to which most Hindu recruits were subject made them an instant success. The British called them 'Gurkhas' and, as such, they were added to the list of India's recruit-worthy 'martial races'. The name, an alternative spelling for 'Gorkha', was however misleading. There was neither a Gurkha race nor a Gurkha tribe. Although many Gurkhas were familiar with the Gorkhali language, few were natives of Gorkha. Those recruited in Nepal itself were more likely to be of less illustrious Gurung, Magar or Tamang parentage. The first British recruits came not from what was now Nepal but mostly from Garhwal and Kumaon (Uttarkhund), whose hereditary rulers had been dislodged by the Gorkha conquests and were now reinstated under British protection. In 1815 the first battalion of what would become the second regiment of British Indian Gurkhas was recruited by Lieutenant Frederick Young in Sirmur, which is nowadays not even in Uttarkhund but in Himachal.

The success of the experiment was never in doubt. In the armies of Nepal itself, of British India, the Republic of India and the United Kingdom, a proliferation of Gurkha units would serve with distinction wherever the tides of war and the eddies of peacekeeping took them. In Himālaya, detachments of Gurkhas provided diplomatic expeditions with armed escorts and gave to undiplomatic ones, like the Younghusband expedition, muscle and firepower. Survey parties in the mountains relied on them as porters; early mountaineers took advantage of their high-altitude survival skills and agility.

In this last context, Gurkhas anticipated the role played in the twentieth century by the Sherpas of the Solu Khumbu region in eastern Nepal. 'The natives whom we found the most plucky were Nepalese Tibetans, the so-called Sherpahs,' reported the leader of a Norwegian expedition to Kabru, a satellite of Kangchenjunga, as early as 1907.

Altitude and ice seemed to suit the Sherpas. As both guides and porters they just needed more experience with ice axes and ropes. 'If they get attached to you, they will do anything.'[27] While exploring other peaks accessible from Darjeeling, in the years before the First World War the maverick climber Alexander Kellas also relied heavily on Sherpas, and when, after the war, the British turned their attention to Mount Everest, it was Dr Kellas who urged their employment, so ensuring their rightful share in the eventual conquest of that mountain in 1953.

By profession shepherds from the high pastures of eastern Nepal, Sherpas were readily distinguishable from the 'Gurkha' peasant-farmers and infantrymen of the central lowlands. Until the Sherpas' extraordinary abilities were acknowledged, they were more usually numbered among another group of Nepali peoples whom the British knew as *bhotia* (*botia*, *bhutea*, *bhotiya* and so on). *Bod/Bhot* being the ancient name for Tibet, *bhotia* simply meant 'Tibetan' and was applicable to any of the peoples living in the high valleys along the Tibetan frontiers of Sikkim, Nepal, Kumaon and Garhwal. Some *bhotia* looked Tibetan, spoke Tibetan, spun prayer wheels and had migrated from Tibet within living memory. Others claimed descent from fugitive settlers from India and had adopted the cults and caste practices of Hinduism. But what distinguished them all as *bhotia* was a peripatetic lifestyle that straddled the frontier. As pastoralists, Bhotias moved their flocks up into the mountains in summer, and as inveterate traders they frequented the great seasonal markets beyond the mountains in Tibet. Their contacts were as extensive, and their credit as good, in Purang and Gartok as in Almora or Srinagar (the Garhwal capital, not to be confused with Kashmir's Srinagar).

Moorcroft seems to have been the first to appreciate the unique status of the Bhotia. Entering Tibet from Garhwal in 1812, he and his Anglo-Indian companion had been halted just short of the 5,070-metre Niti Pass. The Tibetans were unconvinced by their disguise and the local Bhotia resented the commercial competition they posed. Negotiations were getting nowhere – until, that is, Moorcroft, a physician by training though a vet by profession, opened a field surgery. Among his first patients was a boy who was 'successfully tapped for dropsy'. The boy turned out to be the son of 'one of the wealthiest and most influential Bhotias', and the gratitude of the boy's father Deb Singh Rawat and of his uncle Bir Singh, both from the neighbouring Johar valley, knew no bounds. Moorcroft and Hearsey were sent on their way with a guide, porters and a pair of riding yaks. In Tibet they received further Bhotia assistance in the purchase of Tibetan goats plus a consignment of the precious *pashm*; and on recrossing the mountains they were again indebted to the Bhotia brothers when they secured Moorcroft's release from detention by the Nepali authorities.[28]

Thus began what Charles Allen calls 'a remarkable alliance between the Bhotias and the British that still survives to this day'. Bhotia support provided by the Singhs of Johar featured in nearly all the mid-century missions and expeditions into western Tibet, including several hunting trips made by Edmund Smyth, Kumaon's first education officer.

> [The Bhotias] pass their lives in trade with Tibet and they are the only people allowed by the Tibetan authorities to enter the country for the purposes of trade [wrote Smyth]. From June till November they are constantly going over the passes, bringing the produce of Tibet (borax, salt, wool, gold dust, also ponies) and taking back grain of all kinds, English goods, chiefly woollens and other things. The goods are carried on the backs of sheep, goats, ponies, yaks and *jhoopoos* (a cross between the Tibetan yak and the hill cow).

When in 1862 Smyth was asked by Captain Montgomerie of the Survey of India to recommend 'some trustworthy men to train as explorers', he unhesitatingly chose Mani Singh Rawat, the son of Moorcroft's Deb Singh Rawat, and Mani Singh's cousin, Nain Singh Rawat. The latter, being one of Smyth's schoolteachers, was already addressed as *pandit*. In the Survey he became 'No. 1' or 'the Pundit', an alias which was soon adopted as a generic term for all the Bhotia subsequently recruited by the Survey. Equipped with prayer wheels and a string of beads adapted for recording their progress, and with a compass and sextant for basic triangulation and a boiling-point thermometer for altitude measurement, each pundit operated as an inexpensive one-man survey team.

Nain Singh showed what could be achieved. In January 1866 he reached Lhasa, the first surveyor from India ever to do so and the first to record the city's exact location and altitude. A year later he was the first to report on the singing sands of Thok Jalung and confirm the existence of the gold mines. His 1866–7 route survey of around 3,320 kilometres was exceeded by his next from Ladakh to Lhasa in 1874 and by the 4,450 kilometres covered by his cousin 'AK' (aka Kishen Singh) in 1878–82. And there were many other odysseys extending deep into Xinjiang, the Pamirs and the Hindu Kush and lasting for two or three years. Only the indomitable Kintup in his toing and froing between Lhasa and the Tsangpo gorges was absent for longer and may have covered more ground.

Along with the Gurkhas and the Sherpas, the Bhotias won for Nepal more in the way of international recognition than was forthcoming for Himālaya's other satellite states. Yet not even military recruitment and two world wars, let alone clandestine survey work and the seasonal market for mountaineering guides, did much to resolve Nepal's demographic problem. For its chronic lack of employment opportunities the twentieth century's only solution was emigration. Nepalis were drawn in such numbers to Sikkim and Bhutan and to the tea plantations of West Bengal and Assam that Gangtok and Darjeeling began to feel like Nepali cities. A Gorkha National Liberation Front has represented the interests of these incomers since 1980. 'Demanding the recognition of Gorkhali as one of India's official languages, the GNLF strives, not without occasional violence, for an autonomous Nepali enclave within West Bengal or even a separate Nepali state within the Republic of India.'[29]

One argument for resisting this demand for a Nepali-majority state in India is that it already exists. It's called Sikkim. It might even have been called Bhutan. In 1975 it was agitation by the large Nepali community settled in Sikkim that had prepared the ground for Indira Gandhi's intervention. This had a knock-on effect in Bhutan, where fear of a similar fate persuaded a country famed for its 'gross national happiness index' to eject into stateless misery the one-sixth of its population who had migrated there from Nepal. Some of the displaced were no doubt recent illegal immigrants, but others had been constructing Bhutan's vestigial infrastructure for decades.

Nepal, of course, protested over the Bhutanese expulsion of its citizens. On the understanding that the international community would help resettle these so-called Lhotshampa ('Southerners'), Katmandu set up temporary camps, but it withheld citizenship. The displaced were now refugees in their ancestral homeland – and so, as late as 2016, some remained. Meanwhile over 100,000 had been resettled in the

US, Canada and the Netherlands. Others joined the exodus of Nepali contract labour to construction sites in Delhi and Mumbai, the Gulf, Malaysia, Singapore and Hong Kong. Unwanted in their native land and liable to be laid off in their place of work, the Nepalis' search for their own share of 'gross national happiness' looks like being a long one.

13

Shawl Wars

On the last fatal day not one-half could handle their arms, and when a few fled, the rush became general. But death was waiting for them all.

Alexander Cunningham (1854)[1]

Insurgencies, earthquakes and constitutional upheavals notwithstanding, twenty-first-century Nepal retains its sovereignty by playing off against one another the ambitions of its giant neighbours, India and China. Kashmir, having no sovereignty with which to juggle, fares less well. In the time taken to write this book the former Jammu and Kashmir princely state, already rent by three Indo-Pak wars, has been shorn of its less Muslim extremities, stripped of any residual autonomy, isolated by all manner of restrictions and subjected to indefinite occupation by up to half a million Indian troops. What was once the largest of British India's princely states has been pared down to the bitterly contested Kashmir valley plus Jammu, and this in turn faces an uncertain fate as the only Muslim-majority state within a Modi-fied Indian republic bent on asserting the primacy of its Hindu majority. Kashmir's 'troubles' may not be over.

Nor are they new. They date back to the state's nineteenth-century inception and the commercial attractions of the 'cashmere' wool to which the valley gives its name. The history of the shawl-wool trade, the most valuable in all Himālaya, is actually much older and argues strongly for amicable relations between the semi-nomadic wool-producers of western Tibet and the sedentary wool-spinners and weavers of the 'vale of Kashmir'. But between the two were interposed the Great Himālaya and the Tibetan frontier. Snowfall closed the passes for months at a time. Brigands found ready sanctuary in the frontier's frigid no man's land. Losses were heavy, as were tariffs. Demand habitually exceeded supply, which pleased the upland goatherds and was resented by the downstream traders and processors.

Bogle had been instructed to report on how the trade might be managed to the advantage of British India. For William Moorcroft, relocating the shawl-wool traffic turned into something of a personal crusade; it was the main reason for his 1812 journey to Kailas–Manasarowar, and during the two years (1820–2) he spent in Ladakh on his second journey, he compiled what may be the most extensive report ever written on highland goat husbandry and the techniques of shawl-production.

Moorcroft entertained such expectations of the industry that he considered cutting out the Tibetans altogether; instead of trans-Himalayan bartering for the wool (*pashm*), why not acquire the goats that produced it? Introducing Tibet's goats to the warmer climes and better grazing of, say, Garhwal or even Scotland was expected to increase the yield of *pashm* and secure its future supply. In practice it would probably have done neither. But this remained to be shown; Bogle's specimen goats had died in the heat of Bengal and half Moorcroft's herd was lost at sea. (They were the nannies; the billies, on a different ship, reached England but were as useless for breeding purposes as Hastings's bull yak.)

Moorcroft's researches revealed almost too much about *pashm*. Ten years previous to his visit an epidemic had decimated the herds of western Tibet. As a result the value of Kashmir's shawl output at around £300,000 per annum had been halved and the wool merchants of Ladakh had to import *pashm* from the Kyrghyz of the Pamirs and Tien Shan. This back-up supply reached Leh by yak, camel and pony from Xinjiang and Turkestan over the world's highest trade route with a succession of formidable passes that climaxed in the 5,847-metre Karakoram Pass. As shawls, some of the finished *pashm* then went back the same way to the markets of central Asia and even Russia.

Somewhat ahead of his age, Moorcroft was deeply suspicious of any Russian interest in British India. At a time when over 3,000 kilometres still separated the British and Russian empires – and half a century before the Great Game would directly impinge on the Tibetan borderlands – he supposed himself engaged in a battle of wits with a Kashmiri Jew in the employ of the tsar. This Aga Mehdi appeared to be using the trade in *pashm* as a cover for winning influence and allies in Kashmir and Lahore. Moorcroft, 'convinced that nothing less than the safety of India lay in his hands', prepared to expose the impostor – only to be denied that satisfaction when Aga Mehdi died 'of a sudden and violent disorder' (probably pulmonary oedema) on the Karakoram trail. 'Had he lived, he might have produced scenes in Asia which would have astonished some of the cabinets in Europe,' reported Moorcroft with cheerful hyperbole.[2]

Riding yaks provided by their Bhotia friends (on horseback), the be-turbaned William Moorcroft and Hyder Young Hearsey approach Manasarowar and Mount Kailas in 1812

Well informed about *pashm*, Moorcroft was less sure about *tus*. He knew of its exceptional quality; it sold for double the price paid for *pashm* and it came mainly from the Changthang rather than western Tibet or Ladakh itself. He understood it was the underwool of some species of wild goat, perhaps the markhor, and this may indeed have sometimes been the case. But there seems no reason to doubt that, then as since, *tus*/shahtoosh was mostly obtained from the chiru antelope.

Fleet of foot and slender-horned, the chiru was unknown to zoology at the time. But in 1824, from his lonely posting at the British residency in Katmandu, Brian Hodgson took a break from his Buddhist studies to investigate the habits of a 'Himalayan unicorn'. Better known as the one-horned Indian rhinoceros, there were several specimens in Katmandu's royal menagerie, and it was Hodgson's good fortune to witness a pair mating. He was there again, eighteen months later, to attest the safe delivery of a male calf and to record in some detail its infancy. His 'Remarks on the Procreation of the Rhinoceros' as published in one of Calcutta's learned journals was the first account of how rhinos mated and the first of Hodgson's innumerable contributions on Himalayan mammals and birds.

One unicorn led to another. Hodgson had no sooner revealed the mechanics of rhino sex than he was tempted to pontificate on the authenticity of a Tibetan unicorn. 'Is it not singular that the Bhotias continue most steadfastly and universally to maintain the existence of the Unicorn?' he asked in a letter to a fellow naturalist. The creatures in question were not found in Nepal but were known from heads obtained in central Tibet. From one such hanging in Katmandu's Swayambhunath temple, Hodgson was soon convinced they in fact belonged to a bicornate (two-horned) deer or antelope which had shed one of its horns. He even obtained a live specimen of the bicornate version. It had been reared at Tashilunpo by the Panchen Lama and was procured for Hodgson by Bhimsen Thapa, the bloodstained minister who had headed the Nepal government throughout the Anglo-Nepal War. Sadly, the antelope had no sooner disposed of misunderstandings concerning its genus than it succumbed to the heat. Hodgson was not surprised. Katmandu is a stifling 2,500 metres lower than Shigatse, added to which the antelope, beneath a stiff outer coat of fur, was found to be encased in 'a spare fleece of the softest wool'.[3] Tibet knew the animal as the chiru and science now knows it as the *Pantholops hodgsonii*. But Hodgson, like Moorcroft, seems to have been unaware that its thermal vest was in fact the priceless *tus*.

According to Moorcroft, 500–1,000 horse-loads of shawl wool, each of nearly 300 pounds (135 kilos), were sent from Ladakh down to Kashmir every year. The cost of carriage was 33 rupees per horse whether the wool was *pashm* or *tus* and whether it had come from Tibet or Turkestan. 'The duties collected at various places in Ladakh and Kashmir' were more variable and, as estimated by Moorcroft, 'amounted to ninety-five rupees [per horse-load]'.[4] In other words the import trade yielded an annual revenue of 50,000–100,000 rupees. Naturally every stage in the production process attracted more taxes, with the heaviest tariffs of all being reserved for the finished shawls. Kashmir's labour market relied on the shawl industry; Ladakh's entire economy depended on the carrying trade; and the duties realised on this traffic in *pashm* were what made Ladakh's otherwise bleak and stony scarps of abiding interest to its neighbours.

At the time of Moorcroft's visit, Ladakh's political status was open to doubt. The Kashmir valley had just been wrenched from Afghan control by Ranjit Singh's Sikh kingdom of Lahore. As a result the Sikhs were demanding from Ladakh the tribute the Ladakhis had previously paid to the Afghans. But according to the Ladakhis the tribute had originally been exacted by the Mughal emperor Aurangzeb in payment for Mughal assistance in expelling a seventeenth-century invasion of Ladakh by 'Kalmak Tartars'. Hence the Afghan, and now Sikh, demand for tribute was valid only if exercised on behalf of the emperor in Delhi; and since the current emperor was a pensioner of the British, it fell to the now paramount East India Company to decide the matter. Better still, since according to Moorcroft the Ladakhis were ready to tender allegiance to the Company, they were quite entitled to deny the tribute to anyone else.

Such at least was the reasoning of the incorrigibly optimistic Moorcroft. He liked the easy-going Ladakhis and could see nothing but advantage – commercial, political and strategic – in securing a British toehold in the heart of western Himālaya. That was not, though, how Calcutta saw it. Moorcroft had been denied any diplomatic status and only grudgingly authorised to undertake his journey in a private capacity. The news that he was now offering to accept the Ladakhis' tender of allegiance to the Company and had therefore advised Maharajah Ranjit Singh to withdraw his demand for tribute brought only stinging rebukes from Calcutta and Delhi. His intervention was denounced, payment of his salary withheld and a grovelling apology sent to Lahore. Luckily the maharajah chose not to take offence. With a swipe at the Company's 'misplaced squeamishness and unnecessary timidity', a chastened Moorcroft dragged himself away from Ladakh. Effectively disowned by his employer, in 1823 he departed Kashmir heading east for Afghanistan, the Oxus and mortal oblivion.[5]

Moorcroft's fear was that Ranjit Singh, the founder of the Sikh kingdom of Lahore, would add Ladakh to his 1819 conquest of the

Kashmir valley. Six years earlier Sikh troops had halted Nepal's western advance into what is now Himachal Pradesh by aiding the besieged ruler of Kangra. But that was a one-off; for the Sikhs the mountains were a distraction, more often a sanctuary than a source of income. When Ladakh resumed its tribute payments to the Sikh governor in Kashmir, Ranjit Singh let matters rest. It was not the misshapen maharajah whose one-eyed gaze was fixed on Ladakh (the other eye having been lost to smallpox); it was the hawkish attentions of three Dogra (Hindu) brothers from Jammu, a small fief in the foothills south of Kashmir. Basking in the maharajah's favour, one of these brothers served Ranjit Singh as his all-powerful chief minister at court, another as his ruthless enforcer in the field.

The last was the spade-bearded rajah Gulab Singh who, in the 1830s, oversaw a succession of campaigns ostensibly intended to establish Sikh control of Kashmir's hill regions. In retrospect, Gulab Singh's real objective appears to have been that of engrossing the revenues of these regions and carving out a Himalayan kingdom for himself; but perhaps to allay any such suspicions he entrusted the conduct of his most ambitious design to Zorawar Singh, governor of the recently acquired township of Kishtwar tucked into the Chenab valley between Jammu and Zanskar. Despite being another Singh, Zorawar Singh was neither a Sikh nor a Dogra but, like some of Nepal's gentry, claimed Rajput descent. Confusingly, then, it was this supposedly Rajput general Zorawar Singh, under the direction of the Dogra rajah Gulab Singh, himself a feudatory of the Sikh maharajah Ranjit Singh, who in 1834 responded to a minor dispute in Ladakh by embarking on the conquest of *la-dvags*, the 'land of the high passes' – a country and a people whom geography, race, religion and most of history declared to be indisputably Tibetan.

The sporadic campaigning that ensued would be notable if only because, *pace* Younghusband, invasions of Himālaya mounted from the plains of India are almost unknown. This is not to say that Zorawar Singh's achievement would be ignored. To South Asian schoolchildren his invasion of Tibet may be as well known today as once was Younghusband's incursion to British schoolchildren. Noting how the enterprise resulted in Ladakh being attached to Kashmir and so, ultimately, to the Republic of India, Indian scholars proudly claim the invasion as a combined Dogra–Sikh enterprise. Pakistani scholars, on the other hand, stress the prominent role played by Muslims in the army and government of the kingdom of Lahore, a city which is itself now in Pakistan.

Zorawar Singh would be in the field for eight years. It was testimony not so much to the stout resistance offered by the ancient matchlocks and steel-tipped lances of the Ladakhi militia as to Zorawar's breathtaking ambition, plus some unexpected distractions. His ultimate goal is said to have been either Lhasa or Yarkand, and he might actually have achieved one of them had he taken the Tibetan winter more seriously and not been easily sidetracked; on one occasion he set off to besiege Skardu and dethrone Ahmed Shah, the ruler of Baltistan who had lately played host to the wayfaring Godfrey Thomas Vigne; on another it was to deal with the machinations of a rival who ought to have been an ally. This was yet another Singh, in fact Mihan Singh, the Sikh governor of Kashmir. Appointed by the maharajah in Lahore, Mihan Singh might have been expected to aid and abet Zorawar's invasion of Ladakh. Both men, after all, were serving the same government. Yet the involvement of Kashmir's governor cast the whole affair in a new light: evidently it had very little to do with territory, tribute or even Tibet and everything to do with shawl wool.

Approaching Ladakh from Kishtwar in the south rather than by the usual route from Kashmir in the west, Zorawar Singh's hardened campaigners had made short work of the Ladakhi forts until they reached the preposterously sited monastery of Lamayuru on the main Srinagar–Leh road. By now winter was closing in. Unaccustomed to ice and altitude, the Dogra army was feeling the cold. After an exchange of hostages, Zorawar welcomed an invitation to repair to Leh for peace talks.

These dragged on for four months and despite mutual suspicions were eventually successful. But by then opposition to the invaders was breaking out anew in Suru to the west and in Zanskar to the south. Zorawar and his Dogra troops were no sooner scrambling into Zanskar to deal with this uprising than 'intelligence arrived that an insurrection had broken out in Leh'. It seemed that, at the instigation of the Kashmir governor Mihan Singh, Ladakh's titular ruler had repudiated the peace treaty, authorised reprisals against those Ladakhis who had collaborated with the invader and, as reported by one of Zorawar's commanders, 'closed the roads to the merchants'. The merchants affected were the traders in *pashm*, and it took a footnote by Major Alexander Cunningham, the next British visitor to Leh, to interpret this information. Evidently the Kashmir governor, in acting in defiance of his Sikh overlord, was acting in defence of the vital traffic in *pashm* on which Kashmir's shawl industry depended. 'Mihan Singh's intention was undoubtedly to force the whole trade through Kashmir, which otherwise, owing to the Dogra occupation of Ladakh, would have been turned into other channels leading through Kishtwar and the Dogra territories to India.'[6] Hostilities resumed. The Ladakhis were again quashed and another treaty was signed. The war indemnity now dwarfed the original tribute and would take years to pay off. A puppet ruler was installed in Leh but soon deposed in favour of the original incumbent. Zorawar Singh returned to the warmth of Kishtwar for the winter of 1838, and he did so again in 1840 after his conquest of Baltistan.

'Elated with his success, Zorawar Singh now threatened the neighbouring states and even talked of invading Yarkand,' says Cunningham. But western Tibet, and especially those districts which had been attached to Ladakh when King Singge Namgyal had annexed Gugé soon after Father Andrade's stay, seemed 'more accessible'. Gugé's well-endowed monasteries would reward the Dogras with rich pickings, while 'their country produced the finest shawl-wool'. To anyone keen on controlling the *pashm* trade, western Tibet was a more attractive proposition than Lhasa or Yarkand. Gulab Singh, for one, 'would be highly pleased' by Tibet's subjection: 'it would throw the whole trade in shawl-wool into [his] hands'.[7] In effect, the wool-producing regions would at last be opened to the wool-traders of Ladakh, and the revenue from both would become a monopoly of Rajah Gulab Singh of Jammu.

At the head of an army of 6,000 Dogras and Sikhs plus a few turncoat Ladakhis, in May 1841 Zorawar struck south-east from Leh into Tibet. He followed the Indus upstream towards Kailas–Manasarowar. As Sven Hedin would discover, the distances were vast but the opposition was negligible. Even the temperature remained bearable. Rudok and then Gartok surrendered without a shot being fired. Headquarters were established west of Manasarowar near the infant Satluj. Zorawar himself is said to have made the circuit of the holy lake while his men, under the direction of two suitably zealous Muslim commanders, ransacked temples and monasteries as far away as Spiti, now in Himachal, and the Kumaon frontier. 'This work was executed with iconoclastic fury,'

Leh, the capital of Ladakh, is dominated by its own Potala-like palace. The city changed hands repeatedly during the Dogra encroachments of the 1830s and '40s

notes Cunningham. The Tibetans were few and appeared powerless to prevent the desecration. But when word reached Lhasa it was a different matter. An expeditionary force of some 10,000 Tibetan and Chinese troops was hastily assembled and in late summer began the long march up the Tsangpo–Brahmaputra valley.

By the time Zorawar got word of this approaching army it was November and getting colder. Winters of heavy snow are not typical of western Tibet; neither, though, are they particularly unusual. And whatever the snowfall, the frost commonly lasts for months. A temperature at which spit hits the ground as a gob of ice may have seemed low enough to cool even military ardour. Somewhat contemptuously, therefore, Zorawar despatched 300 men, and then another 600, to waylay the enemy. Each of these parties was 'surrounded and cut to pieces'. Their commanders, two brothers, were taken captive. Both actions seem to have taken place south of Rakas Tal, the lesser of the two great lakes somewhat west of Manasarowar.

Bowing to the exigencies of the moment, Zorawar now took the field in person. At the head of his combined force of about 6,000, he marched south to Purang (Burang), the Tibetan district north of Kumaon. He was unopposed until he drew near to Taklakot, site of a great market where Moorcroft had bought *pashm* in 1812. Here on 10 December both sides opened fire in desultory fashion. It continued for two days with neither side establishing an advantage. Then on 12 December Zorawar Singh was 'struck by a ball in the shoulder'. What sounds like a lucky potshot proved decisive. 'As he fell from his horse, the Chinese made a rush, and he was surrounded and slain.'

> His troops were soon thrown into disorder and fled on all sides, and his reserve of 600 men gave themselves up as prisoners. All of the principal officers were captured and out of the whole army ... not more than 1000 escaped alive, and of these some 700 were prisoners of war.[8]

'The Indian soldiers of Zorawar Singh fought under very great disadvantages,' adds Cunningham. As leader, five years later, of a commission tasked with fixing the Ladakh–Tibet border, Cunningham knew something of the terrain and had interviewed some of the survivors.

> The battlefield was upwards of 15,000 feet [4,572 metres] above the sea, and the time mid-winter when even the day temperature never rises above freezing point and the intense cold of night can only be borne by people well covered with sheepskins and surrounded by fires. For several nights the Indian troops had been exposed to all the bitterness of the climate. Many had lost the use of their toes and fingers; and all were more or less frost-bitten. The only fuel available was the Tibetan furze which yields more smoke than fire; and the more reckless soldiers had actually burned the stocks of their rifles to obtain a little temporary warmth. On the last fatal day not one half of the men could handle their arms; and when a few fled, the rush became general. But death was waiting for them all; and the Chinese gave up the pursuit to secure their prisoners and plunder the dead, well knowing that the unrelenting frost would spare no one.[9]

A handful of survivors found their way over the passes into Kumaon, from where news of the disaster spread to Katmandu and Simla and so

to Lahore and Jammu. But, of these fugitives, half were said to have
succumbed to the cold and nearly all had lost digits to frostbite. 'These
few, and the prisoners, form the whole number that escaped with their
lives,' reported Cunningham.

The rout was not on the scale of Napoleon's 1812 retreat from Russia,
nor has this deterred some latter-day enthusiasts from hailing Zorawar
as 'the Indian Napoleon'. Recalling another wintry fiasco, Cunningham
drew a more timely and apposite comparison. 'In this very month, and
in the same year, 1841, the British army, of about the same strength, was
destroyed at Kabul.'[10]

The death of Zorawar Singh and the annihilation of his entire army left
Ladakh wide open to the victorious Chinese and Tibetans from Lhasa.
While overwintering in western Tibet they reclaimed Gugé and the
Gartok region, and in the spring of 1842 entered Ladakh. Leh's Dogra
garrison was besieged and the country's titular ruler restored. But the
Ladakhis' delight in so unexpectedly regaining their independence from
the combined attentions of Jammu, Kashmir and Lahore was short-
lived. Within six weeks, according to Cunningham, they were 'rudely
awakened from their dream of liberty by the musketry of their old
enemies'. Another Dogra army had crossed into 'the land of the high
passes' as soon as the snows melted. With distressing ease it now drove
the Lhasa army back into Tibet and captured its commander. Without
the faithful Zorawar Singh, Rajah Gulab Singh was still as determined
as ever to assert his authority in Ladakh and exercise control of the wool
trade.

A peace treaty of September 1842 appeared to give him just that.
The four signatories included representatives of the governments of
Lhasa and Beijing on one side and on the other of the Sikh *khalsa*
(roughly 'brotherhood') and the newly styled 'Maharajah Bahadur
Gulab Singh' of Jammu. Ladakh's status was barely mentioned and
no Ladakhis are shown as signatories. The place of signing was about
as remote as could be. It was on the southern shore of the Pangong
Lake in the Chushul region of eastern Ladakh – in fact exactly where,
in late 2020, Indian and Chinese negotiators backed by tanks and
heavy artillery were striving to defuse another explosive Sino-Indian
confrontation.

In 1842, the Chushul treaty's terms for 'peace, friendship and unity' were agreed on condition that the old boundary between Ladakh and Tibet was re-established and no hindrance offered to Ladakhi traders entering any of the signatories' territories. As the treaty put it, 'we shall not even to the extent of a hair's breadth act in contravention of the terms agreed ... [regarding] the fixed boundaries of Ladakh and the keeping open of the route for wool, shawls and tea'. This was all very well, but it solved nothing. It left uncontested both the Sikh claim to Ladakh and the Dogra occupation of the country, which in turn ensured Gulab Singh's access to the *pashm* trade.

Four years later, in 1846, Major Alexander Cunningham and his border commissioners found little changed. Though the Ladakhi government was still functioning, it was Gulab Singh who ruled and his Dogra troops who, from various fortified positions, ensured compliance. The Ladakhis appreciated the safer roads and the more predictable duties. Indeed trade was flourishing. But as Cunningham noted, 'the maharajah himself [Gulab Singh] is the chief trader in his own dominions, more particularly in the two staple articles, the import of saffron, and the export of wool'.[11] The saffron came from the crocus fields of Kashmir, Himālaya's main source for this coveted spice. But of the shawl wool only half was now going to Kashmir; the other half was being despatched down-country to Kashmiri weavers who had settled in the towns and cities of the adjacent plains.

There, in the Punjab, everything had indeed changed. In 1839 Maharajah Ranjit Singh had breathed his last. The succession had then been contested, repeatedly, as the Sikh empire based on Lahore began to unravel. At the signing of the 1842 Chushul treaty in Tibet, the Sikh interest had been jointly credited to Sher Singh, already the third maharajah in succession to the great Ranjit, and to the *khalsa*, the Sikh component of Lahore's massive army which many Sikh *sardars* (leaders) regarded as a more legitimate sovereign than any short-lived pretender to the throne. A year later heads were rolling in Lahore, including those of Maharajah Sher Singh and of Dhyan Singh, the Dogra brother who had been Ranjit's chief minister and Gulab Singh's trusted champion at court. The army was now in control – if anyone was. Courtiers curried favour with the troops; contenders for the throne were touted, then toppled. Gulab Singh kept his distance and bided his time.

When his time came it was courtesy of the British. Still operating under the commercial alias of the East India Company, in 1816 the British had acquired by the Sagauli treaty with Nepal not only Kumaon and Garhwal but most of what is now Himachal. The last included a village on a ridge that, according to the then governor-general, made 'a very agreeable refuge from the burning plains of Hindoostaun'. By 1830 this Simla (Shimla) had become the favoured resort of British India's top brass. Policies were now being framed and decisions taken within a day or two's ride of the Himalayan barrier and an easy march of the Satluj and the Anglo-Sikh frontier. It was thus, from close range at Simla, that in the early 1840s the British watched the mounting chaos in Lahore with a mixture of genuine concern and an eye to the main chance.

The First Anglo-Sikh War, when it broke out in 1845, seemed like a case of spontaneous combustion. British minds were no doubt exercised by the need to redeem national pride after the shameful defeat at Kabul in the recent Anglo-Afghan War. There is also ample evidence of British agents intriguing with leading contenders in Lahore and even with Gulab Singh in Jammu. Rumours of these treacherous contacts were rife within the Sikh ranks and, coupled with grievances over arrears of pay and the lack of opportunities for plunder, fanned the army's belligerence and ignited a precipitate advance to the Satluj.

The British responded in force and with almost indecent alacrity. Compared to the Anglo-Nepal War, let alone Zorawar's campaigns in Ladakh and Tibet, the First Anglo-Sikh War promised a struggle to the death between near-equals, with high honours and substantial rewards awaiting the victors. Well supplied with artillery, Lahore's forces were estimated at around 150,000 men, while those available to the British were comparably armed and even more numerous. Only a fraction of both were involved in the war's three setpiece battles yet these were still titanic clashes between the most formidable military powers in mid-nineteenth-century India.

The result, however, was scarcely in doubt. The army of the *khalsa*, having already dispensed with many of its officers and being now unsupported, if not betrayed, by the court in Lahore, was ill-led and uncoordinated. The British were methodical, well served by informants and had as their second-in-command the governor-general himself. In due course they called up massive reinforcements. The back-up which the Sikhs had every reason to expect from Jammu's hardened Dogra regiments simply failed to materialise. They never crossed the Satluj. Evidently Gulab Singh had renewed his overtures to the British.

The price paid for the Dogras' neutrality was revealed in the second of two treaties of 1846. The first, that of Lahore signed with the kingdom's regency council (because the latest maharajah was a minor), included a war indemnity of 15 million rupees. This colossal sum, even when reduced to 5 million, was beyond the means of Lahore's depleted treasury. Only a fraction of it was forthcoming, and hence, as surety for the remainder, the Sikhs were prevailed on to hand over to the East India Company 'all the hilly or mountainous country, with its dependencies, situated to the eastwards of the river Indus and the westward of the river Ravi'.[12] This was hopelessly vague; but as per the second treaty, it was taken to mean Kashmir plus all those 'dependencies', like Ladakh and Baltistan, to which Kashmir's former rulers had laid claim.

The second treaty, that of Amritsar, was exclusively with Gulab Singh. In return for the man now acknowledged as 'Maharajah of Jammu and Kashmir' paying 750,000 rupees towards the outstanding indemnity, he and his heirs were to enjoy independent sovereignty under British oversight of all the 'hilly or mountainous country *et cetera*' just transferred to the Company. Gulab Singh had finally got his way, 'not wholly or in full measure', as Pandit Nehru would put it in his Independence Day oration, 'but very substantially'. Since the payment was in part settlement of the war indemnity, the British had not technically 'sold Kashmir to the highest bidder'. It just looked that way. There were no other takers; the British themselves had no desire to assume the burden of responsibility for this vast, rugged and often disturbed chunk of Himālaya; and as with Nepal, it suited them strategically that it remain a buffer, insulating India from whatever ructions the imperial surges of China or Russia might occasion beyond the mountains.

It remained only to discover what the buffer comprised and how far it extended. This was a matter of urgency. Since Gulab Singh had been responsible for Zorawar's Tibetan incursion, the British feared he might try again; and so airily had the treaty defined his new domain, he might suppose he was within his rights in doing so.

The border commission headed by Cunningham, the first official British intervention in western Himālaya, was on its way within eight weeks of the treaties being signed. But in 1846 the commission confined itself to the Tibetan and Ladakhi frontiers of Kinnaur, Kulu, Lahul and Spiti. These districts, all today in Himachal, had been detached from Sikh and Dogra rule so that, as Cunningham pointed out, the *pashm* from western Tibet could be funnelled through them to weaving centres in the Punjab outside Gulab Singh's jurisdiction. In effect, the tug-of-war over the *pashm* trade was dispersing the weavers and dissipating the shawl industry. Soon more cashmere was being woven in Amritsar and Jalandar than in Kashmir. By the end of the century shawls were being produced in Paisley and Manchester; and by the end of the next century the eastward routing of *pashm* and *tus* handed to communist China and Mongolia a near monopoly of pashmina and shahtoosh.

Neither Tibetan nor Dogra representatives attended Cunningham's 1846 demarcation; and it was the same the next year when Cunningham, now accompanied by Lieutenant Henry Strachey and Dr Thomas Thomson, returned to Ladakh proper; the Chinese and Dogras being either uninformed or uninterested, they simply failed to show up. As a result, the longer the border commission remained in the field, the less it was bothered with borders. That between Ladakh and Tibet had anyway been settled by the Chushul treaty, while those to the north and west of Ladakh were so remote and inaccessible as to defy regulation. It had always been assumed that the commission would be as much about fact-finding as frontier-making and so it now proved. As an exercise in scientific exploration the 1847–8 Ladakh border commission was the most thorough yet undertaken anywhere in Himālaya.

Each of the three commissioners was a specialist in his own field, and to each was allocated an area roughly the size of the British Isles. Each, too, would present his findings in a solid work of scholarship, all of them informative if indigestible. Strachey, the most perceptive of Himālaya's nineteenth-century geographers, took on western Tibet. With Bhotia help he had crossed from Kumaon to Kailas–Manasarowar in 1846 and now began systematically criss-crossing the whole region as he mapped his way from Leh to Pangong Tso to Gartok and Gugé and then back again. 'The result, in so far as it inevitably tends to emphasise the integrity of the region, is a curious memorial to one who was supposed to be defining a boundary across it.'¹³

Cunningham himself concentrated on Ladakh. To Calcutta's scholars Alexander Cunningham was best known as the young engineer who in the 1830s had devoted fourteen months to cutting a shaft down through the iron-clamped rock of the great Dhamekh stupa at Sarnath (near Varanasi). It cost him 500 rupees and he found nothing of value; but he correctly identified the stupa as Buddhist and would devote the rest of his life to Buddhist researches. For someone so curious about the Buddhist faith Ladakh was of course the ideal posting. But while Hodgson in Katmandu was preoccupied with discovering and interpreting texts, Cunningham in Ladakh was more excited by statuary and monuments. On his return journey to India he logged Buddhist sites in Kashmir and then Swat, the valley thought to have been the ancient Buddhist centre of Udyana (Uddiyana) from where Padmasambhava flew via Rewalsar to sow the first seeds of Vajrayana Buddhism in Tibet.

When Cunningham arrived back in Simla in 1848 he was accompanied by a camel loaded with Buddhist statuary, three unknown Sanskrit dramas and 'the oldest dated inscription hitherto found in India'. This was not what was expected of a military commissioner, but his report on *Ladák, Physical, Statistical, and Historical* made ample amends and was more exhaustive even than Moorcroft's. It was also his last word on Himālaya. In 1861, after the suppression of the Indian Mutiny/National Uprising, Cunningham resigned from the army with the rank of brigadier-general and spent the next twenty-five years touring, inventorying and occasionally conserving those mainly Buddhist sites of north India that Curzon would hail as 'the greatest galaxy of monuments in the world'. Cunningham is celebrated today as the founder of the Archaeological Survey of India, indeed the father of Indian archaeology.

And finally there was Thomas Thomson. The inclusion of a botanist in a commission to somewhere as lacking in vegetation as Ladakh might have seemed an extravagance. He was no surveyor, had to borrow a sextant and compass and got Cunningham to show him how surveyors used them. But Thomson was a doctor by profession, a chemist and mineralogist like his father, and a most methodical observer who, even when captured by the Afghans during the recent war, had persevered with his *flora afghanica*. He also developed an affinity for the mountains, which may explain why it fell to Thomson to tackle the northern boundary of Gulab Singh's new domain.

This northern boundary was taken to follow a mountain range running parallel to the upper Indus but beyond it. Thomson called the range 'the Muztagh or Kouen Lun', Cunningham 'the Bolor or Karakoram'. No European had yet crossed it, though Moorcroft and Vigne had collected native reports of the main routes. The commissioners were aware of these reports but failed to realise that the four names they were using for the range were in fact the names of four distinct mountain systems. South to north, from the Indus to the Tarim basin on the other side, the mountain barrier was nearly 300 kilometres deep. Moreover from east to west it was around 500 kilometres long; after forming the northern ramparts of Ladakh and then Baltistan, the barrier continued on to Gilgit before merging with the Pamirs and the Hindu Kush. Under the terms of the ill-drafted Amritsar treaty this whole arena could be claimed – and was being claimed – as part of Gulab Singh's Jammu and Kashmir state.

The enormity of Thomson's assignment dawned on him only slowly. After a gentle ride up through Spiti to Ladakh, he headed north from Leh for the Nubra valley in October 1846. Nubra, the last oasis of cultivation on the calvary that was the Karakoram route from Leh to Yarkand, was where Moorcroft had had to turn back. Thomson too shied away; October was too late in the year for exploration at altitude. Instead he followed the Shyok tributary of the Indus down to Baltistan. A plan to continue downriver to Gilgit for a bit of boundary work had to be abandoned because the area 'was not in a fit state for scientific investigation' – that is, the Dogras and the native Dards were fighting. And a third plan for wintering in Kashmir fared no better; the passes were already snowbound. There was nothing for it but to overwinter in Baltistan. Thomson arrived back in Skardu on Christmas Day and stayed put for six weeks.

In February 1847 he tried again to reach Gilgit, only to find the fighting had restarted. Turning back for the umpteenth time, he retreated up the Indus to the Dras tributary and, through waist-deep snow, finally escaped down to Kashmir and springtime. It might be supposed that by now the good doctor had had enough. For six months he hadn't seen a fellow countryman or slept in a bed. Yet he was no sooner in reach of both than he turned back again. Permission had been given for him to explore a new route to Ladakh through Zanskar and then make another attempt on the Karakoram trail from Leh to Yarkand.

Sven Hedin would rate this 1848 journey of Thomson 'one of the most important and successful ever undertaken against the secrets of the highest mountainland on the earth'.[14] The new route by way of Chamba and the River Chenab (Chandrabhaga) proved a plantsman's delight, but the onward trail through Zanskar became a botanist's despair: 'I had nowhere before seen a country so utterly waste.'

Strachey was in Leh to offer encouragement. Thomson rested up for a week but left again in mid-July. Of his travelling arrangements he says next to nothing. He evidently hired 'cattle', presumably yaks, and retained his horse for crossing rivers. The Khardung La, the first of the five great passes on the world's highest trade route, took him to the Shyok River and the Nubra valley where he had turned back the previous year. Here he stocked up with twenty days' supplies and veered off to the north-east up a trail well advertised by the bleached skeletons of long-dead pack animals. The going was 'exceedingly steep'. A ridge measured at 15,300 feet (4,660 metres) by his boiling-point thermometer didn't even have a name, and the view of the onward trail confirmed reports of 'its extreme difficulty'. The only traffic was 'a party of merchants' travelling from Yarkand to Leh. They were not carrying *pashm*. Their ponies were too emaciated to be carrying anything other than a little grain.

The next pass was the Sasser, which he measured as 17,600 feet (5,364 metres). It was reached by crossing glaciers and it descended on to a stony plain. Looking back, he was surprised to see that it formed part of 'a grand snowy range ... with a direction from south-east to north-west'.

> Many very lofty peaks rose above the others at intervals. The height of the more distant ones I could not venture to estimate, but I felt at the time fully convinced that a very high peak, just opposite to me, and distant ... about ten miles from the edge of the plain, was 6000 or 7000 feet higher than the ground on which I stood, or at least 24,000 feet [7,315 metres] above the level of the sea.[15]

He couldn't be sure of course, but Thomson being Thomson, he 'thought it right that I should state my impression at the time, formed without any wish to exaggerate'.

The range was in fact that now known as the Saltoro, and Thomson's peak, assuming it was Saltoro Kangri (25,400 feet or 7,742 metres asl), overlooks what is today the world's most elevated battlefield. Saltoro Kangri straddles the absurdly named 'Actual Ground Position Line' from behind which the mountain troops of Pakistan and India have been eyeballing one another for four decades.

The trouble here began in 1957 when the British climber Eric Shipton received permission from the Pakistani authorities to lead an attempt on Saltoro Kangri. To avoid Indian-held territory this meant venturing into the no man's land occupied by the 76-kilometre Siachen glacier, a move which India saw as an infringement of the status quo. Protests were lodged, but other expeditions kept coming till in 1984 Indian forces made a pre-emptive strike. The entire Siachen glacier totalling around 2,500 square kilometres was captured and has been held by India ever since. But the provocations went on; in what's called oropolitics, further mountaineering expeditions served only to rack up the tension. And the fighting went on too. Since the casualties were more often the result of avalanches and frostbite than of enemy fire, even a 2003 ceasefire failed to staunch an appalling loss of life which now stands at over 2,000.

Blissfully oblivious of entering what would become not just 'the world's highest battleground' but surely its least desirable, on 19 August 1848 Thomson left his tent standing at 17,200 feet (5,242 metres). He had already left his horse behind. Now, like a mountaineer approaching the summit, he discarded anything not essential to a final push on to the Karakoram Pass.

But if the pass was meant to be the climax of his two years in the mountains, it failed miserably to live up to expectation. He was already 'suffering very considerably from the rarefaction of the air' with a headache that increased with the slightest exertion. The approach to the pass was uninspiring, a gently sloping gradient of snow-free gravel; not so much as a lichen grew there and the only signs of life were 'a bird the size of a sparrow, a bright metallic-coloured carrion-fly and a small dusky butterfly'.

Most disappointing of all, there was no view. By now he was resigned to seeing nothing beyond the mountains but more mountains, yet it was hard to abandon all hope of a distant prospect of the sands of Xinjiang and, through the haze, perhaps a glimpse of a minaret or a stand of poplars. Instead, after a final scramble up loose fragments of rock for 500 metres, 'I found myself on top of the Karakoram Pass – a rounded ridge connecting two hills which rose somewhat abruptly to perhaps 1000 feet above me.' Using his boiling-point thermometer, he got a water temperature of 180.8 degrees, giving a height of 18,200 feet (5,547 metres), which is pretty much what trigonometry would confirm.

Returning by the way he'd come, Thomson paused near the Sasser to assess the glaciers passed on the ascent. He could only guess at their length and it's unclear whether any of them was the upper reaches of the Siachen. He was certainly within easy reach of Himālaya's longest glacier, but being quite unequipped for ice or snow he was not inclined to stray far from terra firma.

Thomson's account of his journey often reads more like a catalogue of the region's few flora than a narrative of exploration; mishaps and dramas play no part in it, and one can only wonder who were his informants and where he found guides and porters – if indeed he did. The plants are everything; and where there is no flora, the prose is spare to the point of threadbare. The mid-nineteenth century was still very much the 'age of the passes'; the 'age of the peaks', of belays and bivouacs and pipe-smoking summiteers, had yet to dawn.

14

Mountains of Destiny

An emulous man on a mountain is a profanity. He should be there
as a worshipper, impersonal, a pilgrim without a name, lost in the
quest. A boast is unthinkable ... Shepherds and goatherds should
have the naming of mountains.

Edmund Candler (1919)[1]

'Because it's there' – George Mallory's answer to the question of why he
wanted to climb Everest – seems not to have been the deeply existential
revelation many have supposed. Just back from the abortive 1922 assault
on the mountain (the one during which Henry Morshead had lost
three fingers) Mallory's curt reply may have had nothing to do with
his 'having in his wisdom distilled the perfect notion of emptiness and
pure purpose'. More probably he just said the first thing that came into
his head. At the time he was nearing the end of a speaking tour in the
US. It was getting late and his presentation was over. According to a
friend, the throwaway line was aimed at 'a bore who stood between him
and a much needed drink'. It got picked up by the *New York Times* and
entered the realms of perceived wisdom without undue scrutiny.[2]

Certainly, neither Mallory nor anyone else attempting an ascent
of Everest did so in order to establish its height. You didn't need to
climb a mountain to measure it. By the end of the nineteenth century
the heights of most of southern Himālaya's more prominent peaks
had been calculated from chains of triangulation conducted across
the length and breadth of India as part of the Survey of India's Great
Trigonometrical Survey (GTS). Thanks to the GTS, India became

more accurately mapped than anywhere outside Europe, with fingers of this triangulation extending deep into the Himalayan piedmont from Assam to Kashmir and the Hindu Kush.

Mount Everest had been so named in honour of Colonel George Everest, the famously irascible supervisor of the GTS. The colonel's completion of the measurement of a 2,500-kilometre Great Arc of the Meridian from the southern tip of India to Dehra Dun had provided a spine for the skeleton of unassailably located trig points from which the heights of the central Himalayan peaks had been calculated.[3] But once a mountain had been measured, once a name and a space on the map had been found for it and its profile sketched or photographed, there it indubitably was. If remotely feasible, mountaineers like Mallory felt compelled to climb it.

Those for whom the mountains were home did not share this compulsion. Until foreign expeditions became a recognised source of income and advancement, Himālaya's peoples regarded scaling inaccessible peaks as an eccentric and unnecessarily provocative obsession. Did one need to swim the Ganges, clamber over the dome of the Taj Mahal? Such feats were disrespectful. Pilgrims usually processed round mountains, not up them. In their search for hills on which to mount their theodolites, India's surveyors had been repeatedly challenged by locals insisting on the inviolable sanctity of a favoured eminence. Icier peaks sacred to the gods, or peaks that were themselves gods, needed to be approached with placatory caution or avoided altogether. Tsari's sodden slopes and Kawa Karpo's misty profile remained unclimbed out of respect for local beliefs. Even today, of the world's fourteen peaks of over 8,000 metres, only Kangchenjunga – and then only its last few steps – remain virgin territory. A British expedition in 1955 would turn back less than 2 metres from Kangchenjunga's summit to oblige Sikkim's ruler; subsequent summiteers, of which there have been many, claim to have followed suit.

Like other Himalayan giants, Kangchenjunga, the world's third highest peak, already had a name. This was not the case with remoter peaks like Everest and K2. Naming what the GTS had previously called 'Peak XV' as 'Mont [sic] Everest' placed the name of the cantankerous colonel 'just a little nearer the stars than any other lover of the eternal glory of the mountains', enthused one of his Survey successors; yet it was not uncontested. Sven Hedin was among those who claimed that Everest did have a local name. It was Chomolungma, or to the Chinese Qomolungma. 'Everest' should therefore have been dropped. It was retained because 'Chomolungma' was thought to apply to the whole cluster of peaks in the vicinity of Everest, not just to the highest. Conversely K2 was eventually denied to Godwin-Austen on the grounds that condoning foreign names was not Survey of India policy. Other even remoter peaks in the Pamirs have been named and renamed so often that their identity is still a source of confusion.

To clarify such matters, and to give Himalayan mountaineering some institutional respectability, in the late 1920s Sir Geoffrey Corbett, a keen walker and member of both the British Alpine Club and the elite Indian Civil Service, alongwith Major Kenneth Mason of the Survey of India, founded the India-based Himalayan Club. In the first issue of the Club's *Himalayan Journal* it was Corbett who addressed the thorny matter of how to pronounce 'Himalaya' ('superior folk' knew to say 'Hĭmālĭyă'), but it was Mason as editor of the *Journal* and later India's surveyor-general who upheld the club's role as mediator and facilitator of all Himalayan travel and exploration, not just mountaineering. A network of club correspondents and local secretaries provided on-the-spot data for visiting travellers; reports were circulated, maps commissioned, advice dispensed, claims assessed.

Mason was not himself an alpinist, but he was learning. In 1909 he had been engaged in resurveying parts of Kashmir and what he calls the 'Punjab Himālaya'. 'None of us had climbed on ice or snow, and I alone had done some climbing on a rope on rock.' Ice axes proved unobtainable in India but he managed to borrow one until more arrived from England. He also ordered skis. 'I ... took them to my survey stations and learnt the art by trial and error and with the help of Caulfield's book *How to Ski* ... It was all great fun.'[4]

It was also useful experience for Mason's next assignment. In 1908, by agreement with the tsarist authorities and as if to mark the end of the Great Game, it had been resolved that, for both scientific and political reasons, the GTS's web of Indian triangulation should be linked to that being pushed across Central Asia by Russia's surveyors. To that end, by way of Gilgit, Yasin and the Darkot Pass over the Hindu Kush, in 1911 British surveyors reached the Wakhan Pamir and the upper Oxus, but they found hauling their instruments up the peaks beyond their means. The following year was worse. Two men were killed when lightning hit their tent near Gilgit and their colleagues were 'more or less incapacitated for the rest of the year' by the same storm. Their commanding officer was then stricken with acute appendicitis 'and died six days later'.[5]

He was succeeded by Mason. In 1913, with two fellow surveyors plus a doctor, some eager Gurkhas and a team of Hunzakut porters, Mason advanced through the mountains and glaciers on the north side of the Hunza valley. The Russian survey had by now reached the easternmost limit of the Pamirs in the no man's land between the territories of British-controlled Kashmir and Chinese Xinjiang. Mason therefore aimed to intercept the Russian triangulation 'where three empires meet' (those of Russia, China and Great Britain). 'We did not attempt difficult climbs if an easier ascent was possible ... but it would be wrong to give the impression that this country is easy to triangulate.' Trig stations could be as high as 5,000 metres and might have to be occupied for a week or more while waiting for the weather to clear. New peaks in the western Karakoram on the other side of the Hunza declivity were plotted, one of 7,885 metres, and Mason gave his skis another try. He thought he might be the first ever to use them in the Himalayas.

In August 1913 the desired Anglo-Russian rendezvous was effected amid deep snow at a spot called Sarbulak on the Taghdumbash (Tukhtamish) Pamir. Twenty years earlier, while passing through on his way to meeting with Younghusband in Chitral, Lord Curzon had counted eight Pamirs. He rightly described them as neither steppes nor plateaux but elevated valleys so choked with glacial alluvium as to be little more than troughs separated from one another by heavily eroded mountain ranges. All ran east to west, and the annual precipitation of just 2 inches (12 millimetres), most of it snow, was drained by the headwaters of the Oxus. The only exception was the Taghdumbash Pamir. This had a north–south alignment as if to block off the eastern extremities of the other Pamirs and was drained by a feeder of the Yarkand River, itself a tributary of the Tarim of Xinjiang. The Taghdumbash was therefore a natural, if almost featureless and as yet undefined, frontier. On it the Russian triangulation had duly terminated with a triangle, one of whose sides – Mason called it 'the junction side' – was a sight line between Sarbulak and another terminal trig station to the north called Kukhtek.

Ideally the 7 kilometres that separated these two stations might have been physically measured by laying out a baseline. But in blizzard conditions at 4,500 metres asl, clearing obstructions and levelling the ground, then fiddling about with micrometers and spirit levels, was out of the question. Instead, the success of the combined operation was to be judged by how closely the length of the 'junction side' as calculated by the Russian triangulation agreed with that arrived at by the British triangulation. The Russian triangulation, which had been carried south-east from a baseline at Osh in what is now Kyrgyzstan, came up with a value for the Kukhtek–Sarbulak line of 7,134.9 metres. The British triangulation, which had been carried north from a baseline in Kashmir, made it 7,133.4 metres. The discrepancy arising from a combined total of over 700 kilometres of triangulation across some of the most inhospitable terrain on earth was just 150 centimetres.

For this satisfying achievement, and for a subsequent survey of the Shaksgam River north of the Karakorams, Mason was awarded the Royal Geographical Society's Founder's Medal in 1927. He would not return to the Pamirs, but in the following year, just as he and Corbett were enrolling the Himalayan Club's first members, the Pamir region was yielding more of its secrets to an expedition that included genuine mountaineers. Ole Olufsen's Danish expedition of 1896–7 had concentrated on the archaeology and ethnology of the southern Pamirs. This Russo-German Alai-Pamirs Expedition of 1928 included Alpine climbers and concentrated on Himalaya's north-western extremity. Here the Pamirs interlock with the Alai and Tien Shan ranges of Kyrgyzstan amid a congress of glaciers and gelid peaks. To British Himalayanists it was all unknown territory. Indeed the Russo-German expedition's eleven 'crack climbers' were themselves newcomers to the region.

A mixture of German and Austrian nationals, these non-Russians of the Alai-Pamirs Expedition spearheaded a resurgence of Berlin's scientific ambitions after the cultural and financial ruin resulting from the First World War. To fund and administer the revival, a Notgemeinschaft der Deutschen Wissenschaft (Emergency Association for German Science) had been set up, and in the late 1920s had despatched three ambitious 'community expeditions'.

The first was a maritime affair aimed at charting the ocean floor of the Atlantic. A ship called the *Meteor* had begun tacking back and forth between the continental shelves of Africa, South America and Antarctica, and since its soundings were of special interest to an advocate of continental displacement, the great Alfred Wegener had been asked to join it. He had declined. Ensconced in the University of Graz at the time, and busy with the much expanded fourth edition of his *The Origin of Continents and Oceans*, Wegener was more drawn to the Notgemeinschaft's second project – an attempt to gauge the thickness of the Greenland ice cap. He was duly chosen as that expedition's leader; and thus it was that in November 1930 the enigmatic genius who had come up with the most convincing explanation of how the Himalayas had been formed met his death trekking across an Arctic ice cap from the expedition's 'mid-ice' station to the coast.

The Happily the Notgemeinschaft's third venture fared better. Nikolai Petrovich Gorbunov and Willi Rickmer Rickmers, the joint leaders of the Russo-German Alai-Pamirs Expedition, were suitably circumspect. They actively discouraged their colleagues in anything that smacked of 'adventure' and, 'fearing to arouse suspicion' in such a politically sensitive region, had made no mention of the Pamirs at all in the expedition's original prospectus. 'All the same the Chinese governor of Kashgar concentrated troops on the border, and many Kirghiz fled into Chinese territory, when they heard of our coming.'[6] With thirty Europeans and a caravan of 180 horses and 70 camels, the expedition might well have been taken for a small army. But, having left Osh in June 1928, it soon split up into scientific study groups as surveyors, mineralogists, geologists, botanists and zoologists went their separate ways.

The heights seem to have been reserved for the Germans, three of whom scaled a colossus of 7,134 metres. With none of the Himalayan giants yet conquered, they claimed it as the highest peak so far climbed. Formerly known as Mount Kaufmann (after the general who had conducted the tsarist conquests in Central Asia in the 1860s), it had just become Lenin Peak in the belief that it was the highest in the Soviet Union. In 2006 it would be renamed Avicenna (or Ibn Sina) Peak after the eleventh-century philosopher who is acclaimed as a son of Tajikistan. The Kyrghyz, however, have their own name, indeed names, for a mountain that straddles the frontier. And, under whatever dispensation, Lenin Peak was never the highest in the Soviet Union, nor is it the highest in the Republic of Tajikistan. That honour belongs to the nearby 7,495-metre massif of Stalin Peak – which became Peak Communism in the Krushchev era and Ismoil Somoni Peak after the break-up of the Soviet Union.

For Rickmers, the leader of the German team, the great heights in this corner of the Pamirs were more than matched by the great length of the glaciers. 'The biggest discovery, that of the Fedchenko glacier, came as a complete surprise,' says Rickmers. A narrow and sinuous river of ice, it flowed in a northerly direction for all of 77 kilometres and was reckoned the longest in the world outside the polar regions. It was rivalled, and perhaps still is, only by the then 76 kilometres of the Siachen glacier in the eastern Karakorams. Mercifully the Fedchenko was allowed to retain the name of the Russian explorer Alexei Fedchenko, so leaving a neighbouring cascade of ice to be lumbered with the uninspiring title of the Notgemeinschaft glacier (the Emergency Association glacier).

Returning to Europe buoyed by the success on Lenin Peak, Rickmers immediately got in touch with the Himalayan Club in India. Corbett had just resigned on health grounds to be succeeded as Hon. Secretary by Gerard Mackworth-Young, another keen hillwalker and senior member of the Indian Civil Service who, seventeen years earlier, had stumbled on the ruins of Tsaparang and Toling amid the Sutlej canyons of western Tibet. But it was to Kenneth Mason as a fellow Pamirs traveller that Rickmers addressed himself. He was acting, he explained, on behalf of a climbing colleague, Paul Bauer of the Munich-based Alpine Academy. Bauer was hoping to pit the skills of a party of young Austrian and German climbers against the central Himalayas. 'They want to test themselves against something difficult,' explained Rickmers, 'some mountain that will call out everything they've got in terms of courage, perseverance and endurance.'[7]

Mason, notwithstanding the legacy of war, was happy to oblige. By July 1929 Bauer and his eight-man team, including one of the Lenin Peak conquerors, were being briefed in Mason's Calcutta office. Three days later they were in Darjeeling where, courtesy of the Himalayan Club's local representative, ninety porters, some of whom had been with Mallory and Morshead on Everest, awaited them. Undeterred by the onset of the monsoon, Bauer and his team, his *Mannschaft*, set off immediately. The Fatherland's fraught and redemptive affair with the Himalayan heights had begun.

Thanks to their Indian raj, the British enjoyed a political and geographical advantage in the race to claim the first of the Himalayan giants. Along with two Gurkha porters, Alfred Mummery, a legendary British alpinist in a field awash with 'legendaries (and not to be confused with George Mallory of Everest fame), had disappeared during an attempt on Nanga Parbat as early as 1895. Ten years later, in the course of a bad-tempered Anglo-Franco-Swiss expedition, a French climber and three porters had been buried by an avalanche on Kangchenjunga. Soon after, during numerous often solo expeditions between 1907 and 1921, Alexander Kellas, the Nobel prize-winning chemist from Aberdeen who had championed the employment of Sherpas, 'enjoyed himself' (as Mason put it) conducting self-experiments into the effects of altitude while climbing to 7,000 metres among Kangchenjunga's subsidiary peaks. Kellas, however, showed little interest in 'summit-bagging'. He rarely wrote up his exploits and seemed to relish climbing for the sheer hell of it.

But if the British had had a head start in the Himalayas, by 1929 when Paul Bauer and his *Mannschaft* began their first Himalayan assault, Albion's alpinists appeared to be resting on their laurels – or perhaps it was their wreaths. Though they had already climbed past the 28,000-foot (8,5344-metre) mark on Everest, the loss there in 1924 of the revered pairing of George Mallory and Sandy Irvine had triggered something of a national catharsis. Flags had been flown at half-mast and the only climbers' memorial service ever to have been held in St Paul's Cathedral saw the king and the prince of Wales joining the massed mourners. It was followed by another commemorative event in the Royal Albert Hall attended by almost every climber who'd ever glimpsed Everest, along with thousands who hadn't, like the now Sir Francis Younghusband, chairman of the organising Mount Everest Committee.

Yet the usual response to the death of a climber – the announcement of a follow-up assault of the same mountain ('we owe it to his memory') – was muted. For one thing there was, and still is, the possibility that, before they disappeared 'into the silence', one or both climbers had actually reached the summit. In that case an unseemly rush to duplicate their feat would be neither respectful nor necessary. And anyway, who could possibly take over from the immortal Mallory? According to another legendary alpinist, Mallory had attempted Everest not 'because it was there' but 'because Mallory was Mallory'. Without Mallory, why bother?

The inimitable George Mallory inviting sun-stroke on the approach to Mount Everest

Nor, more to the point, was a new assault actually possible. Everest stands on the frontier between Nepal and Tibet. At the time, Nepal was firmly closed to foreigners and had been for over a century. The Everest reconnaissances and assaults of the 1920s had therefore been made from Tibet under permits issued by the Tibetan government. By the terms of the 1905 Younghusband treaty, these were usually issued only to official British expeditions or on advice from the British authorities. But in the case of Mallory's 1924 expedition, a film crew had also been admitted, and though the reform-minded thirteenth Dalai Lama was agreeable, conservative elements among Tibet's monastic establishment objected. They took exception to the growing number of Europeans in Lhasa (earlier in the year David-Néel and Yongden had been there) and especially to 'British expeditions marching across the southern frontier of the nation, disturbing the deities amd corrupting the people'. A gung-ho Tibetanist like 'Hatter' Bailey, then serving as the British political officer in Sikkim, thought these reactionaries needed to be confronted by more Anglo-friendly Tibetans lest they invite in the Chinese.

Such was the situation that brought Bailey back to Lhasa in June 1924. While decrying the Mallory expedition's tendency to provoke hostility, Bailey seems to have attempted to silence the conservative opposition to the Dalai Lama by masterminding what Alex McKay calls a coup d'état. If so, it failed. Anglo-Tibetan relations remained under strain and would go into freefall when the film of the expedition was released. It was bad enough that *The Epic of Everest* contained unflattering footage of Tibet. 'Aristocratic Lhasa did not take kindly to scenes of local men and women delousing their children and eating the lice.' Worse, though, were the theatrical effects with which the film's producers saw fit to introduce its London screenings. In a fancifully designed set representing a Tibetan courtyard, a troupe of seven berobed lamas specially shipped from Gyantse for the occasion performed a temple dance to the tinkling of bells and the blare of Tibetan wind instruments. 'High Dignitaries of Tibetan Church Reach London', announced the *Daily Sketch*; 'Bishop to Dance on Stage; Music from Skulls'.

'What became known as the "Affair of the Dancing Lamas" had profound and lasting consequences,' notes the Everest chronicler Wade Davis. The Dalai Lama himself was horrified. An order for the arrest of the offending monks was issued and Bailey was informed that no further permits for entering Tibet to climb mountains would be issued. 'Reinforcing the strength of the traditionalists, [the affair] undercut the reforms of the Thirteenth Dalai Lama,' writes Davis. Had the reforms been pursued, Tibet 'would no doubt have [been] in a much stronger position to cope ... with the Chinese onslaught of 1949 and the subsequent invasion that led, a decade later, to the death of a free nation'.[8]

More immediately, the ban meant a hiatus in Everest expeditions. Not till 1933 would the Everest Committee resume its programme of laboured reconnaissances and abortive assaults, and then only to limited effect. In quick succession four British Everest expeditions would be mounted in the run-up to the Second World War. Bottled oxygen began to find favour and the new generation of climbers were better equipped than Mallory and Irvine. But they got no higher, and though they did find Mallory's ice axe, they shed no new light on Mallory and Irvine's disappearance.

To the British in 1929, therefore, Everest remained something of a shrine, more like an imperial war grave than an international challenge. Paul Bauer was accordingly given to understand that a German assault was quite out of the question. Polite as ever, Bauer didn't object. He had anticipated having to look elsewhere and had even made a list of his preferences. It included Kangchenjunga and Nanga Parbat, in that order. As the world's third and ninth highest peaks, both were over the unattained peak height of 8,000 metres and, even if they proved quite unclimbable, would be the perfect testing ground for Bauer's youthful team of mainly Bavarian alpinists.

The first German mountaineering expedition to Himālaya concentrated on Kangchenjunga. Like the subsequent attempts on Nanga Parbat it was organised by Paul Bauer (standing, centre). Peter Aufschnaiter (standing left) would accompany Heinrich Harrer on their escape into Tibet in 1944

News that Kangchenjunga was available had gone down well. It was much the most convenient of the Germans' options, being easily reachable from Darjeeling with its railhead and its bazaars thronged with potential porters. Eighty-six hillmen, plus fifteen Nepali Sherpas provided by the Himalayan Club, were thought sufficient for an expedition which, though planned with military efficiency, was positively Spartan by Everest standards. 'Everything, down to the level of boot studs and coffee beans, was minutely counted, measured, weighed, and discarded if it did not meet the test of strictest necessity.'[9] From behind rimless spectacles and a beard as fair as his complexion, Bauer, ex-army and a post-war law graduate of Munich University, had overseen the arrangements. He was methodical, judicious and disciplined. He was also a passionate patriot who bitterly resented the crippling reparations imposed on Germany by the Great War allies. For Bauer, pride of race and love of Fatherland were now to be redeemed in a quest for unspoilt horizons and indisputable triumphs. As he explained it:

> The mountains gave us back what the urban environment threatened to take from us. They proved to us that courage, the love of fighting, and toughness provide us with eternal values. In the despair of those days ... we had to prove – and prove again and again in defiant opposition to the spirit of the times – what could be achieved with those virtues even though everything stood against us.
>
> Out of this, the German Himalaya idea was born, and it was in this spirit that the first German Himalayan expedition had left [for India] in 1929.[10]

The expedition was gone for six months. When it returned to Munich it had claimed no summits and broken no records. Neither, on the other hand, had anyone been lost. Skills honed on the rockfaces of the Alps had been transferred to the great wall of Kanchenjunga's North Ridge; the latest in lightweight climbing tackle and weatherproof clothing had stood the test of the monsoon blizzards; and that redemptive camaraderie born of national humiliation in the committee rooms of Versailles had survived long nights in icy dugouts and exhausting days of tunnelling through frozen obstructions. The *Alpine Journal* hailed the expedition's partial ascent as 'a feat without parallel'. Mason concluded that there was now 'no reason to believe that the summit of Kangchenjunga at 28,146 feet was unattainable'.

Bauer agreed; and in 1931 he returned to the fray. This time a German climber fell to his death, as did one of the porters. Then, in appalling conditions, the thirty-five-year-old Bauer 'strained his heart' ... and had to descend. It sounds like the prelude to the fate that had overtaken Alfred Wegener on the Greenland ice cap a year earlier. Unlike Wegener, Bauer survived. After spending a night in an ice cave without sleeping bag or blanket, he somehow 'dragged himself down to Camp IX'.[11] Other nations were queueing up to try their luck on Kangchenjunga, and monsoonal Sikkim was turning out to be a tough testing ground.

'The German Himalayan idea' would have to be realised somewhere else in Himālaya.

Back in Munich the climbing fraternity had come to the same conclusion. Dr Willo Welzenbach, the strongest of Germany's inter-war climbers, had been an obvious choice for the Notgemeinschaft's Pamirs expedition, yet it was on the Great Himālaya that Welzenbach's sights were set. After studying reports of Mummery's fatal attempt on Nanga Parbat in 1895 Welzenbach became convinced that 'Nanga' was in fact 'the easiest of the eight-thousanders'.[12] With the help of Mason in India and friends in England he had actually organised an expedition for 1930. But he had had to cry off at the last minute, just as he had for the Pamirs expedition. Bauer says it was for health reasons. The record shows that ill-feeling also played a part: Bauer, recalling that he too had toyed with the idea of Nanga Parbat, feared that another German expedition would strain his good relations with the British; Welzenbach 'thought Kangchenjunga an impossible waste of time' and remained wedded to the idea of Nanga Parbat.

Postponed in 1930 and then 1931, Germany's first Nanga Parbat expedition finally got under way following Bauer's return from his near-death experience on Kangchenjunga. But when the team left for India in early 1932 it was without the disgruntled Welzenbach. Willy Merkl, Welzenbach's frequent climbing companion, had 'taken up the idea' and it was therefore under Merkl's leadership, though with thanks to Welzenbach's groundwork and subject to Bauer's scrutiny, that Germany's finest alpinists approached their introduction to what would become known as *der Berg des Schicksals*, the nation's 'Mountain of Destiny'.

To Bauer, who would act as both chronicler and impresario for what he called *The Siege of Nanga Parbat*, it was as if German interest in Nanga Parbat had been preordained. Back in the 1850s Adolf Schlagintweit, one of three brothers commissioned by the East India Company to conduct magnetic surveys in Himālaya, had correctly identified the massive ramparts of what the locals called Diamir or Diarmul as those of Nanga Parbat. Unaware of Godfrey Thomas Vigne's sightings during his Baltistan travels in the 1830s, Bauer therefore claimed the mountain as Schlagintweit's – and so Germany's – discovery.

He was of course better informed about the ill-fated Mummery attempt of 1895. He thought it strange, though, that in the thirty-five years since 1895 no one had attempted to follow it up. 'Was it due to Mummery's eminence among his contemporaries?' Bauer wondered. 'Had the mountain gained an evil reputation? Did people fear its precipices and avalanches? Or was it because a golden age had disappeared with the advent of the First World War?'[13] Just before the war, the indefatigable Dr Kellas had 'caused a sensation' by surveying the approaches to the mountain; he reported on them in the *Himalayan Journal*. And in the same year, 1913, Bauer heard tell that a certain 'E. Candler [had] made a circuit of the peak without attempting its higher regions'. No doubt E. Candler shied away from the heights because he had only one hand; he was of course Edmund Candler, once the *Daily Mail*'s man on the Younghusband expedition and now the footloose author of several Kiplingesque novels and entertaining memoirs.

Candler and a friend had approached the mountain from Kashmir. It was August. They had just returned from accompanying the Hindu faithful on the annual pilgrimage to Lord Shiva's cave at Amarnath. In the same spirit of pilgrimage, and minus any hint of the presumption of the Younghusband expedition, they had determined on a 'walking tour' of Nanga Parbat. Like pilgrims they circumnavigated the mountain clockwise; they made do with minimal comforts, most of these being books; and they were accompanied by two old-timers who had been with Mummery. One of them, a Buddha-like dwarf, had broken a leg on that occasion and had still not fully recovered its use. Limping along beside the one-handed Candler, this Lor Khan tumbled repeatedly yet always emerged upright. It was 'as if he had a gyroscope inside him', wrote Candler.

Candler delighted in the whole experience. In meadows blanketed in wild flowers, all of which he logged, he puffed on his pipe contentedly. After the quartet had plunged to the depths of the desert canyon that is the valley of the Indus, the temperature soared to 110 degrees Fahrenheit (43 degrees Celsius). Where a chunk of the mountain had dammed the river to such disastrous effect in 1841, they waded across a mud slide. They scrambled over glaciers, most of them much advanced since Mummery's day, and they climbed round the snow-pillowed heads of others. The entire circuit took five weeks with scarcely a cloudy day. It had been as if the mountain was on its best behaviour for 'a true worshipper' who decried the 'emulous' boastings of self-serving explorers and empire-builders. The shepherds and goatherds, the yak-herders and the long-distance traders in salt and *kuth*, already had names for all the mountains. They should be respected. Candler shuddered to recall that, 'but for the grace of God', his companions in 1904 would have renamed Chomolhari, the tutelary mountain of Bhutan, as Mount Younghusband.

When dropping from the snow line to the Sahara of the Indus valley, Candler had followed 'an ordered ice-stream curling white and smooth through the pine trees in its clear-cut trough'.[14] This was the north-flowing Rakhiot glacier up which Merkl and his team toiled towards their proposed base camp in 1932. Things were not going well for the first German expedition to attempt Nanga Parbat. None of the eight principals had ever been in Himālaya before, let alone in the politically unstable Kashmir borderlands. They had arrived in Srinagar without British authorisation, without porters and without adequate funds. In fact, it was no longer a German expedition but a German–American one. Originally part-financed by the contributions of German railway workers (Merkl himself worked for the railways), it had been obliged to secure additional funding from the US. In return, an invitation to join the expedition had been extended to two American climbers: Rand Herron, a globe-trotting alpinist with a useful command of German, and Elizabeth Knowlton, a journalist 'who nevertheless showed vigour and endurance' especially when, says Bauer, she 'climbed to Camp V to keep Herron company'.

It seems to have been Herron who persuaded the British authorities to let the expedition proceed and recruit whatever porters it could. The result, a mixed column of around 150 Kashmiris, Baltis and Hunzakuts but no Sherpas, proved disastrous. Strikes and mass desertions dogged the expedition's painful approach and continued during the gradual advance towards the subsidiary Rakhiot Peak. Unfamiliar with the ways of Asian labour, the Germans supposed their men were infected by Mahatma Gandhi's non-cooperation movement. But this didn't explain the theft of eight loads of high-altitude gear; the gear was intended for the porters anyway but, without it, they would suffer at the higher camps. Thanks to the settled weather, the chances of success still looked good, but not without adequate support.

Like Candler, Merkl marvelled at the forests of birches and conifers that clung to the mountain's lower flanks. It was just like home – or rather, thought Merkl, just like *Heimat*, the 'homeland':

> only here the feelings and impressions of the Alps were magnified into the unimaginable massiveness of the Himalaya, homeland for the Indo-Germanic peoples, who were born in these mountains and fell under their power. Here we raised our second Rakhiot camp and to the strains of 'Dear Homeland, We Greet You' cut into our first Bavarian smoked ham.[15]

The Americans Knowlton and Herron, both slim and dark, were somewhat bemused by their shorter, fairer companions in their roll-collared jerseys bursting into song at the slightest opportunity. For many Germans, 1932 was an exciting time. Back home the Nazi party was nearing a critical number of seats in the Reichstag. With the promise of national redemption and a quick recovery from the Great Depression, Adolf Hitler was about to be appointed chancellor and start disbanding the discredited Weimar Republic. Future Himalayan expeditions might not need to depend on the generosity of railwaymen. In Munich, Bauer was already exploring the level of official support on offer from Aryan supremacists for forays into 'the Indo-Germanic homeland'.

Bauer's task would have been easier if Merkl's 1932 expedition had notched up more in the way of achievements. Mason would rate it only a useful reconnaissance and scoff at Merkl's claim to have got 'within a day or two of the top'. It did survive a couple of avalanche scares, and it reached about 22,000 feet (6,705 metres) asl. But with the summit of the Naked Mountain still 'a discouraging two miles and 3,600 vertical feet [1,100 metres] away', the weather suddenly broke, whereupon the expedition 'expired'.[16] In the end it couldn't even be credited with having lost no lives. Outside Cairo on the voyage home, Rand Herron fell head first to his death when, in a cruel nemesis for a fine climber, he either slipped or tripped from the apex of one of the Giza pyramids.

Merkl, nothing if not determined, vowed to try again. With more substantial support from the railwaymen, plus a contribution from the Notgemeinschaft and some as yet tentative backing from Hitler's newly appointed *Führer* of Reichsport, the new expedition, that of 1934, was much the best equipped to date. It also attracted much the most experienced team of climbers. All seven had participated in one or more of the previous expeditions to Lenin Peak, Kangchenjunga or Nanga Parbat, while the eighth, Willo Welzenbach, the original advocate of Nanga Parbat, had consented to act as Merkl's deputy in succession to Herron. Better still, Merkl secured the aid of Kenneth Mason and the Himalayan Club in enlisting the services of thirty-five Sherpa and Bhotia porters from Darjeeling. All these men were accustomed to altitude; indeed half had just been on the 1933 expedition that relaunched the new British assault on Everest. If determination, funding and experience were the essentials, the second German–Austrian expedition to Nanga Parbat looked set to succeed.

Augmented by an army of Baltis, the expedition wound over the passes from Kashmir. 'It was an imposing and almost military spectacle,' wrote Fritz Bechtold, Merkl's closest associate. 'On the afternoon of 6 May 1934 we prepared our skis for the coming day of battle.' The metaphors came straight from the battlefield. Virtues internalised over five years of war, explained Bauer, were 'being repurposed on expeditions to the highest mountain ranges in the world'. Every ascent was now accounted *ein Kampf*, a struggle, and every Merkl was *ein Führer*. Mountaineering epitomised that single-minded teamwork in devotion to a higher cause that National Socialism expected of every true-born German.[17]

The unexpected death, from what was probably high-altitude pulmonary oedema, of the popular Alfred Drexel delayed the provisioning of the climbers' staged camps yet prompted no second thoughts. They followed the Rakhiot glacier in the footsteps of the 1932 expedition. Drexel's grave within sight of base camp gave them another claim on the peak that was already being called Germany's Mountain of Destiny. Already, too, Merkl and Welzenbach were trading the barbed criticisms which, in the event of success, would no doubt have been forgotten.

But as yet the climb was going well. By early July six climbers and sixteen Sherpas had reached Camp VI (7,000 metres asl), while from Camp VII the Austrians Erwin Schneider and Peter Aschenbrenner had reported the summit to be within manageable reach. That night, with an eye to posterity, they packed just a camera and a flag. (Photos taken down at Camp II show the flag to have been the first Nazi swastika flown in Himālaya.) And they waited for their colleagues to join them. None materialised. All were either too ill or too exhausted. The last chance of success was let slip. Later that night, over their flimsy bivouacs and within a day of their goal, the storm broke.

Mason would call it a hurricane. It buried under soft waist-deep snow the intermediate camps, along with the steps and ropeways painstakingly crafted between them. And it raged on for five days, making movement both impossible and increasingly imperative. By the time the evacuation got under way, all the men were suffering from frostbite, altitude or both, and judgements were impaired. Out of the nineteen in the upper camps, eight perished in the white-out descent. Five of the fatalities were Sherpas, whose bravery was universally applauded; three were Germans, among them Merkl and Welzenbach with whom, as leaders, must lie much of the blame.

Merkl, a passionate believer in teamwork, had aimed to get as many climbers on the summit as possible. It was to be a mass demonstration of German solidarity. As a result, when the storm struck, the intermediate camps had been abandoned, communication interrupted, and far too many personnel were marooned on the cramped ledges of the uppermost camps. Lessons needed to be learned. Mason called the tragedy 'the greatest mountain disaster of our time'; and so it was – briefly.

The conclusion drawn in Berlin was that the climbing fratenity must try again, and this time harder. Third time lucky, surely. They owed it to the 'martyrs' of 1934 and above all they owed it to Deutschland. National honour was at stake. At a time when climbing qualified as an Olympic sport, mountaineering was being brought within the brief of the Reich's *Sportsführer*. 'The conquest of the peak is expected for the glory of Germany,' he had told Merkl.[18] One or two fatalities changed nothing. Bauer's role as expedition impresario and go-between with the new regime was officially recognised by his heading his own newly formed German Himalayan Foundation. Discreetly backed by the Nazi government, it was thus Bauer and this supposedly non-political foundation that set about mounting Germany's third assault on Nanga Parbat in 1937.

Even now, in 1937, no peak of over 8,000 metres had been successfully climbed. From Tibet the British were persevering with their Everest encroachments and being repeatedly frustrated by the weather while, courtesy of their Tibetan friends, access was being denied to other nationals. But there was progress elsewhere. In 1934, as Merkl and the Germans came to grief on Nanga Parbat, an Anglo-American team led by Charles Houston and the incorrigible Bill Tilman had surprised itself by putting two men on the summit of Nanda Devi. Much mythologised, Nanda Devi was the highest peak wholly within India (Kangchenjunga being half in Sikkim and Nanga Parbat in Kashmir) and was certainly higher than Lenin Peak. It was therefore the highest conquered to date. But at 7,816 metres it was not quite an *Achttausander*, an eight-thousander like Nanga Parbat.

Of the world's twenty highest peaks, three-quarters were known to be either in Nepal or in the Karakoram; and with Nepal being still closed to foreigners, it was those in the Karakoram that were of special interest. There, in a knot of glaciers and peaks claimed as part of Kashmir's – and subsequently Pakistan's – territory, at least six eight-thousanders snag the clouds, including K2 whose 8,612 metres is not far shy of Everest's 8,848. But tackling a mountain tucked away among other giants at the head of the 60-kilometre Baltoro glacier – and further still from anywhere that could be reached by mules (yaks being not much found west of Ladakh) – called for elaborate logistics as well as climbing skills. As a result the several expeditions that had addressed the Karakorams since Godwin-Austen's survey work in the 1860s had tended to be on a generous scale.

In 1892 Martin Conway, an English academic with wealthy American in-laws and a love of the great outdoors, had set his sights on the Karakorams and put together what is often called the first typically Himalayan mountaineering expedition. It certainly included climbers. Mummery himself had nearly joined it. It was well served by a Swiss guide and Gurkha orderlies; and it ticked all the right boxes in terms of prestigious endorsements, reputable sponsorship, press contracts and, in due course, a lavishly illustrated 700-page acount of the expedition.

> Conway's journey of discovery in 1892 was, for better or worse, the prototype on which so many of the big twentieth century expeditions were based. In particular the first British expeditions to Everest after the First World War owed much to Conway.[19]

But to the disappointment of his climbers Conway preferred exploration and survey work to summitry. Cold-shouldering K2 as 'a fraud' (just 'an ugly mass of rock, without nobility of form or grandeur') and dismissing the other giants as 'terrifying', Conway made much of a lesser peak which he called the 'Golden Throne'. Now known as Baltoro Kangri, this 7,274-metre mountain recommended itself simply because it looked climbable. But appearances were deceptive. The ridge selected for the attack turned out to be separated from the rest of the mountain by a void. Instead of seating themselves atop the Golden Throne, Conway and his colleagues had to be content with the view from one of the Throne's footstools. They called this unremarkable excrescence Pioneer Peak for no obvious reason other than that a climb-worthy feature needed a name. Their barometer gave the height as 22,600 feet (6,888 metres), which Conway would round up to 'over 23,000 feet' (7,010 metres) and claim as a world record. In truth it was considerably less, and so probably nothing of the sort.

Exaggerating elevations and overestimating distances had been carried to greater lengths (and heights) by Conway's Karakoram successors. Like Conway, the suffragist and heiress Fanny Bullock Workman and her obliging husband Dr William Hunter Workman had seen themselves as explorers rather than mountaineers. She wore a stiff tweed skirt, he baggy plus-fours. A Swiss guide cut steps and secured ropes for them. Mishaps or reverses were always down to the incompetence of their porters or the wiles of the locals. As Americans they were uncomfortable in the obsequious Orient and no more at ease on the margins of the raj. They didn't expect to be loved – and they often weren't.

Nonetheless, in the course of some seven expeditions between 1899 and 1912, the Workmans covered more of the Karakoram region than either Godwin-Austen or Conway. For intrepid travellers serving an avid readership, nowhere else short of the poles offered such certain renown. As sheer spectacle, the peaks and glaciers exceeded anything the Workmans had reported on in Scandinavia, Spain or the Alps, all of which they had explored thanks to the mobility on offer from the chain-driven velocipede now known as a bicycle. They often rode a tandem version, though from the south of India to the Himalayas they had pedalled their way on separate machines, which they sensibly abandoned on entering the mountains.

In the Karakorams their pedestrian itineraries were convoluted and would prove hard to fathom, so keeping the next generation of travellers busy offering corrections. Likewise the Workmans' maps, which Mason dismissed as 'attractive ... deceptive and not always reliable'; they may well have been the maps that prompted Edmund Candler's outburst against the 'emulous' boastings of men on mountains. Among the jarring toponyms introduced by the Workmans were a Queen Mary Range, the inevitable Mount Bullock Workman (supposedly 5,930 metres) and a Siegfriedhorn (5,700 metres) for Siegfried, their son.

Oddly, neither the dismissive Conway nor the roving Bullock Workmans paid much attention to the challenge posed by K2. A 1902 expedition led by Oscar Eckenstein, formerly one of Conway's team, had hoped to rectify this but, 'a fractured and sullen party' from the start, it had soon 'dissolved in rancorous disorder'.[20] Though its Swiss–Austrian climbers reached about 6,500 metres, the expedition is mostly remembered for the unsettling presence of Aleister Crowley, later a self-proclaimed Satanist and deviant who would be described as 'the wickedest man in the world', yet withal a capable cragsman.

K2 had to wait till 1909 and the arrival by way of Baltistan and the Baltoro glacier of an expedition so well organised and lavishly

One of Vittorio Sella's incomparable photos of the Duke of Abruzzi's 1909 expedition to the Karakorams. Ascending the Baltoro glacier, the group were searching for their first sight of K2

appointed that it upstaged Conway's exercise and set a whole new standard in Himalayan logistics. For one thing, and twenty years before Germany made its presence felt in Himālaya, it was unequivocally a national enterprise. All of its twelve European personnel were Italians, and none more so than its leader, the thirty-six-year-old Prince Luigi Amedeo di Savoia, Duke of the Abruzzi and grandson of united Italy's first king, Victor Emmanuel II. This meant that *il principe*, the prince, as his entourage called him, belonged to the ancient House of Savoie-Aosta in the Italian Alps, from where most of his seven-strong climbing team hailed. The prince himself had somehow managed to combine a naval commission with a busy climbing career which already included notable ascents in the Alps, in the Ruwenzori of Uganda and of Mount Saint Elias in Alaska. Mountains were in the deep blue of *il principe*'s blood and he knew how to mount an expedition.

With 360 porters and several tonnes of supplies the Italians were on the Baltoro glacier by late May. For days there was no sign of K2. The Prince and his Savoyard climbers consulted with the expedition's three savants: Federico Negrotto, a *marchese* or marquis, was really the Prince's adc though now acting as topographer and paymaster; Dr Filippo de Filippi combined the role of physician with that of naturalist and chronicler; and Vittorio Sella, Filippi's nephew and at fifty the oldest of the party, was simply the greatest mountain photographer of the age. Had the expedition achieved nothing else, it would be remembered today for Sella's icy panoramas.

On May 24 they were nearing the head of the Baltoro when, to the north, a vista up the Godwin-Austen glacier opened. 'Suddenly, without warning, as if a veil had been lifted from our eyes ... down at the end, alone detached from other mountains, soared up K2, the indisputable sovereign of the region, gigantic and solitary, jealously defended by a vast throng of vassal peaks, protected from invasion by miles and miles of glaciers.' De Filippi was transfixed.

> For a whole hour we stood absorbed. We gazed, we minutely
> inspected, we examined with our glasses the incredible rock
> wall. All the time our minds were assailed with increasing doubt,
> culminating almost in certainty, that this side of the mountain was
> not accessible and did not offer even a reasonable point of attack.[21]

The other sides would prove no better. For two months the prince's men probed, mapped and photographed the various possibilities with a thoroughness that would spare subsequent expeditions a lot of reconnaissance. On the mountain they may have reached 6,700 metres, but hopes of further progress were repeatedly dashed. It was not, for once, entirely because of the weather. It was the scale. K2 was just too gigantic, too demanding.

Instead they laid down a marker. In an age of intense competition between Europe's colonial powers, mountains were as subject to casual adoption as fishing grounds or offshore islands. Anyone who could reach the top of a chunk of Himālaya felt entitled to fly a flag on it. The peaks were there for the taking. Everest came to be seen as a British prerogative, Nanga Parbat as quintessentially 'the German mountain', and thanks to *il principe*'s 1909 expedition, K2 would be widely acknowledged as 'the Italian mountain'. Like Candler, the prince disapproved of applying personal names to natural features. Yet maps of the mountain were quickly festooned with an Abruzzi Ridge, Savoia Pass, de Filippi Glacier and Negrotto Saddle. There could be no doubt about who claimed it.

Before withdrawing from the Baltoro region, other peaks were assessed and one, Chogolisa, was climbed to 24,278 feet (7,400 metres). This, according to de Filippi, 'exceeds by 700 feet the greatest altitude up to then achieved by men upon the mountains'.[22] After which the expedition headed for the eastern (or Ladakh) Karakoram and the maze of glaciers from which flows the Shyok tributary of the Indus.

To this same region of fewer peaks but higher passes de Filippi had returned with an enormous multi-disciplinary scientific expedition in 1913–14; and he had returned yet again in 1928–9, this time under the leadership of another prince of the House of Savoy, Aimone, Duke of Spoleto. On neither occasion was K2 reattempted, but more glaciers and watersheds were explored and more Italian names imprinted on the Karakoram map.

A year later, in 1930, Professor Giotto Dainelli, the geologist of the 1913 expedition, and his 'ski-ing partner' Miss Elly Kalau had taken territorial appropriation a stage further. For Dainelli, the Siachen glacier became 'my glacier', only to be followed by a pass on to the Rimu glacier that was 'my pass'. The professor could hardly contain himself.

I got Miss Kalau to make me a little tricolour flag. And next day
I had the unspeakable joy – I, the only Italian, in the presence of
the highest mountains on earth and in the midst of the grandest
region of ice that exists – of shouting a 'Viva l'Italia!' which went
straight up to a sky of marvellous purity, while my little flag floated
triumphantly over a pass which had been crossed for the first time
by an Italian caravan ... I named it the 'Italia Pass'.[23]

Italy's near-monopoly of K2 would not be seriously contested
until the late 1930s. In 1938 a small American team led by Charles
Houston of Nanda Devi fame conducted what Mason calls 'a
successful reconnaissance'. In fact it was almost a successful assault.
The Abruzzi ridge was mastered, an altitude of 26,000 feet (7,925
metres) attained, and not a life was lost. It was followed in 1939 by
its antithesis. Commanded by Fritz Wiessner, a member of Merkl's
1932 Nanga Parbat expedition and now a naturalised American,
the K2 expedition of 1939 would provide a textbook example of
how not to climb a mountain. Ropes fouled, crampons got lost,
communications broke down and the weather did its worst.
Wiessner, unlike Houston, knew not when to turn back and would
be accused of the unpardonable – leaving a colleague to die, and
losing three Sherpas in an abortive attempt to rescue him. 'It is
difficult to record in temperate language the folly of this enterprise,'
sighed Mason.[24] All that could be said was that 'the tragedy was not
quite on the scale of Nanga Parbat 1934'.[25]

Nor indeed on the scale of Nanga Parbat 1937. Germany's tryst with
its Mountain of Destiny had still to be realised; an *Achttausender* had
still to be conquered; national pride had still to be assuaged. It could be
argued that the disaster that had overtaken Merkl and his men in 1934
was unlikely to be repeated. Lightning didn't strike twice. Valuable
experience had been gained and, faced with the need to replace the
deceased climbers, the meticulous Bauer had taken the new recruits
on a bonding exercise on Kangchenjunga in 1936. For leader he had
selected Dr Karl Wien, a veteran of the 1928 Pamirs and the 1931
Kangchenjunga expeditions. Wien headed seven other climbers who
in India would be joined by the usual mix of experienced Sherpas and
dogged Baltis.

Two months after leaving Munich, this third Nanga Parbat expedition was reported to be making good progress. Following the usual Rakhiot approach, by 14 June they had reached Camp IV at 20,280 feet (6,181 metres). From there a group of porters was sent back, leaving seven of the eight German climbers and nine remaining Sherpas at IV. There was no further news of them for four days, and by then it was all over. Camp IV was no more, reported Uli Luft, the remaining German climber, after returning from the site. As later determined, disaster had struck soon after midnight on 15 June when the slumbering camp had been buried under a gigantic avalanche. 'Of the tents and their occupants there was no sign. All had been destroyed.'[26] 'Sixteen men lay entombed in thousands of cubic feet of snow and ice.'[27]

The tragedy was world news. Troops from Gilgit and Chilas were rushed to the scene, a team of Himalayan Club members prepared to double-march from Srinagar, and Bauer himself along with colleagues from the German Himalayan Foundation flew immediately to Karachi and on to Gilgit. Within three weeks of the accident, Bauer was on the mountain with steel probes and 300 metres of rope. 'For a time we had thought it likely that there would be survivors in the upper camp,' he explained.[28]

It was wishful thinking. Just reaching the scene of the tragedy proved difficult enough. Hopelessly unacclimatised, Bauer could barely stand, yet recovered enough to oversee the digging. In the thin air, locating and excavating the tents and the board-stiff corpses with their part-written diaries and personal belongings was a feat in itself. All were compacted within a solid mass of rock and ice 3 metres thick and as hard as concrete. Two of the German climbers were never found, while all the porters were left as they lay. Their surviving colleagues preferred it that way, insisting the bodies be consigned to the peris and djinns of the Naked Mountain among whom they now belonged.

Back at base camp, Bauer's last night on the mountain coincided with a clear sky and a full moon. The campfire flared and the glaciers shone with luminescence. 'One could not quarrel with fate on a night such as this,' he mused. 'Melancholy was out of the question.'

Forward thinking was a different matter. 'Do I have to explain why that night we resolved to continue the work that our friends had begun, and return to Nanga Parbat during the following year?'[29] Return, of course, he would – and within the year. In 1938 Bauer himself led the fourth expedition to confront 'Germany's Mountain of Destiny'. A new team was assembled, again they followed the Rakhiot glacier, and this time it was the weather that obliterated their hopes. Naturally somewhat cautious, Bauer called off the assault when a final push to get within striking distance of the summit was beaten back. They had done just about enough for their 'rescue mission' to qualify as the fourth attempt.

Still there could be no admission of defeat – and certainly not now in 1939. Nazi troops were already rolling into Bohemia as the Munich Pact frayed. Bauer would have been reckoned a traitor had he not tried again. A smaller party this time, and minus Bauer himself whom age and uncertain health precluded, the fifth expedition consisted of just four climbers. It was to explore an alternative approach from the north-west by way of the Diamir glacier which had defeated Mummery in 1895 and was thought by Candler in 1913 to have since been made even more impassable by glacial encroachment. But it was worth investigating.

The 1939 team led by Peter Aufschnaiter (a veteran of Bauer's 1931 Kangchenjunga expedition) and Heinrich Harrer (a twenty-seven-year-old who had just been part of the first successful ascent of the north face of the Eiger) thought the Diamir approach might even be possible; but any attempt would require at least seven climbers, all of them skilled in tackling perpendicular ice and soft rock, and it was quite beyond the capabilities of load-carrying porters. The sixth expedition consoled itself with scaling a minor peak and then dispersed.

Aufschnaiter headed for Gilgit to sound out the British about a possible seventh expedition via the Diamir glacier. Harrer and their two companions travelled to Karachi incognito. It was by now late August 1939. The German army was already advancing into Poland; war was looking inevitable. Reunited with Aufschnaiter in Karachi, Harrer and colleagues piled into what Harrer calls 'a ramshackle vehicle' and drove out of town, heading north and west. The plan was 'to break through to Persia and find our way home from there'. Eluding one British 'tail', they managed about 200 kilometres before an alfresco lunch break was interrupted by a detachment of Indian troops. The war in Europe had begun. Undesirable aliens faced internment. Under guard they were driven to a transit camp and eventually to a detention facility within sight of the mountains at Dehra Dun.

In his *Seven Years in Tibet*, perhaps the most widely read of all books on Tibet, Harrer tells of his and Aufschnaiter's subsequent attempts to escape into neutral Tibet. They eventually crossed the Great Himālaya in disguise, but by then it was April 1944. Eighteen months later, after Tibetan wanderings that reversed Hedin's itinerary from the Tsangpo–Brahmaputra to Kailas–Manasarowar and Toling–Tsaparang, they reached Lhasa. The war had long since ended, yet Harrer seems to have had reason to suppose they might still be handed over to the British for reinternment in India. Tibetan hospitality helped reassure them. Their skills were put to good use: Aufschnaiter undertook water-management projects; Harrer designed gardens, made home movies and tutored the young Dalai Lama. They were discouraged from venturing afield to climb mountains but were under no kind of detention.

Nor do the British appear to have demanded their extradition. In his highly rated bestseller, Harrer offers little by way of explanation for his seven long years in Himālaya beyond the satisfaction he got from being of use to the Tibetans. It seems to have taken him three years just to inform his family he was there, after which the news that his wife had borne him a son must have been eclipsed by the news that she had had their marriage dissolved, presumably on the grounds of Harrer's supposed desertion or death.

That Harrer had a family at all is barely acknowledged in the book. An Austrian by birth, he had studied geography at Graz University and had there met, and in December 1938 married, the eighteen-year-old Lotte, daughter of Alfred Wegener, the university's revered (and, in death, continentally displaced) geophysicist. The wedding had been rushed. At the time Harrer was getting ready for Nanga Parbat in between basking in the Führer's congratulations on his Eiger climb and chasing Himmler's approval of his marriage. The approval was in the form of an affidavit that both bride and groom were of Aryan stock; and it was necessary because Harrer, already a member of Austria's pro-Nazi stormtroopers, had just been enrolled in Himmler's Schutzstaffel, aka the dreaded Nazi SS.

This was not something to be proud of in later life. Harrer preferred not to mention it and anyway maintained that he had joined the SS simply to get a place on the sixth Nanga Parbat expedition. The only time he had worn the SS uniform, he said, was at his wedding. Given that there was a photo of the bridal pair, there was no point in denying it.

Much of this information only came to light in the 1990s when, in connection with a Hollywood version of *Seven Years in Tibet*, an Austrian radio journalist came upon documentary proof of Harrer's SS links. Belatedly substantiated, such links could well explain why Aufschnaiter and Harrer had been in no hurry to leave Tibet: the pair stayed put from the fear, it would seem, not of wartime internment but of a possible war-crimes tribunal. Furthermore their Nazi connections may hint at why they had headed for Tibet in the first place.

Members of the sixth Nanga Parbat expedition were not the only Germans to have been overtaken in Himālaya by the approach of the Second World War. The early days of September 1939 which saw Aufschnaiter and Harrer trying to evade British internment by driving from Karachi to Persia also witnessed the hasty exodus from Calcutta, in this case by flying boat, of another group of German Himalayanists. The five-member Ernst Schäfer German Tibet Expedition of 1938–9 had shown no interest in mountaineering. It was a scientific exercise. The impressive Schäfer was a zoologist and ornithologist whose previous specimen-collecting had taken him from China to the Tibetan borderlands. There he had learned some Tibetan and liked the people, but he had not got beyond Amdo.

The present expedition had extended Schäfer's enquiries by entering Tibet via Sikkim and pressing on to Lhasa. His colleagues were a geologist, a geophysicist, an anthropologist–ethnologist and an entomologist cum cameraman. It all looked innocent enough. From London, Younghusband had offered encouragement, and in India the then viceroy took an active interest. A note of caution had indeed been sounded, but only in retrospect would the expedition be recognised as audacious, even sinister.

Suspicions that Schäfer and his men 'prepared maps and surveyed passes for the possible use of Tibet as a staging ground for guerilla assaults on British India' sound far-fetched.[30] They recall Aghvan Dorzhiev and the Russophobic paranoia that had prompted the Younghusband expedition. Nor was it cause for alarm that Schäfer and his expeditionaries flew the Nazi swastika. It had been Germany's national ensign since 1935 and, as a propitious Hindu–Bon–Buddhist symbol of great antiquity, the swastika's adoption by the Germans could be explained as evidence of German–Tibetan affinity.

Even the cranial casts and measurements of Himalayan physiognomies made by Bruno Beger, the expedition's ethnologist, were not especially suspect. Such techniques 'were within the bounds of international scientific practice at the time'; the de Filippi expedition had adopted similar procedures in the Karakorams.[31] It was not even clear that Heinrich Himmler's Ahnenerbe, an 'Ancestral Heritage' research unit concerned with tracing and promoting the supposed superiority of the Nordic–Aryan race, was funding the expedition. But the mystically gullible Himmler was indeed drawn to the eponymous-sounding Himālaya. Convinced that the roots of the pan-Aryan peoples were to be found on the 'roof of the world' he strongly supported Schäfer's research programme into 'the Indo-Germanic homeland' and took a close personal interest in the whole enterprise. All five of the expedition's scientists were in fact officers in his SS, this being as much a prerequisite for their participation as it had been for Harrer's in the Nanga Parbat expedition.

Whether Harrer had any contact with the Schäfer expedition is not clear. He certainly knew of it. At one point he pauses his own narrative to deplore the insensitive behaviour at the 1939 Tibetan New Year festivities of 'the members of the only German expedition that ever came to Tibet'.[32] Conversely, when escaping from detention in Dehra Dun, he mentions the helpful information obtained from another Dehra Dun escapee, one Hans Kopp. Kopp, 'a Berlin fellow', had previously been expelled from Tibet, explains Harrer without revealing why or when.

Suffice it to say that, even if the war had been prolonged, Harrer and Aufschnaiter could have been reasonably sure of being well received in Lhasa, and that they may have stayed on there to pursue the relationship initiated by Schäfer's men. A letter forwarded by Schäfer from the Tibetan regent to 'his Majesty Führer Adolph [sic] Hitler, Berlin, Germany' had specifically mentioned Tibet's willingness 'to do anything that will help to improve the friendly tie of [this] relationship'.[33] But note also that the disquieting features of the Schäfer expedition probably appeared quite innocent at the time. Membership of the SS was not in itself a crime. Only one member of either expedition would face trial after the war, and that was Bruno Beger, the ethnologist. The charge related to Beger's later involvement in the selection of specimens for a collection of Jewish skeletons. He was convicted, but on appeal the three-year sentence was commuted to one of probation. Nazi Germany stood accused of infinitely worse crimes than uncertain interference in Himālaya.

Epilogue

There comes a day in the life of every Tibetan traveller when he stands on the crest of the last range, and gazes across the foot hills to the plains below ... Behind him rise in awful and paralysing grandeur the most desperate mountains in the world. Below him rise spirals of blue smoke from the hearths of men; and as he looks, and dusk slinks down the sky, he sees as it were men and the children of men, and families gathered into villages, and villages into towns, and towns into cities; and hears the dull roar of transport and industry, as man tries to inhabit the whole earth. But behind the mountains lies the garden of God.

F. Kingdon Ward (1924)[1]

If a proliferation of 'protected areas' is evidence of effective environmental conservation, today's Himālaya would not appear to be at risk. As well as hosting what is easily the world's largest high-altitude ecosystem, Himālaya harbours a unique variety of the more specialised habitats and ecozones that interest bodies like the World Wide Fund for Nature (WWF) and the International Union for Conservation of Nature (IUCN). Both these organisations subscribe to a bewildering classification of sites whose status is to some degree endangered. To the layman the difference between a biodiversity hotspot (BDH), a high-biodiversity wilderness area (HBDWA), a crisis ecoregion (CER), a key biodiversity area (KBDA), a Global 200 ecoregion (G200ER) and anywhere identified by the Alliance for Zero Extinction (AZE) may not be obvious, yet there can be few pockets of Himalayan biodiversity that qualify under none of the categories in this acronym soup.

Even Tibet's inhospitable Changthang cries out for protection; it's characterised by the WWF as 'a high-altitude Serengeti'. Oddly, though, only that part of Tibet designated as 'Eastern Himalayas' (objective: 'empowering communities to protect sacred lands') is listed among the WWF's thirty-five 'Global Priority Places'. If environmental fragility and respect for cultural associations count for as much as species diversity, eastern Himālaya is a prime candidate; and of nowhere is this truer than where the Tsangpo–Brahmaputra snakes through the dripping declivities of Pemako in Tibet's south-eastern extremity.

Since 1913 when Eric 'Hatter' Bailey and Henry Morshead scrambled downriver in the footholds of the indomitable Kintup, the Pemako gorges and their 80 kilometres of still-uncharted river had attracted few visitors, most of them botanists. Best known was Frank Kingdon-Ward whose 1926 *Riddle of the Tsangpo Gorges* proved an instant success. Bailey's *No Passport to Tibet* was still unpublished at the time, and so it was Kingdon-Ward's account which, despite its botanical emphasis, first alerted the general public to one of the twentieth century's greatest geographical conundrums. This was not whether the Yarlung–Tsangpo of Tibet became the Dihang–Brahmaputra of Assam; that much was now certain, Kintup and Bailey having disposed of other theories. The unresolved question was how it actually did so: how did one of Asia's major rivers plummet nearly 3,000 metres in a mere 250 kilometres? And in the process, how did it cut through the world's loftiest mountain range? Did it slither down the steepest of water slides, gurgle into some gargantuan culvert or plunge over an almighty waterfall? Anything seemed possible.

Before heading east across southern Tibet, Kingdon-Ward had consulted Bailey. As political officer in Sikkim, Bailey was as near as the British had yet got to having a resident in Lhasa itself. It was March 1924. Anglo-Tibetan relations were under strain. David-Néel and Yongden were mingling with the pilgrims attending the New Year festival in Lhasa, the personnel of George Mallory's last Everest expedition (plus film crew) were gathering in Darjeeling prior to crossing into Tibet, and according to Kingdon-Ward the Tibetan government had recently 'been embarrassed by the antics' of certain other unnamed English travellers. More foreign visitors were not welcome. It was only thanks

British residency at Gangtok

to Bailey's excellent relations with the Dalai Lama that passports had been obtained for Kingdon-Ward's party.

Leaving Bailey's comfortable Gangtok residence, Kingdon-Ward crossed the Nathu La into Tibet. Compared to 'the small army' of 300 pack animals, 150 porters and 10 climbers heading for Everest, the Pemako expedition was a modest affair – just Kingdon-Ward and Jack Cawdor (the undergraduate earl), plus a nameless terrier, a few mules and a Tibetan-speaking threesome of 'Tom, Dick and Sunny Jim', respectively muleteer, cook and cheerful dogsbody.[2] Their route followed Younghusband's of 1904 to Gyantse, then reversed Bailey and Morshead's of 1913 from Tsetsang down the Tsangpo valley. Pemako's twin portals of Namcha Barwa (7,756 metres) and Gyala Peri (7,150 metres) were dead ahead but hidden by cloud.

The Kingdon-Ward expedition would be gone for a year. Most of it was spent collecting botanical specimens in the vicinity of the gorges and exploring Pomé to the north with its puzzling array of stone watchtowers. Only in November, as winter closed the higher passes and the rhododendrons clenched their leaves against the frost, did they descend from the arctic heights into the subtropical world of the river's gorges. Forest here turned to jungle. Broadleaves and bamboo thickets competed with the cliff-clinging tangles of rhododenron. The only tracks were those left by the zoological oddity that is the moose-nosed takin. As they descended to around 2,500 metres asl, Kingdon-Ward sighted what he called 'baboons' but which may have been stump-tailed macaques.

Once in the gorge they retraced Bailey's route downriver. They paused at Pemakochung, noted Kintup's disappointing falls, sighted Morshead's Rainbow Falls and then passed the point at which Bailey had turned back. Obstructions were surmounted by felling trees, cutting notches in their trunks and propping them against the cliffs to serve as ladders.

> Fierce rapids ate hungrily into the core of the mountain. Already we seemed to be far below the level of the ground, going down, down into the interior of the earth; and as though to emphasise the fact, the temperature grew steadily warmer. And the gorge was growing steadily narrower, the gradient steeper, till the power behind the maddened river was terrific. Its blows fell on rock and cliff with frightful force; and at every turn a huge cavernous mouth seemed to open, and gulp it down faster and faster.[3]

They were halted by the sheer walls of rock closing in from both sides. Boulder-hopping alongside the riverbed was no longer possible. Instead, they scaled the cascade-swept walls to track the river's course from high on the spurs of Namcha Barwa. Some 'Hidden Falls' familiar to the locals stayed hidden, thundering out of sight beneath a cloud of spray. 'Could these be the falls of romance,' wondered Kingdon-Ward, 'the goal of so many explorers'? They couldn't; their reported drop was too modest.

With food running low and the river rampaging west, then north, east and south in a suspiciously wide detour, they opted to replenish supplies and cut across this loop to rejoin the river downstream. Now, therefore, they would be tracking back upstream in a final bid to close the remaining 30-kilometre gap of unexplored gorge. Snow turned to rain. Their last night in the gorge was spent huddled under rocks 'like a lot of rabbits'. Excitement plus the roar of the river precluded sleep. 'The final solution – falls? Or no falls? – was now within our grasp.'[4]

Next day a perilous descent landed them in a white-water grotto. The river had here 'blown a hole 25 feet [4.5 metres] wide clean through the middle of a wall of rock'. With water levels at their lowest (it was by now December) the flood was thundering through this breech; but as the water rose, it would overtop the wall to crash down in a fall of perhaps 12 metres. That was just the beginning of what Kingdon-Ward was now calling the Falls of Brahmaputra. From a vantage point high above this obstruction they sighted two more falls of similar height as the river charged towards them. 'We had now discovered the narrowest and most profound depths of the gorge, where the river, only 30 yards [27 metres] in width, descends in falls and rapids over 130 feet in a mile [40 metres in 1, 600 metres]; and that was something.' Something certainly, but not exactly a Niagara. It was less a waterfall than a water staircase. But, from above the gorge, Kingdon-Ward was in no doubt: the Yarlung–Tsangpo got reborn as the Dihang–Brahmaputra not in a single cascade but in a succession of cataracts. 'I was unable to believe that there is any likelihood of a greater fall in the remaining five miles [8 kilometres] which we did not see.'

So be it. The gorges concealed no watery prodigy. But there was still at least one short section that could be regarded as virgin territory. Enough of the riddle remained. That, plus the grandeur of the setting and the surprises of Himālaya's least 'Tibetan-like' extremity, would be enough to tempt a later generation of explorers.

More botanists came first. Kingdon-Ward returned to Pomé in 1935, and Frank Ludlow and George Sherriff resumed the quest for Pemako's exotic flora later in the 1930s. They visited the region again as soon as the war was over, but by then the days of British access to Tibet were numbered. The advance of the People's Liberation Army into Kham in 1949–50 brought the closure of the Tibetan frontier.

Not till the 1990s, and then largely thanks to American enterprise, was there further news of the gorges. In the meantime Chinese scientists had taken an interest in the region. Calculating that the Tsangpo's gorge was longer and deeper than that of the Colorado River in Arizona, the authorities in Beijing sensed tourist potential. The region was declared the Yarlung–Tsangpo Grand Canyon National Reserve. Among its listed attractions were a 'wealth of medicinal plants as well as Tibet's last remaining tigers and other rare mammals including black muntjac and capped leaf monkeys'.[5] Unfortunately this conservational awareness has not deterred the plans to construct one or more colossal dams. When built – if built – they will divert the river's waters, generate three times more electricity than the Three Gorges Dam on the Yangtse and turn China's Grand Canyon National Reserve into a species graveyard.

Permits for leisure activities were also sparingly issued. Namcha Barwa was first scaled by a Japanese-led team in 1992; in 1993 a small American expedition was forced to turn back at Pemakochung; and later an international attempt to kayak through the gorges ended in disaster. Also in 1993 another US team including Buddhist scholar Ian Baker and climber Ken Storm Jr reached Morshead's Rainbow Falls. Baker and Storm were as captivated by the Bailey/Kingdon-Ward narratives as by the gorges. With other enthusiasts they would return to the area as often as permits were forthcoming. By the late 1990s, amid fierce competition with Chinese explorers, the unexplored gap left by Kingdon-Ward was effectively closed and yet more falls were discovered, the highest being claimed as an impressive 30 metres.

Less contentious discoveries resulted when the adventurers and the botanists combined. In 1996, the British rhododendrologist Kenneth Cox led an expedition to Pomé east of the gorges. Here alpine meadows high above the river afforded 'wonderful plant-hunting' as Cox ticked off the flora recorded by Sherriff and Ludlow in 1938. But something was missing. 'The one great disappointment had been that there was no sign of the red lily,' recalled Cox. Otherwise *Lilium paradoxum*, this unusual and extremely rare plant with its deep-red flower had also been noted by Kingdon-Ward; but there was never more than a single specimen and the seed obtained had failed to germinate.

Miraculously, though, even as Cox rued this failure while he waited to bid farewell to Pomé, June Ross, the party's mountaineer, rejoined her colleagues 'clutching a little bundle'.

> I unwrapped June's specimen, and there it was. Unmistakably this was the red lily ... On a treacherous scree, she had spotted a splash of red. Carefully clambering over to investigate, she'd photographed the flower and picked the single bloom for identification purposes when the unstable scree caused her to make a faster and less comfortable descent than she had intended.[6]

There was no time to climb back up. Their pre-booked slot in Pemako was expiring. But back in Britain a journalist friend of Cox's relished the lily story and managed to interest her newspaper's editor in it.

When Cox and colleagues headed back to Tibet in 1997, this 'Jenny' went with them.

They now concentrated on the uplands south and east of the gorges as *paradoxum*'s most likely habitat. They were disappointed: no lilies at all were reported. But once again compensation was at hand. As they ate their lunch in a forest of rhododendron, journalist 'Jenny' (for some reason Cox withholds her surname) spied another rufous rarity. 'Moving through the undergrowth was a red panda: about the size of an average dog but with a characteristic long striped tail.' Cox simply noted that 'these rather timid animals are a rare sight'. They are indeed. They had been rare in Kingdon-Ward's day and are now on the IUCN's Red List of endangered species. They are not, though, strictly speaking pandas. Despite the appeal of their panda-like facial markings, they are in fact the sole surviving member of an otherwise extinct family of carnivores known as ailurids. Related to the raccoon, they had first been identified by the great Georges Cuvier at the Paris Museum of Natural History in the early 1800s.

Cox's party withdrew from Tibet under pressure from Chinese guards and far from reconciled to having found a rare panda instead of a rare lily. They headed for Katmandu and there, by chance, met Ian Baker and Ken Storm. The Americans were preparing for another foray into the gorges; it would involve crossing a pass in Pomé; 'Jenny' was welcome to join them; but it would be early in the year – early for a ramble on the snowfields and probably too early for lilies in flower.

Storm, the climber, nevertheless entered into the spirit of the thing. 'I examined the ground, probed every plant, and scanned the slopes for a splash of red.' Minute scrutiny of such wild terrain he found strangely exhilarating.

> Then I saw them – not one but several blooms, nodding in the breeze on a steep slope beyond the track I was climbing. The deep burgundy of the corollas burned brightly into the grey day as I carefully … lowered myself to eye-level with them. How delicate they seemed, spotted with drops of rain. I feared my breath might scatter the petals.[7]

A shout brought 'Jenny' to the spot. Her quest was finally over; she had her story. But the journalist was no plantswoman. It was too early for seed anyway. Hence, 'as far as we know', says Cox, 'the red lily is still out there and no one has yet managed to collect it'.

> It is somehow curiously satisfying to know that, early in the twenty-
> first century, there are still a few plants out there in the mountains
> which have managed to avoid being tamed, domesticated and
> preened, plants which only a tiny handful of explorers have ever
> seen.[8]

And perhaps ever will see. Twenty years later the status of the red lily is as uncertain as that of the red panda. Like the monkeys, the takins and the tigers of Himālaya's extreme south-east they will surely be lost for ever when the mega-dams materialise.

Besides reconfiguring much of Himālaya's geography, rule from Beijing continues to sinicise the region's toponyms: Tibet is Xizang; most of south-east Tibet is the prefecture of Nyingchi; and Nyingchi includes, as well as much of Indian-held Arunachal, all of Pemako, which is now called Motuo (or Medok, Medog). Hence official announcements about the world's largest hydro-project may be camouflaged by referring to it as the Motuo Dam, which of course has a protected area of its own, the Motuo State Natural Reserve.

The Tibetans are becoming accustomed to this piecemeal appropriation of their country. In western Tibet, what was once Gugé and then Ngari is now called Ali. Charmless highways connect pop-up townships across the empty Changthang; instant cityscapes dwarf the old monastic towns. Sacred landscapes are bulldozed, rivers rerouted, peaks profaned.

Writing of Pemako, Kingdon-Ward had noted what he called 'a quaint prophecy among the Kongbo Tibetans'. It was recorded in a book by 'some fabulous person' whose image was revered in the little monastery of Payi in Pomé. The 'fabulous person' was evidently a *terton*, meaning someone empowered to discover and interpret the hidden treasures (*terma*) left by Padmasambhava, the eighth-century Guru Rinpoche who first brought the Buddha's teachings to Tibet. Padmasambhava is believed to have actually visited Pemako and to have left news not only of the hidden prophecy but also of Pemako's status as a Shambhala, or Shangri-la, a terrestrial paradise and a gateway to nirvana.

The prophecy itself was less encouraging. According to Kingdon-Ward it was to the effect that 'Namcha Barwa will one day fall into the Tsango gorge and block the river, which will then turn east and flow over the Doshong La.'[9] In other words it foretold of a diversion of the river's waters away from the Dihang–Brahmaputra and into water-short China, which is pretty much what the Motuo/Yarlung–Tsangpo dams are designed to achieve – and what the government of India, noting the likely effect on the discharge of the Brahmaputra, has good grounds to fear.

Nor is Padmasambhava's prophecy absurdly far-fetched. Seismic activity resulting from the ongoing collision of the tectonic plates is as commonplace in the central and eastern Himalayas as in the western Himalayas. In 2005 Azad (Pakistani-held) Kashmir was hit by a massive quake, and in 2015 it was Nepal's turn. Whatever the damage to the natural environment from human activities like logging, mining and dam-building, it is worth bearing in mind that these are as nothing compared to the constant scene-shifting and the loss of lives, livings and species occasioned by mountain uplift, erosion and flash flooding.

In August 1950 an earthquake measuring 8.6 on the Richter scale (and so stronger than either of the above) devastated Pemako/Medog and neighbouring Assam. Hillsides turned into landslides, unsuspected glaciers nosed from the vegetation, and chunks of Namcha Barwa were indeed precipitated into the river. The little monastery at Pemakochung was destroyed. Aerial reconnaissance revealed apocalyptic devastation right through the gorges.

The only Westerner on the spot was Kingdon-Ward. On the last of his innumerable expeditions he happened to be in Rima in 1950. This is a small town in Zayul which lies east of Pemako but south of the McMahon Line and is today claimed by both China and India. It was also the earthquake's epicentre. The deafening explosions that accompanied the quake seemed to be coming from above ground rather than beneath it. Kingdon-Ward wondered whether hostilities between the newly independent Asian rivals had already begun.

The shocks were felt in Calcutta and as far away as central Tibet. In Lhasa they were taken as an ill omen by the advisers of the about-to-flee Dalai Lama. Heinrich Harrer, himself about to depart, also recorded 'some forty detonations' plus 'a huge glow' in the east. For news he relied on All India Radio's reports of how 'mountains and valleys were displaced'. 'The Brahmaputra, blocked by a fallen mountain, had caused immense devastation … Hundreds of monks and nuns were buried in their rock monasteries … Human beings, as if snatched by a demon's hand, disappeared into the suddenly gaping earth.'[10]

Namcha Barwa of course survived; but it was rendered more difficult of access than ever, which may explain why it remained for so long unclimbed. Not till 1990 did Tsuneo Shigehiro and his Japanese team begin their reconnaissance of the mountain. It involved overflying its snowy slopes 'in a small jet' and, perhaps as a quid pro quo, led to the inclusion of six Chinese climbers in the party. Next year one team member was lost to an avalanche, but in 1992 no fewer than eleven climbers finally attained Namcha Barwa's summit.

They did so without bottled oxygen or armies of porters. Among purists a swift and less encumbered approach to mountain-climbing known as 'Alpine style' had now found favour. The siege tactics, the prestocked camps, the fixed ropes and elaborate supply chains of the pre-war era were disdained. Instead climbers aimed to be self-sufficient, equipped only with what they could personally carry for making a hasty thrust to the summit. Airstrips at Gilgit, Skardu, Leh, Lhasa, Pokhara and Katmandu made reaching base camps easier; aerial reconnaissance, radio communication and better weather forecasting helped when they got there. Performance was enhanced by nylon ropes, alloy gadgetry, down-filled clothing and high-protein rations. Disclaiming its debt to the battlefield logistics of the inter-war period, mountaineering had gone back to basics by embracing science.

The results were impressive. Although not quite an eight-thousander, Namcha Barwa was claimed by Shigehiro as 'the highest of the world's still unclimbed mountains' – and it probably was. During the previous four decades Himālaya's other peaks had been succumbing to the ropes and grappling irons of the mountaineers as rapidly as the Changthang's wildlife to the trigger-happy roughriders of the PLA. The highest peaks, like the biggest 'heads', made prime trophies and had been the first to submit. Of the world's fourteen peaks over 8,000 metres, in 1950 not one had been conquered; by 1960 nearly all had. The so-called 'golden age' of bagging Himālaya's loftiest summits had lasted little over a decade.

The post-war endeavours had begun where the pre-war endeavours had ended. The Germans planned a return to Nanga Parbat, the Americans and the Italians again directed their attentions to K2, and the British geared up for a renewed assault on Everest. Yet a French expedition to a hitherto barely noticed peak had pipped them all. On 3 June 1950 Maurice Herzog, an engineer from Lyon, and Louis Lachenal, a Chamonix guide, unfurled a *tricolore* on the summit ridge of Annapurna (8,091 metres) and so broke through the 'ice ceiling' associated with summits of over 8,000 metres. It was not just an unprecedented feat but a wholly unexpected one. There had been no time to recce the mountain; the climbers were without bottled oxygen; unusually the summit had been reached at the very first attempt; and all this despite Annapurna not being the expedition's main objective. That had been nearby Dhaulagiri (8,167 metres). Only when frustrated by Dhaulagiri had Herzog and his colleagues looked further afield. Directed to Annapurna, they had decided, *faute de mieux*, to give it a go.

It was also the first of central Nepal's great array of peaks to be attempted. Tibet was already closed to foreigners. The fourteenth Dalai Lama had yet to flee Lhasa, but, with the PLA advancing into Kham, any Tibetan concession to Western climbers would have gone down badly in Maoist Beijing. On the other hand there was something to be said for the 'bamboo curtain'. The Asian extension of the 'iron curtain' followed Tibet's southern frontier, so leaving Nepal and Bhutan outside it and exposed to overtures from the 'free world'. Courtesy of this new world order, plus some Indian-inspired agitation and Katmandu's usual palace upheavals, just when Lhasa was turning climbers away Nepal started extending a cautious welcome to them.

The British saw this as a long-awaited opportunity to access Everest from the south rather than from Tibet in the north. The Swiss were thinking along the same lines. But the French saw it as an excellent opportunity to steal a march on their European rivals elsewhere. They duly applied to Katmandu for permits, and as per the subtitle of Herzog's account of the expedition, Nepal obliged with a chance to claim 'le premier huit mille'.

Perhaps no climbing book has ever been as widely read as Maurice Herzog's *Annapurna, premier 8.000*. Translated into numerous languages, it sold over 11 million copies and is still reckoned 'the most successful expedition book of all times'.[11] It was also one of the first. With

Mallory and Irvine best known for the mystery that surrounded their disappearance, Herzog became 'the first living international climbing celebrity whose name was known to the general public'.

Though not a literary masterpiece, the book established the essential ingredients of the genre: triumph spiced with tragedy, and neither of them relieved by anything as wishy-washy as self-deprecation. The triumph in the case of Annapurna was in the ascent, the tragedy in the descent. Leaving the summit in an oxygen-starved daze, Herzog had lost his gloves, then his companion. Lachenal would be found, but he too had lost a crampon and his ice axe. On feet already numb with frostbite, both men then stumbled into a crevasse, were buried in it by an avalanche and spent the night pinned down in an ice cave without food or water and with only one sleeping bag. There next day they were found half dead by two would-be rescuers. These, though, were also handicapped; both were suffering the agonies of snow-blindness. Amid more cliff-hanging drama and against all odds, the lame climbers made a hobbled descent piloting their sightless colleagues in a grotesque conga. 'We returned to earth in a fearful mix-up of pain and joy, heroism and cowardice, grandeur and meanness,' recalled one of the rescuers.[12] The snow-blind would recover but Lachenal would live the rest of his life without toes and Herzog, after months of hospitalisation, without either toes or fingers.

Two years later in 1952, from Katmandu a Swiss expedition including Tenzing Norgay, already the pick of the Sherpa climbers, resumed the assault on Everest. They made two attempts and got to within 300 metres of the summit. The fact that the Swiss so nearly succeeded alerted the British to the urgency of restaking their own claim. Ever mindful of Mallory and Irvine, London's Himalayan Committee still considered Everest a peculiarly British responsibility. It was dubbed the Third Pole, as if success in Himālaya would compensate for Captain Scott's having been pipped in the Antarctic. Indifferent to the danger of tempting fate, someone actually pencilled in the Everest summit bid for late May (1953); a successful outcome would be the perfect augury for the 2 June coronation of the young Queen Elizabeth II.

Looking back, the Everest triumph seems as if preordained. The exhaustive reconnaissances, the massive fund-raising effort, the meticulous planning, the tonnes of oxygen, the 350 porters, the choice of the self-effacing John Hunt as leader and the inspired pairing of the laconic Hillary and the incomparable Tenzing – all went to plan. The weather behaved. There were no serious mishaps. It was almost boring. Hunt, unlike Scott of the Antarctic, would not be lionised. With impeccable timing the embedded *Times* correspondent (it was the James Morris who later became Jan Morris) sent a coded message by runner to the nearest radio operator who transmitted news of the 29 May ascent to ensure it made the headlines on Coronation Day itself.

London had almost too much to celebrate. Television, itself a novelty, brought the flag-waving crowds into countless living rooms. The so-called 'new Elizabethan Age' was off to a memorable start. Instead of Annapurna's Gallic pathos or Nanga Parbat's Teutonic angst, Everest filled the streets with neo-imperial euphoria. That neither Hillary nor Tenzing was actually British went largely unremarked. Hunt, anyway, was British enough for both of them, and Hillary as a New Zealander belonged to the Commonwealth. So too would Tenzing when he was proudly claimed by India. Born in Tibet, raised in Nepal and about to make his home in Darjeeling, Tenzing had actually been working in Chitral, part of Pakistan, when the subcontinent was partitioned in 1947. Like other mountain people, he cared little about nationality. If he was a true son of anywhere, it was of Himālaya.

National pride still counted for a lot among European climbers. After a US expedition to K2 in 1953 had been forced back by appalling weather and the loss of a climber, in 1954 the Italians returned in strength to reclaim what was still considered 'the Italian mountain' and was now the highest still unclimbed. With eleven climbers and 500 Hunzakut porters, this 'quasi-official national enterprise' was led by Ardito Desio of the 1929 Italian expedition and equipped on a scale to rival that of de Filippi's 1913 K2 expedition. As with Hunt on Everest, the elaborate arrangements paid off. On 31 July 1954, nearly a hundred years since its discovery, Lino Lacedelli and Achille Compagnoni became the first to stand atop the world's second highest eminence.

Of the prized pre-war peaks, that left Nanga Parbat and Kangchenjunga. Kangchenjunga's was the summit on to which a British expedition of 1955 would take care *not* to set foot. And Nanga Parbat, though indeed scaled, was easily ignored on the grounds of controversy. After claiming so many lives, both German and Sherpa, *der Berg des Schicksals*, the Mountain of Destiny, had been under siege by a German–Austrian team even as the British were settling scores with Everest. To Dr Karl Herrligkoffer the world's ninth highest peak was well worth a seventh attempt. He owed it to the mountain and he owed it to his half-brother. This was Willy Merkl, he who had led the ill-fated expedition of 1934 and whose body still lay among those unrecovered somewhere on the Rakhiot Ridge.

Paul Bauer, the impresario of the previous assaults, had a poor opinion of Dr Herrligkoffer; he was 'unknown in mountaineering circles and without experience of the subject'. His Willy Merkl Memorial Expedition would have to manage without Bauer's support.[13] It also had to manage without Sherpas. Nanga Parbat being now in Pakistan, the government in Karachi baulked at issuing entry visas to men they regarded as citizens of a hostile India; Hunzakut porters must be used instead. News received by radio of the British success on Everest may also have disconcerted Herrligkoffer's team. Bauer suggests it 'acted on them like champagne', but by now it was late June, the weather was deteriorating and tempers were fraying.

Camp IV at 6,700 metres was reached by following Merkl's route up the Rakhiot Ridge. To further honour Merkl's memory Herrligkoffer nursed the idea of seeing all nine of his climbers reach the summit. That, he calculated, meant ordering them back to base camp to wait on the weather and recuperate for the final push. The four lead climbers refused to withdraw; defying repeated commands, instead of descending, they climbed. Their final camp was too small for more than two men; and of these, on 3 July 1953, only Hermann Buhl, the stronger and more experienced of the pair, made it to the top.

Buhl's two days alone and tentless at over 8,000 metres on the most demanding of Himālaya's summits was applauded by his supporting colleagues. But none emulated his achievement and, of those lower down the mountain, several resented it. Herrligkoffer in particular saw it as a betrayal of the consolidated teamwork advocated by Merkl and the earlier expeditions. It was self-indulgent and much too unGerman. The ascent of Nanga Parbat proved to be the first of an eight-thousander made in defiance of orders; it would be the only such peak to have been first climbed solo; and though the expedition was German-led, Buhl was in fact an Austrian. No one doubted his achievement, but Herrligkoffer would 'repeatedly downplay Buhl's importance to the expedition's success'. Buhl retaliated. He contravened the terms of his contract and published a rival account of the climb. The dispute went public and 'led to a round of nasty legal proceedings that divided the German and Austrian mountaineering communities'.[14] Nor was it the end of the matter. Merkl had yet to be vindicated. Herrligkoffer would be back.

He returned in 1962 and again in 1970, this time in charge of the ninth German expedition to the world's ninth highest mountain. The objective now was the first ever ascent of the mountain's south face, an awesome wall of some 4,500 metres of ice and snow. Buhl had thought it unclimbable. Two young brothers in Herrligkoffer's new team begged to differ. Natives of the German-speaking Italian Tyrol, Günther and Reinhold Messner had already mastered the main Alpine climbs. 'The Alps had become too small for me,' recalled the twenty-six-year-old Reinhold.[15] Nanga Parbat would be his first taste of the Himalayas, though from the photos it too looked underwhelming: 'The ascents there seemed to me too flat. Snow plodding I did not want.' Abrim with confidence, Reinhold felt equal to any challenge; the difficulty lay in finding a challenge equal to Reinhold.

What followed on the Naked Mountain in 1970 is still hotly debated. Dr Herrligkoffer would again complain of insubordination and betrayal, while Reinhold's version favours a scarcely credible catalogue of misunderstandings. A weather report from the doctor down at base camp was misinterpreted. Regardless, Reinhold, like Buhl, set off for the summit alone, without orders and without rope or tent. To his surprise he was overtaken by brother Günther, also without rope or tent. Both Messners reached the summit ridge and bivouacked there. It was minus 30 degrees Celsius but they were doing OK.

The descent was reminiscent of Herzog and Lachenal's on Annapurna. Now frostbitten and hallucinating, they opted for the unclimbed western face of the mountain. On good evidence it was later claimed that this had been Reinhold's ambition all along – to complete the first traverse of the mountain. Two of their colleagues appeared within hailing distance but misunderstood the Messners' calls for help. Reinhold then lost contact with Günther. He claims to have spent two days and nights looking for him before concluding he'd been swept away by an avalanche. 'I knew now that Günther was dead – yet I waited for him,' he adds as if by way of exoneration. Reinhold himself was by now unable to walk yet crawled to terra firma. He was eventually rescued by local woodsmen.

A spell in hospital left him with a total of four and a half toes. It seemed he'd be lucky ever to walk again, let alone climb. Yet, nothing if not perverse, it was now that 'I became a mountaineer and later a kind of adventurer' (which was 'not a profession, rather a condition'). Nanga Parbat had redefined his life. Herrligkoffer would again pursue his lead climber through the courts, and Messner would return to the mountain to search, unsuccessfully, for his brother's remains. It was widely supposed that he had sacrificed Günther's safety to his own ambition. The loss did indeed dog his days. But what the Nanga Parbat experience had really changed was his attitude towards the mountains and, with it, the whole nature of mountaineering.

Over the next two decades Reinhold Messner broke every record in the book. Everest was climbed without oxygen, then without a companion, then without either. Nanga Parbat was knocked off again. He chalked up firsts in the Karakorams and was the first man to have scaled all fourteen of the world's *Achttausanders*. He also wrote more books, delivered more lectures and made more films (not to mention enemies) than any contemporary 'adventurer'. His Tyrolean network of Messner Mountain Museums memorialises his achievements while serving as an educational facility.

Part faculty, part industry, Messner himself lent glamour to the ongoing commodification of the mountains. Commercial 'trekking', as opposed to casual hiking with a guide, had begun in Nepal when in 1964 Jimmy Roberts, an Everest veteran and then British military attaché in Katmandu, founded the tour agency Mountain Travel. Others soon followed suit elsewhere. Safer and less spartan than mountaineering, trekking grew exponentially throughout Himālaya – from the Karakoram Highway and Kashmir to Sikkim and Bhutan. Visitor numbers in Nepal rocketed from around 250,000 a year in the 1990s to over a million in 2020. So did the number of climbing permits issued by Katmandu – and the fees charged for them. Adventure tourism became one of the country's largest employers and its main source of foreign exchange. Popular routes spawned tea shops and hostelries. Base camps became eyesores.

Messner disapproved. By 2016 he rated Everest merely 'a kindergarten with miles of fixed rope set up for weekend hill walkers to reach the summit with guides and porters'.[16] The mountains were being demeaned.

*

Today, across the Indus from Nanga Parbat, dust devils whirl about the windowless town houses of Chilas. Tucked into a deep trough and invisible even from the 'fairy meadow' where Edmund Candler once smoked his pipe, Chilas feels like it hasn't known rain for centuries. Shimmering in the heat, Cyclopean chunks of rock pecked with petroglyphs dot the encroaching sands. Lizards scuttle for cover and a call to prayer reminds the faithful that Chilas takes its Islam seriously.

Here, a mere 1,265 metres asl, the Indus belies its Tibetan origins to wriggle from its gorge and carve an arabesque across a desert of gravel. On the right bank runs the Chinese-built and occasionally motorable Karakoram Highway. Nearby there's even room for Chilas Airport. 'Not open to commercial aircraft', says the internet – and not open at all in 2020 because it was being upgraded. Basic facilities were being constructed in recognition of the likelihood that, 40 kilometres downstream where the river resumes its tortured descent amid friable hillsides, a 272-metre wall of concrete that will be Pakistan's biggest-ever dam will need to be more VIP-accessible.

Just as Namcha Barwa stands guard over the eastern end of the Himalayas, so does Nanga Parbat over the western end. Geologists ascribe both mountains to the range's terminal syntaxes; guidebooks imagine them as the Great Himalaya's gables or buttresses. Just as in Pemako the Tsangpo twists and turns in an impenetrable gorge round Namcha Barwa, so in this Kohistan region of Gilgit–Baltistan does the Indus slide round the plinth of Nanga Parbat. Both gorges are claimed as the world deepest, both have excited the dam-builders, and both are now being pulverised by an army of earth-moving machinery prior to inundation.

The Chilas dam is properly the Diamer-Bhasha Dam. (Diamer is one of several spellings of an alternative name for Nanga Parbat; Bhasha is a nearby village in what was formerly Pakistan's North West Frontier Province and is now Khyber-Pakhtunkhwa; and all fall within the region loosely known as Kohistan.) The dam has been on the stocks for longer than the Yarlung–Tsangpo/Motuo Dam and is even more controversial. India objects because Gilgit–Baltistan is considered part of Jammu and Kashmir, the former state which India still claims *in toto*. The dam is thus on Indian soil, says New Delhi, and any Sino-Pak exploitation of the region's resources is therefore an infringement of India's territorial integrity.

This Indian objection proved enough to persuade international benefactors like the World Bank to withdraw promised funding for the dam. Pakistan responded by launching a crowd-funding exercise, which also fell short of expectations. It was only when in 2017–18 Beijing picked up much of the tab as part of its investment in a China–Pakistan Economic Corridor (itself somehow incorporated into Xi Jinping's elasticated Belt and Road Initiative) that work on the project finally began. For a stake likely to exceed US$20 billion, a Chinese consortium got the lion's share of the construction contracts and will no doubt enjoy a handsome share of the operating profits.

Other objections still stand. The 1841 flood that overtook the Lahore army of Sham Singh Atarewala 500 kilometres downstream at Attock can be discounted. The blockage that caused that flood would not now go ignored and could anyway be dislodged. But with the astonishing 'average of 300 earthquakes in a single month at the proposed site of the dam', comparable catastrophes can by no means be discounted and could have fatal consequences for downstream Pakistan.[17]

Pemako may have more in the way of endangered species, but Kohistan has a larger and more intractable human population. Relocating hundreds of villages and forcibly acquiring jealously guarded tribal lands has already stirred up opposition. Kohistan ('Land of the Mountains') was formerly Yaghistan ('Land of the Ungoverned') and is still a redoubt of brigands and other fugitives from justice. Here, and more especially in the upper reaches of the river's steep side-valleys, the state is not welcome. Locals carry guns and wear beards; their women are mostly invisible and are best not acknowledged. Village watchtowers are still in use for sanctuary and defence; feuds over 'honour' (that is, women), timber-extraction, hunting rights and now land sales divide family from family, village from village and clan from clan.

Only on matters of faith is there general agreement. Neighbours like the progressive Ismaili Shi'i of Hunza are regarded as infidels and liable to be attacked. Kohistan's preference is for an intolerant and ultra-conservative Sunni Islam more like that of the Afghan Taliban. Elsewhere in Himālaya such extremist tendencies are found mostly among 'freedom-fighters' joining the conflict in Kashmir.

Kalashnikov-toting jihadists are not usually among the hazards of climbing Nanga Parbat; nor was that why Reinhold Messner called it the 'Killer Mountain'. Yet on the night of 22 June 2013 the greatest ever loss of life on the mountain was down to neither storm, avalanche or the unloved Dr Herrligkoffer. The ten climbers, a multinational group mainly from Ukraine, Slovakia and China, were still at base camp with their Pakistani guide. They retired to their tents at about 10 p.m. and the first shots rang out almost immediately. Cries of 'God is Great' and 'Long Live Usama bin Laden' accompanied the firing. Eleven bodies – ten climbers and a cook – were later recovered.

Climbers continue to characterise Nanga Parbat not as the Naked Mountain, a literal translation, but as Messner's Killer Mountain. The ratio of fatalities to survivors among those who have attempted to climb it is believed to be one of the worst. Danger, like elevation, proves irresistible, attracting ever more contenders, many of them inexperienced and ill acclimatised, and inviting ever more mishaps, many of them fatal. In Nepal scarcely a season passes without a major loss of life: in 2014 forty-three died in a snowstorm on Annapurna and sixteen in an avalache on Everest; the Nepal earthquake of 2015 claimed the lives of at least nineteen climbers; in 2016 only six deaths were recorded – it was 'a normal season' reported one database – and in 2017 only twelve. 'Death rates are reducing,' insisted the database. But the reduction is scarcely borne out by the ongoing catalogue of woe – eleven fatalities in 2019, seventeen in an Annapurna avalanche in 2020, and more in 2021 as pandemic restrictions were eased.

Most disasters occur during the pre-monsoon months of April and May. Ice conditions then favour the summiteer and permits are in great demand. But meteorological conditions are not so obliging. Cloud and high winds result in long delays until a 'weather window' reveals the spectacle of climbers bunched nose to tail on the fixed ropeways. In 2019 photos showing a 500-metre crocodile of climbers snaking up Everest's summit ridge went viral. Colourfully attired in high-viz Gore-Tex and down-filled puffers, they looked like visitors awaiting admission to a wintry Disneyland, or skiers queueing for a cable car to the stars. Margret Grebowicz in the *Atlantic* described the mountain as 'crawling' with amateur adventurers 'whose bank accounts often exceed their mountaineering experience'.

More than 600 people summit per year, paying somewhere between $30,000 and $100,000 each. And more than 200 dead bodies, too costly to remove, remain in plain view, a particularly dramatic kind of human waste.[18]

Other forms of human waste foul the camps and disfigure the route, prompting inadequate and often hazardous efforts at rubbish retrieval. In 2017 an eco-expedition to the Khumbu Icefall at Camp 2 collected '1500 kilos of garbage'. Sherpas, whose own lives are as much at risk as those of their clients, grow to resent the menial role assigned to them and may resort to strike action. 'Everest is over,' says Grebowicz. Scaling Himālaya's loftiest peaks is no longer a redemptive experience.

Nor is pilgrimage. Visiting Kailas in 2011, Stephen Alter found the sacred circuit (or *kora*) of western Tibet's most revered mountain 'littered with rubbish – discarded juice packets, biscuit wrappers, aerosol tins, sanitary pads, cigarette butts and plastic pepsi bottles'. He adds, 'Mountains and rivers are revered and worshipped as maternal deities, yet the same streams of holy water are defiled with untreated sewage from "Vedic Resorts" … [and] the eternal silence of Himālaya echoes with digitised hymns set to a Bollywood beat.'[19]

It doesn't need to be this way. They manage things better in that other great realm of snow and ice at the world's southern extremity. In Antarctica environmental pollution is strictly forbidden and closely monitored. Waste, if not readily compostable or suitable for an incinerator, is mandated for removal to whence it came. Personal items – defunct footwear, dog-eared paperbacks, exhausted toothpaste tubes, redundant underwear, broken goggles, discarded batteries – all must be shipped out. Preserving the continent's pristine environment is recognised as being paramount both for the science conducted there and for the health of our planet.

In Antarctica they worry about an escape of head-lice; in Himālaya they tolerate the presence of human cadavers. The difference has a lot to do with our perception of these frigid realms. Antarctica is universally recognised as the most precious and coherent of ecozones, while Himālaya is rarely acknowledged as anything of the sort – or indeed as existing outside the pages of the *Himalayan Journal*. The former is reserved for 'science and peace', protected by numerous binding treaties, excluded from international conflict and sometimes seen as a governmental paradigm for a new world order. The other is seen as hopelessly fragmented, fair game in a world of jostling nations, much disputed over and more environmentally endangered than anywhere.

It could still be different. There being no native Antarcticans, the southern continent was quietly apportioned among the twentieth century's interested nations without contravening any human rights. This would never be possible in Himālaya; its considerable population invariably contests political encroachment and would certainly object to an international carve-up. But the concerns of the indigenous peoples should not be allowed to deter global anxieties. Mosquitoes have lately brought malaria to Lhasa; temperatures across the 'roof of the world' appear to be rising faster than anywhere else in Asia. This has profound implications way beyond the mountains. To address the depletion of what ICIMOD calls 'the water towers of HKH (Hindu Kush Himālaya)' will require regional states to sign up to an HKH treaty. It could be a first step towards recognising Himālaya as a unique global asset. Accelerated climate change may have an upside.

Antarctica attracts a few tourists, mostly to view the wildlife and check out the assorted huts, masts and cairns listed as its 'historic sites and monuments'. Amazonia has great human diversity plus an inexhaustible variety of other species; Oceania boasts something similar, interspersed by vast expanses of water. But of our planet's major ecozones only Himālaya combines rich cultural traditions and palaces, temples and monasteries with the oddities and natural wonders of the most eye-watering mountain scenery on earth. 'The garden of God' deserves recognition as somewhere special. It cries out for whatever conservational safeguards can be devised.

Notes

PROLOGUE

1 Sir Geoffrey Corbett, 'The Word Himalaya', *Himalayan Journal*, I (1929), pp. 84–6

2 Patrick French, *Younghusband: The Last Great Adventurer*, HarperCollins, London, 1994, p. 162

3 Perceval Landon, *The Opening of Tibet*, Hurst & Blackett, London, 1905, vol. 1, pp. 26–7

4 David Gilmour, *The Ruling Caste: Imperial Lives in the Victorian Raj*, John Murray, London, 2005, p. 26

5 Peter Hopkirk, *The Great Game: On Secret Service in High Asia*, John Murray, London, 1990, pp. 4–5

6 Peter Fleming, *Bayonets to Lhasa: The First Full Account of the British Invasion of Tibet in 1904*, Rupert Hart-Davis, London, 1961, p. 95

7 Curzon to Younghusband, 23 January 1904, quoted in Gordon Stewart, *Journeys to Empire: Enlightenment, Imperialism, and the British Encounter with Tibet, 1774–1904*, Cambridge University Press, Cambridge, 2009, p. 154

8 Edmund Candler, *The Unveiling of Lhasa*, Nelson, London, 1905, repr. 1925, pp. 43–4

9 Ibid., p. 48

10 Fleming, *Bayonets to Lhasa*, p. 102

11 Candler, *The Unveiling of Lhasa*, p. 49

12 Ibid., pp. 105–6

13 Ibid., p. 108

14 Ibid., pp. 136, 139

15 Landon, *The Opening of Tibet*, vol. 1, p. 147

16 Candler, *The Unveiling of Lhasa*, p. 141
17 Ibid., pp. 141–2
18 Landon, *The Opening of Tibet*, vol. 1, p. 150
19 Candler, *The Unveiling of Lhasa*, pp. 144–5
20 Lieutenant Arthur Haddow quoted in French, *Younghusband*, p. 223
21 L. Austine Waddell, *Lhasa and its Mysteries, with a Record of the Expedition of 1903–1904*, John Murray, London, 1905, p. 159
22 Candler, *The Unveiling of Lhasa*, p. 146
23 French, *Younghusband*, p. 229
24 Candler, *The Unveiling of Lhasa*, pp. 289–90
25 Waddell, *Lhasa and its Mysteries*, p. 331
26 Landon, *The Opening of Tibet*, vol. 2, p. 164
27 Younghusband quoted in French, *Younghusband*, p. 252
28 Younghusband quoted in Stewart, *Journeys to Empire*, p. 244
29 Candler, *The Unveiling of Lhasa*, pp. 307–8

I AN OROGENOUS ZONE

1 Thomas Carlyle, *Sartor Resartus*, Saunders & Otley, London, 1838, p. 2
2 Mike Searle, in correspondence with the author, 28 January 2021
3 S. G. Burrard and H. H. Hayden, *A Sketch of the Geography and Geology of the Himalaya Mountains and Tibet*, 2nd edn, Survey of India, Delhi, 1933, p. 9
4 James Abbott, 'Inundation of the Indus', *Journal of the Asiatic Society of Bengal*, XVII (1848), pp. 230–2
5 Hugh Falconer, 'On the Recent Cataclysm of the Indus', *Journal of the Asiatic Society of Bengal*, X, pt 2, no. 116 (1841), pp. 615–20
6 Kenneth Mason, 'Indus Floods and Shyok Glaciers', *Himalayan Journal*, I (1929), pp. 10–29
7 Quoted in Catherine Moorehead, *The K2 Man (and his Molluscs): The Extraordinary Life of Haversham Godwin-Austen*, Neil Wilson Publishing, Castle Douglas, 2013, p. 210
8 M. S. Krishnan, *The Geology of India and Burma*, Higginbothams, Madras, 1941, p. 27
9 John McPhee, *Annals of the Former World*, Farrar, Straus & Giroux, New York, 1998, p. 124
10 Mike Searle, *Colliding Continents: A Geological Exploration of the Himalaya, Karakoram, & Tibet*, Oxford University Press, Oxford, 2013, p. 368
11 Dennis R. Dean, *James Hutton and the History of Geology*, Cornell University Press, Ithaca, NY, 1992, p. 95
12 Ibid., pp. 158–9

13 Gavin Flood, *An Introduction to Hinduism*, Cambridge University Press, Cambridge, 1996, p. 113

14 *Saṃyutta Nikāya*, 15.5.6, quoted in Randy Kloetzli and Alf Hiltebeitel, 'Kāla', in Sushin Mittal and Gene Thursby (eds), *The Hindu World*, Routledge, Abingdon, 2004, pp. 569–70

15 McPhee, *Annals of the Former World*, p. 29

16 Ibid., p. 89

17 Martin J. S. Rudwick, *Earth's Deep History: How It Was Discovered and Why It Matters*, University of Chicago Press, Chicago, 2014, p. 242

2 WAR OF THE PLATES

1 *Rig Veda*, x, 9

2 William Montgomery McGovern, *A Manual of Buddhist Philosophy*, vol. 1: *Cosmology*, K. Paul, Trench, Trubner, London, 1923, p. 66

3 Ibid., p. 61

4 Mott T. Greene, *Alfred Wegener: Science, Exploration, and the Theory of Continental Drift*, Johns Hopkins University, Baltimore, Md., 2015, p. 98

5 Quoted in ibid., p. 234

6 Alfred Wegener, *The Origin of Continents and Oceans*, 4th rev. edn (1929), trans. John Biram, Dover Publications, New York, 1966, p. 17

7 Alfred Wegener quoted in Henry R. Frankel, *The Continental Drift Controversy*, Cambridge University Press, Cambridge, 2012, vol. 1: *Wegener and the Early Debate*, p. 33

8 Alfred Wegener quoted in Greene, *Alfred Wegener*, p. 126

9 Ibid., p. 295

10 Wegener, *The Origin of Continents and Oceans*, p. 2

11 John McPhee, *Annals of the Former World*, Farrar, Straus & Giroux, New York, 1998, p. 117

12 Ted Nield, *Supercontinent: Ten Billion Years in the Life of our Planet*, Granta Books, London, 2007, p. 15

13 Mike Searle, *Colliding Continents: A Geological Exploration of the Himalaya, Karakoram, & Tibet*, Oxford University Press, Oxford, 2013, p. 279

14 *Vishnu Purana*, trans. H. H. Wilson, repr. Ganesha Publishing, London, 2001, ch. iv, p. 32

15 Wegener, *The Origin of Continents and Oceans*, pp. 330–1

16 Pierre Dèzes, 'Tectonic and Metamorphic Evolution of the Central Himalayan Domain in South East Zanskar', PhD thesis, University of Lausanne, 1999, https://zanskar.geoheritage.ch/zanskar_geology.pdf

17 Ibid., pp. 38–9

18 McPhee, *Annals of the Former World*, p. 432

19 Kenneth Deffeyes quoted in ibid., p. 132

20 Ibid., p. 115

21 Frankel, *The Continental Drift Controversy*, vol. 2: *Paleomagnetism and Confirmation of Drift*, p. xvii

22 Henry Hess, 'History of Ocean Basins', 1962, https://ww2.odu.edu/~mcarhart/hist386/texts/HessOceanBasins1962.pdf, pp. 608–9

23 Searle, *Colliding Continents*, p. 12

24 McPhee, *Annals of the Former World*, p. 121

25 Cherry Lewis, *The Dating Game: One Man's Search for the Age of the Earth*, Cambridge University Press, Cambridge, 2000, p. 5

26 Searle, *Colliding Continents*, p. 257

27 Ibid., p. 259

28 Ibid., p. 70

29 Ibid., p. 359

3 A DOMAIN OF ANIMALS

1 Matthew T. Kapstein, 'Remarks on the *Mani Kabum* and the Cult of Avalokitesvara in Tibet', in Gray Tuttle and K. R. Schaeffer (eds), *The Tibetan History Reader*, Columbia University Press, New York, 2013, p. 95

2 Adriaan Camper, letter to Georges Cuvier, July 1800, quoted in Martin J. S. Rudwick, *Bursting the Limits of Time: The Reconstruction of Geohistory in the Age of Revolution*, University of Chicago Press, Chicago, 2005, p. 383

3 J. H. Batten, 'Note of a Visit to the Niti Pass of the Grand Himalayan Chain', *Journal of the Asiatic Society of Bengal*, VII (1838), pp. 314–15

4 Henry Mortimer Durand, *Life of Major-General Sir Henry Marion Durand*, W. H. Allen, London, 1983, vol. 1, p. 36

5 Hugh Falconer, *Palaeontological Memoirs and Notes of the Late Hugh Falconer*, ed. Charles Murchison, Robert Hardwicke, London, 1868, vol. 1: *Fauna Antiqua Sivalensis*, p. xxvii

6 Proby Cautley in *Journal of the Asiatic Society of Bengal*, IV (1835), p. 585

7 Ibid., pp. 584–7

8 Hugh Falconer in *Journal of the Asiatic Society of Bengal*, IV (1835), p. 57

9 Charles Murchison in Falconer, *Palaeontological Memoirs*, vol. 1, p. xxviii

10 Rudwick, *Bursting the Limits of Time*, p. 368

11 Georges Cuvier quoted in Elizabeth Kolbert, *The Sixth Extinction: An Unnatural History*, Picador, New York, 2015, p. 42

12 James Prinsep quoted in Falconer, *Palaeontological Memoirs*, vol. 1, p. xxx

13 Claudine Cohen, *The Fate of the Mammoth: Fossils, Myth, and History*, trans. William Rodarmor, University of Chicago Press, Chicago, 2002, pp. 135–9

14 Proby Cautley in *Journal of the Asiatic Society of Bengal*, IV (1835), p. 587

15 Hugh Falconer and Captain P. T. Cautley, 'Sivatherium Giganteum', *Asiatick Researches*, XIX (1836), and Falconer, *Palaeontological Memoirs and Notes*, vol. 1, pp. 247–79

16 Martin J. S. Rudwick, *Worlds before Adam: The Reconstruction of Geohistory in the Age of Reform*, University of Chicago Press, Chicago, 2008, p. 407

17 *Chambers' Edinburgh Journal*, 23 November 1844, quoted in Falconer, *Palaeontological Memoirs and Notes*, vol. 1, p. 372

18 Falconer, *Palaeontological Memoirs and Notes*, vol. 1, p. li

19 Ibid., p. 369

20 Ibid., p. 295

21 James Prinsep quoted in ibid., p. 307

22 Quoted in Rudwick, *Worlds before Adam*, p. 419

23 A. Ghosh (ed.), *An Encyclopaedia of Indian Archaeology*, E. J. Brill, Leiden, 1990, vol. 1, pp. 312–13

24 Charles Darwin, *The Descent of Man, and Selection in Relation to Sex*, John Murray, London, 1871, vol. 1, p. 3

25 A. Bowdoin Van Riper, *Men among the Mammoths: Victorian Science and the Discovery of Human Prehistory*, University of Chicago Press, Chicago, 1993, p. 82

26 Hugh Falconer, 'On the Ossiferous Cave of Brixham near Torquay', in Falconer, *Paleontological Memoirs and Notes*, vol. 2: *Mastodon, Elephant, Rhinoceros, Ossiferous Caves, Primeval Man and his Contemporaries*, p. 496

27 Charles Murchison in ibid., p. 486

28 James Sackett, 'Boucher de Perthes and the Discovery of Human Antiquity', *Bulletin of the History of Archaeology*, 24(2) (2014), pp. 1–11

4 WHEN MEN AND MONKEYS MEET

1 William Blake, 'Gnomic Verses', Epigrams, Verses and Fragments from the Note-book 1808–11, *The Poetical Works of William Blake*, Oxford University Press, London and New York, 1908, p. 193

2 Perceval Landon, *Lhasa: An Account of the Country and People of Central Tibet ...*, Hurst & Blackett, London, 1905, vol. 1, pp. 107–8, 410

3 David L. Snellgrove, *Buddhist Himālaya: Travels and Studies in Search of the Origin of Tibetan Religion*, Bruno Cassirer, Oxford, 1957, pp. 224–6

4 William Woodville Rockhill, *The Land of the Lamas: Notes of a Journey through China, Mongolia and Tibet*, Longmans, Green, London, 1891, pp. 356–60

5 Gabriel Bonvalot, *Across Thibet, being a Translation of 'De Paris au Tonkin à travers le Tibet Inconnu'*, trans. C. B. Pitman, Cassell, London, 1891, vol. 1, p. 210

6 C. G. Rawling, *The Great Plateau, being an Account of Exploration in Central Tibet, 1903, and of the Gartok Expedition, 1904–5*, Edward Arnold, London, 1905, pp. 222–3

7 Colin Barras, 'Secrets of How Primates Can Live at Extreme Altitude Revealed', *New Scientist*, 23 August 2016, https://www.newscientist.com/article/2101954-secrets-of-how-primates-can-live-at-extreme-altitude-revealed/

8 William F. Ruddiman and John E. Kutzbach, 'Plateau Uplift and Climatic Change', *Scientific American*, 264(3) (March 1991), p. 68, https://www.jstor.org/stable/24936829?seq=3#metadata_info_tab_contents

9 Sheila Melvin, 'Archaeology: Peking Man Still Missing and Missed', *New York Times*, 5 October 2005, https://www.nytimes.com/2005/10/11/health/archaeology-peking-man-still-missing-and-missed.html

10 Mark Aldenderfer and Yang Zinong, 'The Prehistory of the Tibetan Plateau in the Seventh Century: Perspectives and Research from China and the West since 1950', *Journal of World Prehistory*, 18(1) (March 2004), p. 15

11 Carl Zimmer, 'High Ceilings and a Lovely View: Denisova Cave Was Home to a Lost Branch of Humanity', *New York Times*, 30 January 2019, https://www.nytimes.com/2019/01/30/science/neanderthals-denisovans-humans.html

12 Bo Li et al., 'How Midnight Digs at a Holy Tibetan Cave Opened a Window to Prehistoric Humans Living on the Roof of the World', University of Wollongong, 29 November 2020, https://www.uow.edu.au/media/2020/how-midnight-digs-at-a-holy-tibetan-cave-opened-a-window-to-prehistoric-humans-living-on-the-roof-of-the-world.php

13 Cynthia M. Beall, 'Adaptation to High Altitude: Phenotypes and Genotypes', *Annual Review of Anthropology*, 43 (2014), p. 261

14 Emilia Huerta-Sanchez et al., 'Altitude Adaptations in Tibetans caused by introgression of Denisovan-like DNA', *Nature*, 512, 2 July 2014, pp. 194–7

15 Beall, 'Adaptation to High Altitude', p. 255

16 Robert Sanders, 'Tibetans Adapted to High Altitude in Less Than 3,000 Years', *Berkeley News*, 1 July 2010, https://news.berkeley.edu/2010/07/01/tibetan_genome/

17 Alex Riley, 'How Tibetans Survive Life on the "Roof of the world"', *BBC Future*, 27 February 2017, https://www.bbc.com/future/article/20170227-how-tibetans-survive-life-on-the-roof-of-the-world

5 OF FLOWERS AND TOWERS

1 Fernand Braudel, *The Mediterranean and the Mediterranean World in the Reign of Philip II*, trans. Siân Reynolds, Collins, London, 1972 (orig. pub. 1949), vol. 1, p. 34

2 Mike Searle, *Colliding Continents*, Oxford University Press, Oxford, 2013, p. 7

3 O. Olufsen, *Through the Unknown Pamirs: The Second Danish Pamir Expedition, 1898–99*, Heinemann, London, 1904, pp. 178–87

4 Ibid., pp. 172–3

5 Sophie Ibbotson, 'A lost city rises: archaeologists have unearthed the "Machu Picchu of Tajikistan"', *Telegraph*, 16 August 2019, https://www.telegraph.co.uk/travel/destinations/asia/tajikistan/machu-picchu-tajikistan-karon/

6 John Keay, *China: A History*, HarperCollins, London, 2008, p. 134

7 John Mock, 'Raising the Alarm: Defensive Communications Networks and the Silk Roads through Wakhan and Chitral', *Silk Road*, 15 (2017), p. 2

8 Captain F. Kingdon Ward, *The Riddle of the Tsangpo Gorges*, 1926, repr. Antique Collectors' Club, Woodbridge, 2001, p. 6

9 Ibid.

10 Ibid., p. 110

11 Ibid., p. 161

12 Ibid., p. 162

13 R. A. Stein, *Tibetan Civilization*, Faber & Faber, London, 1972, pp. 119–20

14 Ibid., p. 29 (photo of Kongbo towers opposite p. 49 attrib. 'Maharajah of Sikkim')

15 Ibid., p. 162

16 Ibid, p. 149

17 C. G. Rawling, *The Great Plateau, being an Account of Exploration in Central Tibet, 1903, and of the Gartok Expedition, 1904–5*, Edward Arnold, London, 1905, pp. 280–1

18 Quoted in Alex McKay, *Tibet and the British Raj: The Frontier Cadre, 1904–1947*, Curzon Press, Richmond, 1997, p. 159

19 G. M. Young, 'A Journey to Toling and Tsaparang in Western Tibet', *Journal of Punjab Historical Society*, VII (1919), p. 108

20 Ibid., p. 116

21 Giuseppe Tucci and Captain E. Ghersi, *Secrets of Tibet, being the Chronicle of the Tucci Scientific Expedition to Western Tibet (1933)*, Blackie & Son, London, 1935, p. vii

6 SCHOLAR, EXPLORER, WRITER, PILGRIM

1 Charles Allen, *A Mountain in Tibet: The Search for Mount Kailas and the Sources of the Great Rivers of India*, André Deutsch, London, 1982, p. 16

2 Alex McKay (ed.), *The History of Tibet*, RoutledgeCurzon, London and New York, 2003, vol. 1, p. 3

3 Giuseppe Tucci and Captain E. Ghersi, *Secrets of Tibet, being the Chronicle of the Tucci Scientific Expedition to Western Tibet (1933)*, Blackie & Son, London, 1935, p. x

4 Ibid., p. 100

5 Ibid., p. 104, quoted in Poncar Jaroslav and John Keay, *Tibet*, Edition Panorama, Mannheim, 2008, p. 47

6 Tucci and Ghersi, *Secrets of Tibet*, p. 136

7 Fosco Maraini, *Secret Tibet*, trans. Eric Mosbacher and Guido Waldman, Harvill Press, London, 2002, pp. 196, 280

8 Tucci and Ghersi, *Secrets of Tibet*, pp. 198–9

9 David Snellgrove, *Asian Commitment: Travels and Studies in the Indian Subcontinent and South-East Asia*, Orchid Press, Bangkok, 2000, pp. 145–6

10 McKay (ed.), *The History of Tibet*, vol. 1, p. 477

11 David L. Snellgrove (ed. and trans.), *The Nine Ways of Bon: Excerpts from gZi-brjid*, Oxford University Press, London, 1967, p. 151

12 R. A. Stein, *Tibetan Civilization*, Faber & Faber, London, 1972, p. 164

13 Snellgrove (ed. and trans.), *The Nine Ways of Bon*, p. 13

14 David L. Snellgrove, *Himalayan Pilgrimage: A Study of Tibetan Religion by a Traveller through Western Nepal*, Prajna Press, Boulder, Col., 1981, p. 43

15 Per Kvaerne, *The Bon Religion of Tibet: The Iconography of a Living Tradition,* Shambhala Publications, Boston, 1995, p. 13

16 John Vincent Bellezza, *The Dawn of Tibet: The Ancient Civilization on the Roof of the World*, Rowman & Littlefield, Lanham, Md., 2014, p. 18

17 Ibid., p. 31

18 Ibid., p. 2

19 Ibid., p. 144

20 Mark Aldenderfer, 'Variations in Mortuary Practice on the Early Tibetan Plateau and the High Himalayas', *Journal of the International Association for Bon Research*, 1 (2013), pp. 293–4

21 Bellezza, *The Dawn of Tibet*, p. 171

22 Ibid., p. 158

23 Tucci and Ghersi, *Secrets of Tibet*, p. 98

7 PILGRIMS' PROGRESS

1 Quoted in Kenneth Cox, 'Frank Kingdon Ward and the Plant Hunters of China and the Himalaya', in Kenneth Cox, Kenneth R. Storm Jr and Ian Baker, *Frank Kingdon Ward's Riddle of the Tsangpo Gorges: Retracing the Epic Journey of 1924–25 in South-East Tibet*, Antique Collectors' Club, Woodbridge, 2001, p. 62

2 Quoted in Alex McKay, *Tibet and the British Raj: The Frontier Cadre, 1904–1947*, Curzon Press, Richmond, 1997, pp. 90–1

3 F. M. Bailey, 'From the Outposts: A Quiet Day in Tibet', *Blackwoods Magazine*, 189 (February 1911), pp. 270–5

4 Toni Huber, *The Cult of Pure Crystal Mountain: Popular Pilgrimage and Visionary Landscapes in Southeast Tibet*, Oxford University Press, New York, 1999, p. 4

5 Ibid., p. 21

6 Derek Waller, *The Pundits: British Exploration of Tibet and Central Asia*, University Press of Kentucky, Lexington, Ky., 1990, p. 236

7 F. M. Bailey, *No Passport to Tibet*, Rupert Hart-Davis, London, 1957, p. 26

8 Ibid., p. 55

9 Ibid., p. 282

10 Ibid., pp. 31–2

11 Ibid., pp. 113–14

12 L. Austine Waddell, *Lhasa and its Mysteries, with a Record of the Expedition of 1903–1904*, John Murray, London, 1905, p. 62

13 Cox, 'Frank Kingdon Ward and the Plant Hunters of China and the Himalaya', p. 33

14 Bailey, *No Passport to Tibet*, p. 125

15 Ibid., p. 136

16 Ibid., pp. 198–9

17 F. M. Bailey, 'Exploration of the Tsangpo or Upper Brahmaputra', *Geographical Journal*, XLIV(4) (October 1914), p. 355

18 McKay, *Tibet and the British Raj*, pp. 111–12

19 Colonel Vinayak Bhat, 'Despite Modi–Xi bonhomie, China moves into Arunachal Pradesh, builds new roads and barracks', *ThePrint*, 22 June 2018, https://theprint.in/defence/south-china-sea-to-the-himalayas-chinas-salami-slicing-now-reaches-arunachal/73316/

8 THE KARAKORAM ANOMALY

1 Godfrey Thomas Vigne, *Travels in Kashmir, Ladak, Iskardo, the Countries Adjoining the Mountain Course of the Indus, and the Himalaya North of the Punjab*, Henry Colburn, London, 1842, vol. 2, p. 285

2 Tobias Bolch et al., 'Status and Change of the Cryosphere in the Extended Hindu Kush Himalaya Region', in Philippus Wester et al. (eds), *The Hindu Kush Himalaya Assessment: Mountains, Climate Change, Sustainability and People*, ICIMOD, 2019, https://doi.org/10.1007/978-3-319-92288-1_7

3 Mark Carey, 'The History of Ice: How Glaciers became an Endangered Species', *Environmental History*, 12(3) (July 2007), p. 510

4 *Climate Change 2007: Impacts, Adaptation and Vulnerability: Contribution of Working Group II to the Fourth Assessment Report of the Inter-governmental Panel on Climate Change*, Cambridge University Press, Cambridge, 2008, p. 493

5 Carey, 'The History of Ice: How Glaciers became an Endangered Species', p. 498

6 Elizabeth Kolbert, *Field Notes from a Catastrophe*, Bloomsbury, London, 2006, pp. 49–50

7 Bolch et al., 'Status and Change of the Cryosphere', pp. 213–14

8 Vigne, *Travels in Kashmir*, vol. 2, p. 204

9 Ibid., pp. 238–9

10 Ibid., p. 286

11 Kenneth Mason, *Abode of Snow: A History of Himalayan Exploration and Mountaineering*, Rupert Hart-Davis, London, 1955, p. 77

12 Catherine Moorehead, *The K2 Man (and his Molluscs): The Extraordinary Life of Haversham Godwin-Austen*, Neil Wilson Publishing, Castle Douglas, 2013, pp. 92, 128

13 H. H. Godwin-Austen, 'The Glaciers of the Mustakh Range', *Proceedings of the Royal Geographical Society*, 34 (1864), p. 35

14 Ibid., p. 36

15 Ibid., p. 40

16 *International Boundary Study No. 85: China–Pakistan Boundary*, https://fall.fsulawrc.com/collection/LimitsinSeas/IBS085.pdf

17 Ibid, p. 51

18 Lionel James Picton, *Thoughts on Feeding*, Faber & Faber, London, 1946, p. 64

19 John Staley, *Words for my Brother: Travels between the Hindu Kush and the Himalayas*, Oxford University Press, Karachi, 1982, p. 24

20 Ibid., pp. 18–19

21 E. O. Lorimer, *Language Hunting in the Karakoram*, Allen & Unwin, London, 1939, p. 20

22 John Mook, 'Dards, Dardistan and Dardic: An Ethnographic, Geographic and Linguistic Conundrum', http://www.mockandoneil.com/dard.htm

23 Staley, *Words for my Brother*, pp. 54–5

24 Ingvar Nøstegård Tveiten, 'Glacier Growing – A Local Response to Water Scarcity in Baltistan and Gilgit, Pakistan', Masters thesis, Norwegian University of Life Sciences, 2007, http://www.umb.no/statisk/noragric/publications/master/2007_ingvar_tveiten.pdf

25 Janaki Lenin, 'Artificial Glaciers could help Ladakh villages adapt to climate change', *Guardian*, 24 February 2015, https://www.theguardian.com/environment/2015/feb/24/artificial-glacier-could-help-ladakh-villagers-adapt-to-climate-change

26 Elizabeth Kolbert, 'The Ice Stupas; Artificial Glaciers at the Edge of the Himalayas', *New Yorker*, 13 May 2019, https://www.newyorker.com/magazine/2019/05/20/the-art-of-building-artificial-glaciers

9 SUBLIME DELIVERANCE

1 Swami Pranavananda, *Kailas–Manasarovar*, S. P. League, Calcutta, 1949, pp. 6–7

2 E. O. Lorimer, *Language Hunting in the Karakoram*, Allen & Unwin, London, 1939, p. 21

3 D. L. R. Lorimer, 'The Supernatural in the Popular Belief of the Gilgit Region', *Journal of the Royal Asiatic Society of Great Britain and Ireland*, no. 3 (July 1929), pp. 512–13

4 Ibid., pp. 519–21

5 Giuseppe Tucci, *The Religions of Tibet*, Routledge & Kegan Paul, London, 1950, p. 216

6 Alexandra David-Néel, *My Journey to Lhasa*, Heinemann, London, 1927, p. 30

7 Ibid., p. 222

8 Ernst Schäfer, quoted in Bianca Horlemann, 'The Goloks through Western Eyes: Fascination and Horror', https://www.academia.edu/6465012/The_Golok_of_Tibet_through_Western_Eyes_Fascination_and_Horror

9 Alexandra David-Néel, *A l'Ouest Barbare de la Vaste Chine*, Plon, Paris, p. 187

10 Francis Younghusband and George Pereira, *Peking to Lhasa: The Narrative of Journeys in the Chinese Empire Made by the Late Brigadier-General George Pereira*, Constable, London, 1925, pp. 116–17

11 Leonard Clark, *The Marching Wind*, Funk & Wagnalls, New York, 1954, p. 29

12 Ibid., p. 195

13 John Town, 'Amné Machin: A Closer Look', *Alpine Journal* (1988–9), p. 78

14 August Gansser in Arnold Heim and August Gansser, *The Throne of the Gods: An Account of the First Swiss Expedition to the Himalayas*, Macmillan, London, 1939, p. 98

15 Ibid., pp. 97–8; and Swami Pranavananda, *Exploration in Tibet*, 2nd edn, University of Calcutta, Calcutta, 1950, p. 172

16 John Snelling, *The Sacred Mountain: Travellers and Pilgrims at Mount Kailas in Western Tibet and the Great Universal Symbol of the Sacred Mountain*, East–West Publications, London, 1983, p. 18

17 Tucci, *The Religions of Tibet*, pp. 217–19

10 SWEDE AND SWAMI

1 Pandit Jawaharlal Nehru, Foreword to Swami Pranavananda, *Kailas–Manasarovar*, S. P. League, Calcutta, 1949, p. xvii
2 Swami Pranavananda, *Exploration in Tibet*, 2nd edn, University of Calcutta, Calcutta, 1950, p. 7
3 S. G. Burrard and H. H. Hayden, *A Sketch of the Geography and Geology of the Himalaya Mountains and Tibet*, 2nd edn, Survey of India, Delhi, 1933, p. 228
4 Pranavananda, *Exploration in Tibet*, p. 7
5 Ibid., p. 65
6 William Moorcroft, 'A Journey to Lake Mánasaróvara in Ún-Dés, a Province of Little Tibet', *Asiatick Researches*, 12 (1816), quoted in Charles Allen, *A Mountain in Tibet: The Search for Mount Kailas and the Sources of the Great Rivers of India*, André Deutsch, London, 1982, p. 91
7 H. Y. Hearsey quoted in Garry Alder, *Beyond Bokhara: The Life of William Moorcroft, Asian Explorer and Pioneer Veterinary Surgeon, 1767–1825*, Century Publishing, London, 1985, p. 155
8 H. H. P. Deasy, *In Tibet and Chinese Turkestan, being the Record of Three Years' Exploration*, Fisher & Unwin, London, 1901, p. 26
9 Allen, *A Mountain in Tibet*, pp. 161, 201
10 Sven Hedin, *Trans-Himalaya: Discoveries and Adventures in Tibet*, Macmillan, London, 1909, vol. 1, pp. 1, 3–4
11 Sven Hedin, *My Life as an Explorer*, Cassell, London, 1926, p. 102
12 Hedin, *Trans-Himalaya*, vol. 1, pp. 31–2
13 Ibid., pp. 97–100
14 Pranavananda, *Exploration in Tibet*, p. 192
15 Hedin, *Trans-Himalaya*, vol. 1, p. 2
16 Hedin, *My Life as an Explorer*, p. 370
17 Hedin, *Trans-Himalaya*, vol. 1, pp. 271, 272
18 Ibid., p. 401
19 Ibid., vol. 2, p. 90
20 Ibid., p. 97
21 Ibid.
22 Ibid., pp. 112, 197
23 Ibid., pp. 213–14
24 Ibid., pp. 217, 218
25 Pranavananda, *Exploration in Tibet*, p. 196

II SAGES AND HEROES

1 Quoted in David L. Snellgrove and Hugh Richardson, *A Cultural History of Tibet*, Orchid Press, Bangkok, 2005, p. 24

2 Amar Kaur Jasbir Singh, *Himalayan Triangle: A Historical Survey of British India's Relations with Tibet, Sikkim and Bhutan, 1765–1950*, British Library, London, 1988, p. 110

3 Alex McKay, *Pilgrimage in Tibet*, Curzon, London, 1998, p. 180

4 David Snellgrove, *Buddhist Himālaya: Travels and Studies in Search of the Origins of Tibet's Religions*, Bruno Cassirer, Oxford, 1957, pp. 174–5

5 Ibid., p. 173

6 Matthew T. Kapstein, *The Tibetan Assimilation of Buddhism: Conversion, Contestation, and Memory*, Oxford University Press, New York, 2000, pp. 133–4

7 Giuseppe Tucci, *The Religions of Tibet*, Routledge & Kegan Paul, London, 1980, p. 39

8 Snellgrove, *Buddhist Himālaya*, p. 123

9 Sam van Schaik, *Tibet: A History*, Yale University Press, New Haven, 2011, p. 4

10 Christopher I. Beckwith, *The Tibetan Empire in Central Asia: A History of the Struggle for Great Power among the Tibetans, Turks, Arabs, and Chinese during the Early Middle Ages*, Princeton University Press, 1987, pp. 19–20

11 Van Schaik, *Tibet: A History*, p. 14

12 John Keay, *China: A History*, HarperCollins, London, 2008, p. 276

13 Christopher I. Beckwith, 'Empire in the West', in Michael Aris and Aung San Suu Kyi (eds), *Tibetan Studies in Honour of Hugh Richardson*, Aris & Phillips, Warminster, 1979, p. 31

14 Matthew Kapstein quoted in van Schaik, *Tibet: A History*, p. 39

15 Tucci, *The Religions of Tibet*, p. 13

16 R. A. Stein, *Tibetan Civilization*, Faber & Faber, London, 1972, p. 68

17 Van Schaik, *Tibet: A History*, p. 44

18 Tucci, *The Religions of Tibet*, p. 16

19 Giuseppe Tucci quoted in Poncar Jaroslav and John Keay, *Tibet*, Edition Panorama, Mannheim, 2008, p. 47

20 Snellgrove, *Buddhist Himālaya*, p. 123

21 L. Austine Waddell, *The Buddhism of Tibet or Lamaism, with its Mystic Cults, Symbolism and Mythology, and its Relation to Indian Buddhism*, W. H. Allen, London, 1895, pp. 15, 33

22 Ibid., p. 30

23 Perceval Landon, *Lhasa; An Account of the Country and People of Central Tibet ...*, Hurst & Blackett, London, 1905, pp. 348–9

24 Snellgrove, *Buddhist Himālaya*, pp. 119–20, 177

25 Edmund Candler, *The Unveiling of Lhasa*, Nelson, London, 1905, repr. 1925, pp. 300–1

26 Ibid., p. 302

27 Van Schaik, *Tibet: A History*, pp. 59–60; Tucci, *The Religions of Tibet*, p. 22

28 Helmut Hofman, 'Early and Medieval Tibet', in Denis Sinor (ed.), *The Cambridge History of Early Inner Asia*, Cambridge University Press, Cambridge, 1990, p. 392

29 Van Schaik, *Tibet: A History*, p. 89

30 Tucci, *The Religions of Tibet*, p. 37

31 Stein, *Tibetan Civilization*, p. 80

32 Ibid., p. 81; Giuseppe Tucci, *Tibet: Land of Snows*, Elek, London, 1973, p. 37

33 Van Schaik, *Tibet: A History*, p. 115

34 Tucci, *Tibet: Land of Snows*, p. 41

35 Stein, *Tibetan Civilization*, p. 84

36 Van Schaik, *Tibet: A History*, p. 123

37 Tucci, *Tibet: Land of Snows*, p. 42

38 Stein, *Tibetan Civilization*, p. 85

12 GOLD DUST AND YAK TAILS

1 Quoted in William Woodville Rockhill, *The Land of the Lamas: Notes of a Journey through China, Mongolia and Tibet*, Longmans, Green, London, 1891, p. 360

2 Ibid., p. 274

3 R. A. Stein, *Tibetan Civilization*, Faber & Faber, London, 1971, p. 149

4 Sam van Schaik, *Tibet: A History*, Yale University Press, New Haven, 2011, p. 67

5 Thomas Thomson, *Western Himalaya and Tibet: A Narrative of a Journey through the Mountains of Northern India during the Years 1847–8*, 1852, repr. Ratna Pustak Bhandar, Katmandu, 1979, pp. 212–13

6 T. G. Montgomerie, 'Report of the Trans-Himalayan Explorations during 1867', *Proceedings of the Royal Geographical Society*, 13(3) (1868–9), pp. 183–98

7 C. G. Rawling, *The Great Plateau, being an Account of Exploration in Central Tibet, 1903, and of the Gartok Expedition, 1904–5*, Edward Arnold, London, 1905, pp. 60–1

8 John Snelling, *The Sacred Mountain: Travellers and Pilgrims at Mount Kailas in Western Tibet, and the Great Universal Symbol of the Mountain*, East–West Publications, London, 1983, pp. 42–3

9 Michael Buckley, *Meltdown in Tibet: China's Reckless Destruction of Ecosystems from the Highlands of Tibet to the Deltas of Asia*, Palgrave Macmillan, London, 2014, pp. 144–5

10 'The price of gold: mining in Tibet', *The Economist*, 6 April 2013, https://www.economist.com/china/2013/04/06/the-price-of-gold

11 Ed Yong, 'The World's Most Valuable Parasite is in Trouble', *The Atlantic*, 22 October 2018, https://www.theatlantic.com/science/archive/2018/10/tibetan-caterpillar-fungus-trouble/573607/

12 Buckley, *Meltdown in Tibet*, p. 9

13 Kate Teltscher, *The High Road to China: George Bogle, the Panchen Lama and the first British Expedition to Tibet*, Bloomsbury, London, 2007, p. 19

14 Ibid., p. 78

15 John Keay, *Midnight's Descendants: South Asia from Partition to the Present Day*, William Collins, London, 2014, p. 185

16 John Keay, *The Great Arc: The Dramatic Tale of How India was Mapped and Everest was Named*, HarperCollins, London, 2000, p. 40

17 Quoted in John Keay, *The Honourable Company: A History of the English East India Company*, HarperCollins, London, 1991, p. 424

18 Teltscher, *The High Road to China*, p. 239

19 Ibid., p. 241

20 John Pemble, *The Invasion of Nepal: John Company at War*, Oxford University Press, 1971, pp. 57–8

21 Ibid., p. 10

22 Teltscher, *The High Road to China*, p. 245

23 Ed Douglas, *Himalaya: A Human History*, Bodley Head, London, 2020, p. 172

24 Charles Allen, *The Prisoner of Kathmandu*, Haus Publishing, London, 2015, p. 40

25 John Keay, *India Discovered*, William Collins, London, 1988, p. 67

26 Pemble, *The Invasion of Nepal*, p. 312

27 C. W. Rubenson quoted in Maurice Isserman and Stewart Wheeler, *Fallen Giants: A History of Himalayan Mountaineering from the Age of Empire to the Age of Extremes*, Yale University Press, New Haven, 2008, p. 78

28 Charles Allen, *A Mountain in Tibet: The Search for Mount Kailas and the Sources of the Great Rivers of India*, André Deutsch, London, 1982, pp. 84–6

29 Keay, *Midnight's Descendants*, p. xxii

13 SHAWL WARS

1 Alexander Cunningham, *Ladák, Physical, Statistical, and Historical; with Notices of the Surrounding Countries*, W. H. Allen, London, 1854, p. 352

2   Quoted in John Keay, *When Men and Mountains Meet: The Explorers of the Western Himalayas, 1820–65*, John Murray, 1977, pp. 35–6

3   Charles Allen, *The Prisoner of Kathmandu: Brian Hodgson in Nepal, 1820–43*, Haus Publishing, London, 2015, p. 70

4   William Moorcroft and George Trebeck, *Travels in the Himalayan Provinces of Hindustan and the Panjab; in Ladakh and Kashmir; in Peshawar, Kabul, Kunduz, and Bokhara*, ed. H. H. Wilson, John Murray, London, 1837, vol. 2, p. 166

5   Ibid., vol. 1, p. 359

6   Alexander Cunningham, *Ladák, Physical, Statistical, and Historical; with Notices of the Surrounding Countries*, W. H. Allen, London, 1854, p. 234

7   Ibid., p. 351

8   Ibid., p. 352

9   Ibid., p. 353

10 Ibid., p. 353n

11 Ibid., p. 274

12 Keay, *When Men and Mountains Meet*, p. 172

13 Ibid., p. 177

14 Quoted in ibid., p. 179

15 Thomas Thomson, *Western Himalaya and Tibet: A Narrative of a Journey through the Mountains of Northern India during the Years 1847–8*, London, 1852, repr. Ratna Pustak Bhandar, Katmandu, 1979, p. 429

14 MOUNTAINS OF DESTINY

1   Edmund Candler, *On the Edge of the World*, Cassell, London, 1919, pp. 59–60

2   Wade Davis, *Into the Silence: The Great War, Mallory and the Conquest of Everest*, Bodley Head, London, 2011, pp. 465–6

3   See John Keay, *The Great Arc: The Dramatic Tale of How India was Mapped and Everest was Named*, HarperCollins, London, 2000

4   Kenneth Mason, *Abode of Snow: A History of Himalayan Exploration and Mountaineering*, Rupert Hart-Davis, London, 1955, pp. 142–3

5   Ibid., p. 144

6   W. Rickmer Rickmers, 'The Alai-Pamirs in 1913 aand 1928', *Geographical Journal*, 74(3) (September 1929), p. 220

7   Mason, *Abode of Snow*, p. 195

8   Davis, *Into the Silence*, pp. 560–4

9   Maurice Isserman and Stewart Weaver, *Fallen Giants: The History of Himalayan Mountaineering from the Age of Empire to the Age of Extremes*, Yale University Press, New Haven, 2008, p. 134

10 Paul Bauer, quoted in Harald Höbusch, *'Mountain of Destiny': Nanga Parbat and its Path into the German Imagination*, Camden House, Rochester, NY, 2016, p. 94

11 Mason, *Abode of Snow*, pp. 199–201

12 Reinhold Messner, *Solo: Nanga Parbat*, Kaye & Ward, London, 1980, p. 22

13 Paul Bauer, *The Siege of Nanga Parbat, 1856–1963*, Rupert Hart-Davis, London, 1963, p. 35

14 Edmund Candler, *On the Edge of the World*, Cassell, London, 1919, p. 80

15 Willy Merkl, *Die Deutsche–Amerikanische himalaja Expedition 1932*, p. 65, quoted in Isserman and Weaver, *Fallen Giants*, p. 153

16 Isserman and Weaver, *Fallen Giants*, p. 154

17 Höbusch, *'Mountain of Destiny'*, pp. 92–3

18 Isserman and Weaver, *Fallen Giants*, p. 179

19 Jim Curran, *K2: The Story of the Savage Mountain*, Hodder & Stoughton, London, 1995, p. 51

20 Isserman and Weaver, *Fallen Giants*, p. 61

21 Filippo de Filippi, *Karakoram and Western Himalaya, 1909: An Account of the Expedition of H.R.H. Prince Luigi Amedeo of Savoy*, Constable, London, 1912, p. 218

22 Ibid., p. 321

23 Giotto Dainelli, *Buddhists and Glaciers of Western Tibet*, Kegan Paul, Trench & Trubner, London, 1933, pp. 200–1

24 Mason, *Abode of Snow*, p. 263

25 Isserman and Weaver, *Fallen Giants*, p. 221

26 Mason, *Abode of Snow*, p. 235

27 Isserman and Weaver, *Fallen Giants*, p. 100

28 Bauer, *The Siege of Nanga Parbat*, p. 111

29 Ibid., p. 143

30 Karl E. Meyer, 'Nazi Trespassers in Tibet', *New York Times*, 7 July 1997, https://www.nytimes.com/1997/07/07/opinion/nazi-trespassers-in-tibet.html?pagewanted=1?pagewanted=1

31 Isrun Engelhardt, 'The Ernst-Schaefer-Tibet-Expedition (1938–1939): New Light on the Political History of Tibet in the First Half of the 20th Century', in Alex McKay (ed.), *Tibet and her Neighbours: A History*, Edition Hansjörg Mayer, London, 2003, p. 190

32 Heinrich Harrer, *Seven Years in Tibet*, Rupert Hart-Davis, London, 1953, Pan repr. 1965, p. 190

33 Engelhardt, 'The Ernst-Schaefer-Tibet-Expedition', p. 193

EPILOGUE

1 F. Kingdon Ward, *The Romance of Plant Hunting*, Edwin Arnold, London, 1924, pp. 252–3

2 Wade Davis, *Into the Silence: The Great War, Mallory and the Conquest of Everest*, Bodley Head, London, 2011, p. 498; Captain F. Kingdon Ward, *The Riddle of the Tsangpo Gorges*, 1926, repr. Antique Collectors' Club, Woodbridge, 2001, p. 73

3 Kingdon Ward, *The Riddle of the Tsangpo Gorges*, pp. 210–11

4 Ibid., p. 232

5 'Paradise found', *Independent*, 6 March 2005, https://www.independent.co.uk/news/world/asia/paradise-found-5349738.html

6 Kenneth Cox, 'Exploring Pomé 1996–97', in Kingdon Ward, *The Riddle of the Tsangpo Gorges*, p. 290

7 Ken Storm Jr quoted in ibid., p. 294

8 Ibid.

9 Kingdon Ward, *The Riddle of the Tsangpo Gorges*, p. 160

10 Heinrich Harrer, *Seven Years in Tibet*, Rupert Hart-Davis, London, 1953, Pan repr. 1965, p. 259

11 Maurice Isserman and Stewart Weaver, *Fallen Giants: The History of Himalayan Mountaineering from the Age of Empire to the Age of Extremes*, Yale University Press, New Haven, 2008, pp. 252–3

12 Lionel Terray, *Conquistadors of the Useless*, Mountaineers Books, Seattle, 2001, pp. 86, 295

13 Paul Bauer, *The Siege of Nanga Parbat, 1856–1953*, Rupert Hart-Davis, London, 1956, p. 184

14 Isserman and Weaver, *Fallen Giants*, p. 304

15 Reinhold Messner, *Free Spirit: A Climber's Life*, Mountaineers Books, Seattle, 1991, p. 121

16 Ed Caesar, 'Reinhold Messner: the man who left his life on the mountain', *GQ*, 30 August 2020, https://www.gq-magazine.co.uk/lifestyle/article/reinhold-messner

17 Dipanjan Roy Chaudhuri, 'China-funded Diamer-Bhasha Dam in PoK an Ecological Disaster in the Making', *Economic Times*, 18 May 2020, https://economictimes.indiatimes.com/news/politics-and-nation/china-funded-diamer-basha-dam-in-pok-is-an-ecological-disaster-in-making/articleshow/75773410.cms

18 Margret Grebowicz, 'The World's Most Difficult Mountain May Soon be Fully Conquered', *The Atlantic*, 1 March 2018, https://www.theatlantic. com/science/archive/2018/03/k2-last-problem-of-the-himalayas/554618/

19 Stephen Alter, *Wild Himalaya: A Natural History of the Greatest Mountain Range on Earth*, Aleph Book Company, New Delhi, 2019, p. 378

Acknowledgments

This book took longer than expected. It draws on the rambles and ruminations of a lifetime and on the insights of so many whose names I've forgotten that the apologies would outnumber the acknowledgements that follow.

For their courtesy and ongoing support my first thanks go to the staff of National Library of Scotland in Edinburgh, where much of the research was done prior to the pandemic closure, and to the London Library whose waiving of postal charges in response to the same closure was a typically generous gesture of incomparable value.

In India the 2018 Mussoorie Mountain Festival brought the encouragement of Stephen Alter, Shekhar Pathak and others and proved the perfect stimulus to get writing. As always, the hospitality and insights of friends Bikram Grewal and Alpana Khare in Dehra Dun and Gurgaon enlivened the research with anecdote and refreshment. A big thank you, too, to other Dun luminaries: to Santanu Sarkar for a host of introductions and a large chunk of his time; to Bill Aitken whose inspirational books and fearless disposition more than justify his status as a living legend; to Kirsty Chakrabharti in her lamaist Shangri-La for helpful hints and hospitality; and to Lokesh Ohri for introducing me to his ground-breaking study of Jaunsar-Bawar, to which I have dismally failed to do justice.

Long ago Christina Noble and I entrusted selves and suitcases to three Ladakhis for a long walk from Kulu to Kashmir. It took about two months. From the experience was born the 'Kulu Trekking Agency', which later became 'West Himalayan Holidays' and whose clients I would occasionally accompany on historically themed treks. Christina and Kranti Singh were old friends, Chamba, Tanjin, Dorze, Prem and the others soon became so, and all were a rich source of Himalayan lore as well as the best of company.

Earlier still there was Kashmir. I was in my twenties, keen to fish and trying to write. Ghulam Mohamed Major, shrewdest of houseboat proprietors, appointed himself my 'Kashmiri father', mentor and rather unorthodox shikari. Three dedicated Scots doctors, Donald Duck and Jean Duck of Rainawari and Marie Mitchell from Anantnag, helped preserve my sanity.

Subsequent Himalayan ventures came courtesy of the BBC and long-suffering Radio 3 producer David Perry. From Sikkim and Kathmandu to Swat and Gilgit we trundled into the hills, and together we also made it to St Andrews to meet Hugh Richardson for what must have been one of the former Lhasa resident's last interviews.

A very big thank you too to David Campbell and the Drue Heinz Charitable Trust in whose castellated retreat on the shores of Lake Como the writing at last got off to a sunny start. Schuyler Jones CBE of Wichita has long been an inspiration and proved a dependable correspondent. Mike Searle at Oxford kindly corrected my understanding of plate tectonics, and Jim Naylor in the US explained the workings of isostasy. James McConnachie generously put aside his work on Kangchenjunga to run an eye over chapter 14.

The doyen of Himalayan photographers Jaroslav Poncar has been a friend and inspiration for at least thirty years. We collaborated on some of his magnificent panoramic books (published in Germany) and he has now provided most of the colour plates that grace *Himālaya*. For his incomparable images and encouragement, I can't thank him enough.

In Argyll son-in-law Amor Aditya was the first to read the text and it was he who discovered how to insert the vital makron in 'Himālaya'. Nell, my daughter, fathomed the apparent inability of Apple and Windows to communicate with one another, while their daughters Coco and Indi left us to get on with it. By way of thanks the book is dedicated to all four of them.

Amanda not only made up for my hopeless ignorance of all things IT but endured that most difficult of domestic presences, a husband with a wayward broadband connection. A word of thanks would be preposterous. The book itself is my tribute.

Andrew Gordon of David Higham and Associates placed the book and Michael Fishwick at Bloomsbury commissioned it. Peter James meticulously edited it. Lauren Whybrow researched the illustrations and oversaw production. Very many thanks to all of them, as also to Mike Athanson for the maps, Kieron Connolly for assisting and David Atkinson for the index.

Image Credits

Page 63: Tundop Sonam, to the left, and the wild yak (a bull) that he has just shot. Camp LV at La Schung Tso, Changtang (Photo: Sven Hedin 7 November 1906); used with kind permission of the Sven Hedin Foundation © Sven Hedin Foundation

Page 70: Group of Ani (nuns), photo taken by J.B. Noel, Mount Everest Expedition 1922. In the collection of Royal Geographical Society via Getty Images © Royal Geographical Society

Page 87: Yamchun Fort, photo by Tiffany Kary/Bloomberg via Getty Images © Bloomberg

Page 93: Mud watchtowers, Tibet, photo by Francis Kingdon-Ward, 1942, image from Royal Geographical Society via Getty images © Royal Geographical Society

Page 103: Giuseppe Tucci with Tibetan official in Chushul, 1948 © IsMEO

Page 105: Vultures at Sky Burial, taken by Rabden Lepcha (?), Sir Charles Bell's Mission to Lhasa 1920-21. The Tibet Album, 05 Dec. 2006. The Pitt Rivers Museum © The Pitt Rivers Museum

Page 122: Gyantse Dzong, photo by Captain W. Macready/Royal Geographical Society via Getty Images © Royal Geographical Society

Page 128: Expedition members, British expedition to Everest, John B. Noel Collection, photo by J.B. Noel/Royal Geographical Society via Getty Images © The Royal Geographical Society

Page 136: Kintup, photo by Gerald Burrard/Royal Geographical Society via Getty Images © The Royal Geographical Society

Page 174: Alexandra David-Néel and Yongden © Maison Alexandra David- Néel

Page 179: Golok man and his wife taken by Rabden Lepcha, Sir Charles Bell's Mission to Lhasa 1920-21. The Tibet Album, 05 Dec. 2006. The Pitt Rivers Museum © The Pitt Rivers Museum

Page 189: Portrait of Swami Prananvananda (of the Holy Kailas and Manasarovar), in *Exploration in Tibet* by Swami Prananvananda (University of Calcutta, Calcutta: 1939)

Page 197: Sven Hedin, in Tibetan attire, with his men and two dogs upon leaving Tibet after his third expedition. (Photo: Rev. Karl Marx, Poo - 29 August 1908); used with kind permission of the Sven Hedin Foundation © Sven Hedin Foundation

Page 215: The 14th Dalai Lama, unknown photographer, from the collection of Harry Staunton/Diana Hughes. The Tibet Album, 05 Dec. 2006. The Pitt Rivers Museum © The Pitt Rivers Museum

Page 227: 'The Editors in their Office', photo credited to *Scribners* magazine, *c.*1900

Page 236: Sera Monastery, Sir Charles Bell's Mission to Lhasa 1920-21. The Tibet Album, 05 Dec. 2006. The Pitt Rivers Museum © The Pitt Rivers Museum

Page 252: Chomolhari from Phari, taken by Charles Bell or Rabden Lepcha, Sir Charles Bell's Mission to Lhasa 1920-21. The Tibet Album, 05 Dec. 2006. The Pitt Rivers Museum © The Pitt Rivers Museum

Page 269: Moorcroft and Hearsey, painted by Hyder Young Hearsey in 1812, image reproduction © The British Library

Page 275: Leh, Ladakh, photo by Haeckel collection/ullstein bild via Getty Images

Page 297: The first German mountaineering expedition to Himālaya concentrated on Kangchenjunga © Archiv des Deutschen Alpenvereins, München

Page 307: Duke of the Abruzzi's expedition to Karakorum, photograph by Vittorio Sella, via Getty Images

PLATE SECTION

All images Copyright © Jaroslav Poncar apart from page 16, top, Copyright © Mike Searle

Index

A Note on the Author

John Keay's involvement with Himālaya dates back to the 1960s when he was a foreign correspondent in Kashmir. In the 1970s he published two standard works on the exploration of the Western Himalayas and in the 1980s wrote and presented a seven-part BBC Radio 3 documentary series on the Himalayan kingdoms. He has continued to specialise in Asian affairs and is probably the only writer in English to have produced 5000-year histories of both India and China. 'Exhaustively researched' and 'exquisitely written' (*Observer*), these have been through numerous editions from single-volume paperbacks (HarperCollins) or two-volume boxed hardbacks (Folio Society). Keay is married, has four children, lives in the West Highlands and travels whenever possible.

johnkeay.com

A Note on the Type

The text of this book is set Adobe Garamond. It is one of several versions of Garamond based on the designs of Claude Garamond. It is thought that Garamond based his font on Bembo, cut in 1495 by Francesco Griffo in collaboration with the Italian printer Aldus Manutius. Garamond types were first used in books printed in Paris around 1532. Many of the present-day versions of this type are based on the *Typi Academiae* of Jean Jannon cut in Sedan in 1615.

Claude Garamond was born in Paris in 1480. He learned how to cut type from his father and by the age of fifteen he was able to fashion steel punches the size of a pica with great precision. At the age of sixty he was commissioned by King Francis I to design a Greek alphabet, and for this he was given the honourable title of royal type founder. He died in 1561.